T0402161

Springer Tracts in Modern Physics

Volume 244

For further volumes:
http://www.springer.com/series/426

Springer Tracts in Modern Physics

Springer Tracts in Modern Physics provides comprehensive and critical reviews of topics of current interest in physics. The following fields are emphasized: elementary particle physics, solid-state physics, complex systems, and fundamental astrophysics.

Suitable reviews of other fields can also be accepted. The editors encourage prospective authors to correspond with them in advance of submitting an article. For reviews of topics belonging to the above mentioned fields, they should address the responsible editor, otherwise the managing editor.

See also springer.com

Martin Greiter

Mapping of Parent Hamiltonians

From Abelian and non-Abelian Quantum Hall States to Exact Models of Critical Spin Chains

Martin Greiter
Institut für Theorie der Kondensierten Materie (TKM)
Karlsruhe Institut für Technologie (KIT)
Campus Süd
Karlsruhe
Germany
e-mail: greiter@tkm.uni-karlsruhe.de

ISSN 0081-3869 e-ISSN 1615-0430
ISBN 978-3-642-24383-7 e-ISBN 978-3-642-24384-4
DOI 10.1007/978-3-642-24384-4
Springer Heidelberg Dordrecht London New York

Library of Congress Control Number: 2011940032

Cover design: eStudio Calamar, Berlin/Figueres

Printed on acid-free paper

Springer is part of Springer Science+Business Media (www.springer.com)

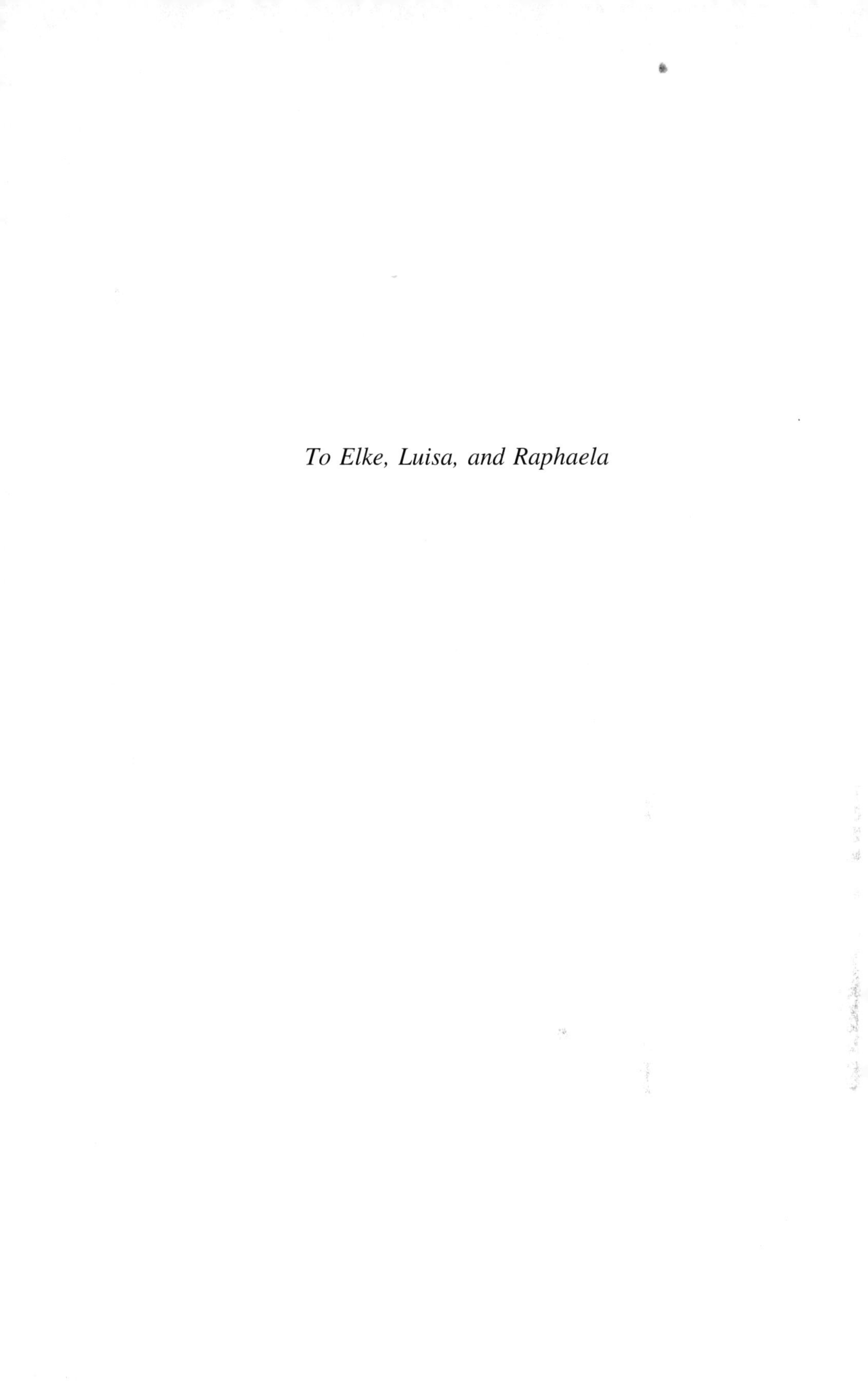

To Elke, Luisa, and Raphaela

Preface

The immediate advance we communicate with this monograph is the discovery of an exact model for a critical spin chain with arbitrary spin S, which includes the Haldane–Shastry model as the special case $S = \frac{1}{2}$. For $S \geq 1$, we propose that the spinon excitations obey a one-dimensional version of non-Abelian statistics, where the topological degeneracies are encoded in the fractional momentum spacings for the spinons. The model and its properties, however, are not the only, and possibly not even the most important thing one can learn from the analysis we present.

The benefit of science may be that it honors the human spirit, gives pleasure to those who immerse themselves in it, and pragmatically, contributes to the improvement of the human condition in the long term. The purpose of the individual scientific work can hence be either a direct contribution to this improvement, or more often an indirect contribution by making an advance which inspires further advances in a field. When we teach Physics, be it in lectures, books, monographs, or research papers, we usually teach what we understand, but rarely spend much effort on teaching how this understanding was obtained. The first volume of the famed course of theoretical physics by L. D. Landau and E. M. Lifshitz [1], for example, begins by stating the principle of least action, but does nothing to motivate how it was discovered historically or how one could be led to discover it from the study of mechanical systems. This reflects that we teach our students how to apply certain principles, but not how to discover or extract such principles from a given body of observations. The reason for this is not that we are truely content to teach students of physics as if they were students of engineering, but that the creative process in physics is usually erratic and messy, if not plainly embarrassing to those actively involved, and hence extremely difficult to recapture. As with most of what happens in reality, the actual paths of discovery are usually highly unlikely. Since we enjoy the comfort of perceiving actions and events as more likely and sensible, our minds subconsciously filter our memory to this effect.

One of the first topics I immersed myself in after completing my graduate coursework was Laughlin's theory of the fractionally quantized Hall effect [2].

I have never completely moved away from it, as this work testifies, and take enormous delight whenever I recognize quantum Hall physics in other domains of physics. More important than the theory itself, however, was to me to understand and learn from the way R. B. Laughlin actually discovered the wave function. He numerically diagonalized a system of three electrons in a magnetic field in an open plane, and observed that the total canonical angular momentum around the origin jumped by a factor of three (from $3\hbar$ to $9\hbar$) when he implemented a Coulomb interaction between the electrons. At the same time, no lesser scientists than D. Yoshioka, B. I. Halperin, and P. A. Lee [3] had, in an heroic effort, diagonalized up to six electrons with periodic boundary conditions, and concluded that their data were "supportive of the idea that the ground state is not crystalline, but a translationally invariant "liquid."" Their analysis was much more distinguished and scholarly, but unfortunately, did not yield the wave function.

The message I learned from this episode is that it is often beneficial to leave the path of scholarly analysis, and play with the simplest system of which one may hope that it might give away natures thoughts. For the Laughlin series of quantized Hall states, this system consisted of three electrons. I spend most of my scientific life adapting this approach to itinerant antiferromagnets in two dimensions, where I needed to go to twelve lattice sites until I could grasp what nature had in mind. But I am digressing. To complete the story about the discovery of the quantum Hall effect, Laughlin gave a public lecture in Amsterdam within a year of having received the Nobel price. He did not mention how he discovered the state, and at first couldn't recall it when I asked him in public after the lecture. As he was answering other questions, he recalled the answer to mine and weaved it into the answer of another question. During the evening in a cafe, a very famous Russian colleague whom I regard with the utmost respect commented the story of the discovery with the words "But this is stupid!".

Maybe it is. If it is so, however, the independent discoveries of the spin $\frac{1}{2}$ model by F. D. M. Haldane [4] and B. S. Shastry [5] may fall into the same category. Unfortunately, I do not know much about these discoveries. Haldane told me that he first observed striking degeneracies when he looked at the model for $N = 6$ sites numerically, motivated by the fact that the $1/r^2$ exchange is the discrete Fourier transform of $\epsilon(k) = k(k - 2\pi)$ in one dimension. Shastry told me that he discovered it "by doing calculations", which is not overly instructive to future generations. If my discovery of the general model I document in this monograph will be perceived in the spirit of my friends comment, I will at least have made no attempt to evade the charge.

In short, what I document on these pages is not just an exact model, but a precise and reproducible account of how I discovered this model. This reflects my belief that the path of discovery can be as instructive to future generations as the model itself. Of course, the analysis I document does not fully reflect the actual path of discovery, but what would have been the path if my thinking had followed a straight line. It took me about four weeks to obtain all the results and about four months to write this monograph. The reason for this discrepancy is not that my writing proceeds slowly, but that I had left out many intermediate steps when I did the calculation. The actual path of discovery must have been highly unlikely. In any event, it is comforting to me that, now that I have written a scholarly and coherent account of it, there is little need to recall what actually might have happened.

I am deeply grateful to Ronny Thomale for countless discussions and his critical reading of the manuscript, to Burkhard Scharfenberger, Dirk Schuricht, and Stephan Rachel for collaborations on various aspects of quantum spin chains, to Rose Schrempp and the members of the Institute for Theory of Condensed Matter at KIT for providing me with a pleasant and highly stimulating atmosphere, and especially to Peter Wölfle for his continued encouragement and support.

I further wish to thank Ms. Ute Heuser from Springer for her highly professional handling of the publication process.

Karlsruhe, April 2011 Martin Greiter

References

[1] L.D. Landau, E.M. Lifshitz, *Mechanics* (Butterworth-Heinemann, Oxford, 1982)

[2] R.B. Laughlin, Anomalous quantum hall effect: an incompressible quantum fluid with fractionally charged excitations. Phys. Rev. Lett. **50**, 1395 (1983)

[3] D. Yoshioka, B.I. Halperin, P.A. Lee, Ground state of two-dimensional electronsin strong magnetic fields and $\frac{1}{3}$ quantized hall effect. Phys. Rev. Lett. **50**,1219 (1983)

[4] F.D.M. Haldane, Exact Jastrow–Gutzwiller resonant-valence-bond ground state of the spin-$\frac{1}{2}$ antiferromagnetic Heisenberg chain with $1/r^2$ exchange. Phys. Rev. Lett. **60**, 635 (1988)

[5] B.S. Shastry, Exact solution of an $S=\frac{1}{2}$ Heisenberg antiferromagnetic chain with long-ranged interactions. Phys. Rev. Lett. **60**, 639 (1988)

Contents

Chapter 1
Introduction and Summary

Fractional quantization, and in particular fractional statistics [1, 2], in two-dimensional quantum liquids is witnessing a renaissance of interest in present times. The field started more than a quarter of a century ago with the discovery of the fractional quantum Hall effect, which was explained by Laughlin [3] in terms of an incompressible quantum liquid supporting fractionally charged (vortex or) quasiparticle excitations. When formulating a hierarchy of quantized Hall states [4–7] to explain the observation of quantized Hall states at other filling fractions fractions, Halperin [5, 6] noted that these excitations obey fractional statistics, and are hence conceptually similar to the charge-flux tube composites introduced by Wilczek two years earlier [8]. Physically, the fractional statistics manifests itself through fractional quantization of the kinematical relative angular momenta of the anyons.

The interest was renewed a few years later, when Anderson [9] proposed that hole-doped Mott insulators, and in particular the t–J model [10, 11] universally believed to describe the CuO planes in high T_c superconductors [12, 13], can be described in terms of a spin liquid (i.e., a state with strong, local antiferromagnetic correlations but without long range order), which would likewise support fractionally quantized excitations. In this proposal, the excitations are spinons and holons, which carry spin $\frac{1}{2}$ and no charge or no spin and charge $+e$, respectively. The fractional quantum number of the spinon is the spin, which is half integer while the Hilbert space (for the undoped system) is built up of spin flips, which carry spin one. One of the earliest proposals for a spin liquid supporting deconfined spinon and holon excitations is the (Abelian) chiral spin liquid [14–17]. Following up on an idea by D.H. Lee, Kalmeyer and Laughlin [14, 15] proposed that a quantized Hall wave function for bosons could be used to describe the amplitudes for spin-flips on a lattice. The chiral spin liquid state did not turn out to be relevant to CuO superconductivity, but remains one of very few examples of two-dimensional spin liquids with fractional statistics. Other established examples of two-dimensional spin liquids include the resonating valence bond (RVB) phases of the Rokhsar–Kivelson model [18] on the triangular lattice identified by Moessner and Sondhi [19], of the Kitaev model [20], and of the Hubbard model on the honeycomb lattice [21].

M. Greiter, *Mapping of Parent Hamiltonians*, Springer Tracts in Modern Physics 244,
DOI: 10.1007/978-3-642-24384-4_1,

While usually associated with two-dimensional systems, fractional statistics is also possible in one dimension. The paradigm for one-dimensional anyons are the spinon excitations in the Haldane–Shastry model [22, 23], a spin chain model with $S = \frac{1}{2}$ and long-ranged Heisenberg interactions. The ground state can be generated by Gutzwiller projection of half-filled bands of free fermions, and is equivalent to a chiral spin liquid in one dimension. The unique feature of the model is that the spinons are free in the sense that they only interact through their fractional statistics [24, 25]. The half-fermi statistics was originally discovered and formulated through a fractional exclusion or generalized Pauli principle [26], according to which the creation of two spinons reduces the number of single particle states available for further spinons by one. It manifests itself physically through fractional shifts in the spacings between the kinematical momenta of the individual spinons [27–29].

The present renaissance of interest in fractional statistics is due to possible applications of states supporting excitations with non-Abelian statistics [30] to the rapidly evolving field of quantum computation and cryptography [31, 32]. The paradigm for this universality class, is the Pfaffian state introduced by Moore and Read [33] in 1991. The state was proposed to be realized at the experimentally observed fraction $\nu = \frac{5}{2}$ [34] (i.e., at $\nu = \frac{1}{2}$ in the second Landau level) by Wen, Wilczek, and ourselves [35, 36], a proposal which recently received experimental support through the direct measurement of the quasiparticle charge [37, 38]. The Moore–Read state possesses $p + \mathrm{i}p$-wave pairing correlations. The flux quantum of the vortices is one half of the Dirac quantum, which implies a quasiparticle charge of $e/4$. Like the vortices in a p-wave superfluid, these quasiparticles possess Majorana-fermion states [39] at zero energy (i.e., one fermion state per pair of vortices, which can be occupied or unoccupied). A Pfaffian state with $2L$ spatially separated quasiparticle excitations is hence 2^L fold degenerate [40], in accordance with the dimension of the internal space spanned by the zero energy states. While adiabatic interchanges of quasiparticles yield only overall phases in Abelian quantized Hall states, braiding of half-vortices of the Pfaffian state will in general yield non-trivial changes in the occupations of the zero energy states [41, 42], which render the interchanges non-commutative or non-Abelian. In particular, the internal state vector is insensitive to local perturbations—it can only be manipulated through non-local operations like braiding of the vortices or measurements involving two or more vortices simultaneously. For a sufficiently large number of vortices, on the other hand, any unitary transformation in this space can be approximated to arbitrary accuracy through successive braiding operations [43]. These properties together render non-Abelions preeminently suited for applications as protected qubits in quantum computation [30, 32, 44–46]. Non-Abelian anyons are further established in certain other quantum Hall states described by Jack polynomials [47–49] including Read–Rezayi states [50], in the non-Abelian phase of the Kitaev model [20], in the Yao–Kivelson model [51], and in the non-Abelian chiral spin liquid proposed by Thomale and ourselves [52]. In this liquid, the amplitudes for renormalized spin-flips on a lattice with spins $S = 1$ are described by a bosonic Pfaffian state.

The connection between the Haldane–Shastry ground state, the chiral spin liquid, and a bosonic Laughlin state at Landau level filling fraction $\nu = \frac{1}{2}$ suggests that one may consider the non-Abelian chiral spin liquid in one dimension as a ground state for a spin chain with $S = 1$. This state is related to a bosonic Moore–Read state at filling fraction $\nu = 1$. In this monograph, we will introduce and elaborate on this one-dimensional spin liquid state, construct a parent Hamiltonian, and generalize the model to arbitrary spin S. We further propose that the spinon excitations of the states for $S \geq 1$ will obey a novel form of "non-Abelian" statistics, where the internal, protected Hilbert space associated with the statistics is spanned by topological shifts in the spacings of the single spinon momenta when spinons are present.

Most of the book will be devoted to the construction of the model Hamiltonian for spin S. In Chap. 2, we introduce three exact models, and the ground state for the $S = 1$ spin chain for which we wish to construct a parent Hamiltonian. The exact models consist of Hamiltonians, their ground states, and the elementary excitations, which are in some cases exact and in others approximate eigenstates of the Hamiltonian. In Sect. 2.1, we review the Laughlin $\nu = \frac{1}{m}$ state for quantized Hall liquids,

$$\psi_0(z_1, z_2, \ldots, z_M) = \prod_{i<j}^{M}(z_i - z_j)^m \prod_{i=1}^{M} e^{-\frac{1}{4}|z_i|^2}, \tag{1.1}$$

where the z_i's are the coordinates of M electrons in the complex plane, and m is odd for fermions and even for bosons. For $m = 2$, its parent Hamiltonian is given by the kinetic term giving rise to Landau level quantization supplemented by a δ-function potential, which excludes the component with relative angular momentum zero between pairs of bosons. The ground state wave function for a bosonic $m = 2$ Laughlin state is similar to the ground state of the Haldane–Shastry model we review in Sect. 2.2,

$$\psi_0^{\mathrm{HS}}(z_1, z_2, \ldots, z_M) = \prod_{i<i}^{M}(z_i - z_j)^2 \prod_{i=1}^{M} z_i, \tag{1.2}$$

where the z_i's are now coordinates of spin flips for a spin chain with N sites on a unit circle embedded in the complex plane, and $M = \frac{N}{2}$. The Haldane–Shastry Hamiltonian,

$$H^{\mathrm{HS}} = \left(\frac{2\pi}{N}\right)^2 \sum_{\alpha<\beta}^{N} \frac{\boldsymbol{S}_\alpha \boldsymbol{S}_\beta}{|\eta_\alpha - \eta_\beta|^2}, \tag{1.3}$$

where $\eta_\alpha = e^{\mathrm{i}\frac{2\pi}{N}\alpha}$ are the coordinates of the N sites on the unit circle, however, bears no resemblance to the δ-function Hamiltonian for the Laughlin states. We will elaborate in Sect. 3.1 that these models are both physically and mathematically sufficiently different to consider them unrelated. Even the ground state wave functions, when adapted as far as any possible by formulating the bosonic Laughlin state on the sphere and by inserting a quasihole at the south pole, differ due to different Hilbert space normalizations. From a scholarly point of view, there just appears to be no connection.

From a pragmatic point of view, however, we may view both Hamiltonians as devices to obtain the coefficients of the polynomial

$$\prod_{i<i}^{N}(z_i - z_j)^2$$

for particle numbers such that the Hamiltonians can be diagonalized numerically. In fact, Haldane [4] introduced the parent Hamiltonian for the Laughlin state in order to obtain the coefficients of all the configurations of the state vector for $N = 6$, which he could then compare numerically to the exact ground state for Coulomb interactions. This raises the question whether the recipes used by both Hamiltonians for obtaining these coefficients are really different. If one wishes to attribute the results we presented to a discovery, this discovery is that they are not.

When we "derive" the Haldane–Shastry model from the bosonic $m = 2$ Laughlin state and its δ-function parent Hamiltonian in Chap. 3, we really first extract this recipe from the quantum Hall Hamiltonian, and then use it to construct a parent Hamiltonian for the quantum spin chain, which has to be Hermitian, local, and invariant under translations, parity, time reversal, and SU(2) spin rotations. Written in the language of the spin system, the recipe is the condition that the Haldane–Shastry ground state is annihilated by the operator

$$\Omega_\alpha^{\text{HS}} = \sum_{\substack{\beta=1\\ \beta\neq\alpha}}^{N} \frac{1}{\eta_\alpha - \eta_\beta} S_\alpha^- S_\beta^-, \qquad \Omega_\alpha^{\text{HS}} \left|\psi_0^{\text{HS}}\right\rangle = 0 \quad \forall\, \alpha. \tag{1.4}$$

The Haldane–Shastry model has been known for more than two decades, but while Haldane and Shastry independently discovered it, we derive it. Unlike the discoveries, this derivation lends itself to a generalization to higher spins. The construction of exact models of critical spin chains following the line of reasoning we use in our derivation of the Haldane–Shastry model is the subject of this monograph.

In Sect. 2.3, we review the properties of the Moore–Read state [33, 35, 36],

$$\psi_0(z_1, z_2, \ldots, z_N) = \text{Pf}\left(\frac{1}{z_i - z_j}\right) \prod_{i<j}^{N}(z_i - z_j)^m \prod_{i=1}^{N} e^{-\frac{1}{4}|z_i|^2}, \tag{1.5}$$

at Landau level filling fraction $\nu = \frac{1}{m}$, where m is even for fermions and odd for bosons, with emphasis on the non-Abelian statistics of the half-vortex quasiparticle excitations. For $m = 1$, the Pfaffian state is the exact ground state of the kinetic Hamiltonian supplemented by the three-body interaction term [36]

$$V = \sum_{i,j<k}^{N} \delta^{(2)}(z_i - z_j)\delta^{(2)}(z_i - z_k). \tag{1.6}$$

The bosonic $m = 1$ ground state is similar to the ground state wave function of the critical $S = 1$ spin liquid state we introduce in Sect. 2.4,

$$\psi_0^{S=1}(z_1, z_2, \ldots, z_N) = \mathrm{Pf}\left(\frac{1}{z_i - z_j}\right) \prod_{i<j}^{N} (z_i - z_j) \prod_{i=1}^{N} z_i, \tag{1.7}$$

which describes the amplitudes of renormalized spin flips

$$\tilde{S}_\alpha^+ = \frac{S_\alpha^z + 1}{2} S_\alpha^+, \tag{1.8}$$

on sites $\eta_\alpha = e^{\mathrm{i}\frac{2\pi}{N}\alpha}$ on a unit circle embedded in the complex plane. These spin flips act on a vacuum where all the N spins are in the $S^z = -1$ state. In Sect. 2.4.5, we propose that the momentum spacings between the individual spinon excitations of this liquid alternate between being odd multiples of $\frac{\pi}{N}$ and being either even or odd multiples of $\frac{\pi}{N}$. (Since the spacings for bosons or fermions are multiples of $\frac{2\pi}{N}$, an odd multiply of $\frac{\pi}{N}$ corresponds to half-fermion, and an even multiple to boson or fermion statistics.) When we have a choice between even and odd, this choice represents a topological quantum number. The momentum spacings hence span an internal or topological Hilbert space of dimension 2^L when $2L$ spinons are present, as appropriate for Ising anyons. These spacings constitute the analog of the Majorana fermion states in the cores of the half-vortex excitations of the Moore–Read state.

In Chap. 4, we derive a parent Hamiltonian for the $S = 1$ spin liquid state (1.5) from the three-body parent Hamiltonian (1.6) of the Moore–Read state. The steps are similar to those taken for the Haldane–Shastry model, but technically more involved. The defining condition for the state, i.e., the recipe used by the quantum Hall Hamiltonian to specify the coefficients of the polynomial

$$\mathrm{Pf}\left(\frac{1}{z_i - z_j}\right) \prod_{i<j}^{N} (z_i - z_j),$$

is in the language of the $S = 1$ spin model given by

$$\Omega_\alpha^{S=1} = \sum_{\substack{\beta=1 \\ \beta \neq \alpha}}^{N} \frac{1}{\eta_\alpha - \eta_\beta} (S_\alpha^-)^2 S_\beta^-, \qquad \Omega_\alpha^{S=1} \left|\psi_0^{S=1}\right\rangle = 0 \ \ \forall\, \alpha. \tag{1.9}$$

As an aside, we also find that the state is annihilated by the operator

$$\Xi_\alpha = \sum_{\substack{\beta,\gamma=1 \\ \beta,\gamma \neq \alpha}}^{N} \frac{S_\alpha^- S_\beta^- S_\gamma^-}{(\eta_\alpha - \eta_\beta)(\eta_\alpha - \eta_\gamma)} - \sum_{\substack{\beta=1 \\ \beta \neq \alpha}}^{N} \frac{(S_\alpha^-)^2 S_\beta^-}{(\eta_\alpha - \eta_\beta)^2}, \quad \Xi_\alpha \left|\psi_0^{S=1}\right\rangle = 0 \ \ \forall\, \alpha, \tag{1.10}$$

which we do not consider further. A Hermitian and translationally invariant annihilation operator for the $S = 1$ spin liquid state (1.5) is given by

$$H_0 = \frac{1}{2} \sum_{\alpha=1}^{N} \Omega_\alpha^{S=1^\dagger} \Omega_\alpha^{S=1}. \tag{1.11}$$

Since the state is a spin singlet, i.e., invariant under SU(2) spin rotations, all the different tensor components of (1.11) must annihilate it individually. In Sect. 4.5, we obtain the desired parent Hamiltonian for the $S = 1$ spin liquid state (1.7),

$$H^{S=1} = \frac{2\pi^2}{N^2} \left[\sum_{\alpha \neq \beta}^{N} \frac{\boldsymbol{S}_\alpha \boldsymbol{S}_\beta}{|\eta_\alpha - \eta_\beta|^2} - \frac{1}{20} \sum_{\substack{\alpha,\beta,\gamma \\ \alpha \neq \beta,\gamma}}^{N} \frac{(\boldsymbol{S}_\alpha \boldsymbol{S}_\beta)(\boldsymbol{S}_\alpha \boldsymbol{S}_\gamma) + (\boldsymbol{S}_\alpha \boldsymbol{S}_\gamma)(\boldsymbol{S}_\alpha \boldsymbol{S}_\beta)}{(\bar{\eta}_\alpha - \bar{\eta}_\beta)(\eta_\alpha - \eta_\gamma)} \right], \tag{1.12}$$

by projecting out the component of H_0 which is invariant under parity, time reversal, and SU(2) spin rotations. The energy of the ground state (1.7) is given by

$$E_0^{S=1} = -\frac{2\pi^2}{N^2} \frac{N(N^2+5)}{15}. \tag{1.13}$$

Finally, we use the same methods to obtain vector annihilation operators for the $S = 1$ spin liquid state in Sect. 4.6.

In Chap. 5, we generalize the model to arbitrary spin S. We do, however, no longer start with a quantum Hall state and its parent Hamiltonian, but generalize the spin liquid states and the defining conditions for $S = \frac{1}{2}$ and $S = 1$, i.e., the conditions (1.4) and (1.9), directly to higher spins. To generalize the state vector, we first recall from Sect. 2.4.4 that the $S = 1$ spin liquid can be obtained by taking two (identical) Gutzwiller or Haldane–Shastry ground states and projecting onto the triplet or $S = 1$ configuration at each site [53]. This projection can be accomplished conveniently if we write the Haldane–Shastry ground state (2.2.3) in terms of Schwinger bosons,

$$\begin{aligned} |\psi_0^{\text{HS}}\rangle &= \sum_{\{z_1,\ldots,z_M;w_1,\ldots,w_M\}} \psi_0^{\text{HS}}(z_1,\ldots,z_M)\, a_{z_1}^{+} \ldots a_{z_M}^{\dagger} b_{w_1}^{+} \ldots b_{w_M}^{\dagger} |0\rangle \\ &\equiv \Psi_0^{\text{HS}}[a^\dagger, b^\dagger]\, |0\rangle, \end{aligned} \tag{1.14}$$

where $M = \frac{N}{2}$ and the w_k's are those coordinates on the unit circle which are not occupied by any of the z_i's. The $S = 1$ spin liquid state (1.7) can then be written

$$|\psi_0^{S=1}\rangle = \left(\Psi_0^{\text{HS}}[a^\dagger, b^\dagger]\right)^2 |0\rangle. \tag{1.15}$$

To generalize the ground state to arbitrary spin S, we just take $2S$ (identical) copies Haldane–Shastry ground state, and project at each site onto the completely symmetric representation with total spin S. In terms of Schwinger bosons,

$$|\psi_0^S\rangle = \left(\Psi_0^{\text{HS}}[a^\dagger, b^\dagger]\right)^{2S} |0\rangle. \tag{1.16}$$

This state is related to bosonic Read–Rezayi states [50] in the quantum Hall system. In Sect. 5.2, we verify that the state is annihilated by the operator

$$\Omega_\alpha^S = \sum_{\substack{\beta=1 \\ \beta\neq\alpha}}^{N} \frac{1}{\eta_\alpha - \eta_\beta} (S_\alpha^-)^{2S} S_\beta^-, \qquad \Omega_\alpha^S |\psi_0^S\rangle = 0 \quad \forall\alpha. \tag{1.17}$$

In Sect. 5.3, we follow the same steps as for the $S=1$ state to construct a parent Hamiltonian for the spin S state (1.16), and obtain

$$H^S = \frac{2\pi^2}{N^2} \left[\sum_{\alpha\neq\beta}^{N} \frac{\boldsymbol{S}_\alpha \boldsymbol{S}_\beta}{|\eta_\alpha - \eta_\beta|^2} - \frac{1}{2(S+1)(2S+3)} \sum_{\substack{\alpha,\beta,\gamma \\ \alpha\neq\beta,\gamma}}^{N} \frac{(\boldsymbol{S}_\alpha \boldsymbol{S}_\beta)(\boldsymbol{S}_\alpha \boldsymbol{S}_\gamma) + (\boldsymbol{S}_\alpha \boldsymbol{S}_\gamma)(\boldsymbol{S}_\alpha \boldsymbol{S}_\beta)}{(\bar{\eta}_\alpha - \bar{\eta}_\beta)(\eta_\alpha - \eta_\gamma)} \right]. \tag{1.18}$$

The energy eigenvalue is given by

$$E_0^S = -\frac{2\pi^2}{N^2} \frac{S(S+1)^2}{2S+3} \frac{N(N^2+5)}{12}. \tag{1.19}$$

This is the main result we present. In Sect. 5.4, we construct the vector annihilation operators

$$\begin{aligned} \boldsymbol{D}_\alpha^S &= \frac{1}{2} \sum_{\substack{\beta \\ \beta\neq\alpha}} \frac{\eta_\alpha + \eta_\beta}{\eta_\alpha - \eta_\beta} \left[\mathrm{i}(\boldsymbol{S}_\alpha \times \boldsymbol{S}_\beta) + (S+1)\boldsymbol{S}_\beta - \frac{1}{S+1} \boldsymbol{S}_\alpha \left(\boldsymbol{S}_\alpha \boldsymbol{S}_\beta\right) \right], \\ \boldsymbol{D}_\alpha^S |\psi_0^S\rangle &= 0 \quad \forall\, \alpha, \end{aligned} \tag{1.20}$$

and

$$\begin{aligned} \boldsymbol{A}_\alpha^S = &\sum_{\substack{\beta \\ \beta\neq\alpha}} \frac{\boldsymbol{S}_\alpha (\boldsymbol{S}_\alpha \boldsymbol{S}_\beta) + (\boldsymbol{S}_\alpha \boldsymbol{S}_\beta) \boldsymbol{S}_\alpha + 2(S+1)\boldsymbol{S}_\beta}{|\eta_\alpha - \eta_\beta|^2} \\ &+ \sum_{\substack{\beta,\gamma \\ \beta,\gamma\neq\alpha}} \frac{1}{(\bar{\eta}_\alpha - \bar{\eta}_\beta)(\eta_\alpha - \eta_\gamma)} \\ &\cdot \left[- \frac{(\boldsymbol{S}_\alpha \boldsymbol{S}_\beta)\boldsymbol{S}_\alpha (\boldsymbol{S}_\alpha \boldsymbol{S}_\gamma) + (\boldsymbol{S}_\alpha \boldsymbol{S}_\gamma)\boldsymbol{S}_\alpha (\boldsymbol{S}_\alpha \boldsymbol{S}_\beta)}{S+1} \right. \\ &\left. + 2(S+2)\boldsymbol{S}_\alpha (\boldsymbol{S}_\beta \boldsymbol{S}_\gamma) - \boldsymbol{S}_\beta (\boldsymbol{S}_\alpha \boldsymbol{S}_\gamma) - (\boldsymbol{S}_\alpha \boldsymbol{S}_\beta)\boldsymbol{S}_\gamma \right], \\ \boldsymbol{A}_\alpha^S |\psi_0^S\rangle = &\ 0 \quad \forall\, \alpha. \end{aligned} \tag{1.21}$$

In Sect. 5.5, we evaluate the parity and time reversal invariant scalar operators

$$\sum_{\alpha} \boldsymbol{D}_{\alpha}^{S\dagger} \boldsymbol{D}_{\alpha}^{S} \quad \text{and} \quad \sum_{\alpha} \boldsymbol{S}_{\alpha} \boldsymbol{A}_{\alpha}^{S}, \tag{1.22}$$

and find that both of them reproduce the model (1.18). The factorization of H^S is terms of $\boldsymbol{D}_{\alpha}^{S\dagger}$ and $\boldsymbol{D}_{\alpha}^{S}$ shows that $\left|\psi_0^S\right\rangle$ is not just an eigenstate of (1.18), but also a ground state. Numerical work [54] indicates that $\left|\psi_0^S\right\rangle$ is the only ground state of $\left|\psi_0^S\right\rangle$. In Sect. 5.6, we show that the model (1.18) reduces to the Haldane–Shastry model if we take $S = \frac{1}{2}$.

We conclude with a brief discussion of several unresolved issues as well as possible generalizations of the model in Chap. 6. These include the quest for integrability, the correctness and universality of our assignments for the SU(2) level $k = 2S$ anyon-type momentum spacings of the spinon excitations and the feasibility of applications as protected cubits in quantum computation. We outline how to generalize the model to symmetric representations of SU(n), where the non-abelian statistics of the spinons appears to have no correspondence in a quantum Hall system.

References

1. F. Wilczek, *Fractional Statistics and Anyon Superconductivity* (World Scientific, Singapore, 1990)
2. A. Stern, Anyons and the quantum Hall effect—a pedagogical review. Ann. Phys. **323**, 204 (2008). January special issue 2008
3. R.B. Laughlin, Anomalous quantum Hall effect: an incompressible quantum fluid with fractionally charged excitations. Phys. Rev. Lett. **50**, 1395 (1983)
4. F.D.M. Haldane, Fractional quantization of the Hall effect: a hierarchy of incompressible quantum fluid states. Phys. Rev. Lett. **51**, 605 (1983)
5. B.I. Halperin, Statistics of quasiparticles and the hierarchy of fractional quantized Hall states. Phys. Rev. Lett. **52**, 1583 (1984)
6. B.I. Halperin, Statistics of quasiparticles and the hierarchy of fractional quantized Hall states. Phys. Rev. Lett. **52**, E2390 (1984)
7. M. Greiter, Microscopic formulation of the hierarchy of quantized Hall states. Phys. Lett. B **336**, 48 (1994)
8. F. Wilczek, Quantum mechanics of fractional-spin particles. Phys. Rev. Lett. **49**, 957 (1982)
9. P.W. Anderson, The resonating valence bond state in La_2CuO_4 and superconductivity. Science **235**, 1196 (1987)
10. F.C. Zhang, T.M. Rice, Effective Hamiltonian for the superconducting Cu oxides. Phys. Rev. B **37**, 3759 (1988)
11. H. Eskes, G.A. Sawatzky, Tendency towards local spin compensation of holes in the high-T_c copper compounds. Phys. Rev. Lett. **61**, 1415 (1988)
12. J. Zaanen, S. Chakravarty, T. Senthil, P. Anderson, P. Lee, J. Schmalian, M. Imada, D. Pines, M. Randeria, C. Varma, M. Vojta, M. Rice, Towards a complete theory of high T_c. Nat. Phys. **2**, 138 (2006)
13. J. Orenstein, A.J. Millis, Advances in the physics of high-temperature superconductivity. Science **288**, 468 (2000)
14. V. Kalmeyer, R.B. Laughlin, Equivalence of the resonating-valence-bond and fractional quantum Hall states. Phys. Rev. Lett. **59**, 2095 (1987)

15. V. Kalmeyer, R.B. Laughlin, Theory of the spin liquid state of the heisenberg antiferromagnet. Phys. Rev. B **39**, 11879 (1989)
16. D.F. Schroeter, E. Kapit, R. Thomale, M. Greiter, Spin Hamiltonian for which the chiral spin liquid is the exact ground state . Phys. Rev. Lett. **99**, 097202 (2007)
17. R. Thomale, E. Kapit, D.F. Schroeter, M. Greiter, Parent Hamiltonian for the chiral spin liquid. Phys. Rev. B **80**, 104406 (2009)
18. S.A. Kivelson, D.S. Rokhsar, J.P. Sethna, Topology of the resonating valence-bond state: solitons and high-t_c superconductivity. Phys. Rev. B **35**, 8865 (1987)
19. R. Moessner, S.L. Sondhi, Resonating valence bond phase in the triangular lattice quantum dimer model. Phys. Rev. Lett. **86**, 1881 (2001)
20. A. Kitaev, Anyons in an exactly solved model and beyond. Ann. Phys. **321**, 2 (2006)
21. Z.Y. Meng, T.C. Lang, S. Wessel, F.F. Assaad, A. Muramatsu, Quantum spin liquid emerging in two-dimensional correlated Dirac fermions. Nature **464**, 847 (2010)
22. F.D.M. Haldane, Exact Jastrow–Gutzwiller resonant-valence-bond ground state of the spin-$\frac{1}{2}$ antiferromagnetic Heisenberg chain with $1/r^2$ exchange. Phys. Rev. Lett. **60**, 635 (1988)
23. B.S. Shastry, Exact solution of an $S = \frac{1}{2}$ Heisenberg antiferromagnetic chain with long-ranged interactions. Phys. Rev. Lett. **60**, 639 (1988)
24. F.D.M. Haldane, *Physics of the Ideal Semion Gas: Spinons and Quantum Symmetries of the Integrable Haldane–shastry Spin Chain*, ed. by A. Okiji, N. Kawakami. Correlation Effects in Low-Dimensional Electron Systems, (Springer, Berlin, 1994)
25. M. Greiter, D. Schuricht, No attraction between spinons in the Haldane–Shastry model. Phys. Rev. B **71**, 224424 (2005)
26. F.D.M. Haldane, Fractional statistics in arbitrary dimensions: a generalization of the Pauli principle. Phys. Rev. Lett. **67**, 937 (1991)
27. M. Greiter, D. Schuricht, Comment on spinon attraction in spin-1/2 antiferromagnetic chains. Phys. Rev. Lett. **96**, 059701 (2006)
28. M. Greiter, D. Schuricht, Many-spinon states and the secret significance of Young tableaux. Phys. Rev. Lett. **98**, 237202 (2007)
29. M. Greiter, Statistical phases and momentum spacings for one-dimesional anyons. Phys. Rev. B **79**, 064409 (2009)
30. A. Stern, Non-Abelian states of matter. Nature **464**, 187 (2010)
31. A.Y. Kitaev, Fault-tolerant quantum computation by anyons. Ann. Phys. **303**, 2 (2002)
32. C. Nayak, S.H. Simon, A. Stern, M. Freedman, S. Das Sarma, Rev. Mod. Phys. **80**, 1083 (2008)
33. G. Moore, N. Read, Nonabelions in the fractional quantum Hall effect. Nucl. Phys. B **360**, 362 (1991)
34. R. Willett, J.P. Eisenstein, H.L. Störmer, D.C. Tsui, A.C. Gossard, J.H. English, Observation of an even-denominator quantum number in the fractional quantum Hall effect. Phys. Rev. Lett. **59**, 1776 (1987)
35. M. Greiter, X.G. Wen, F. Wilczek, Paired Hall state at half filling. Phys. Rev. Lett. **66**, 3205 (1991)
36. M. Greiter, X.G. Wen, F. Wilczek, Paired Hall states. Nucl. Phys. B **374**, 567 (1992)
37. M. Dolev, M. Heiblum, V. Umansky, A. Stern, D. Mahalu, Observation of a quarter of an electron charge at the $\nu = 5/2$ quantum Hall state. Nature **452**, 829 (2008)
38. I.P. Radu, J.B. Miller, C.M. Marcus, M.A. Kastner, L.N. Pfeiffer, K.W. West, Quasi-particle properties from tunneling in the $\nu = 5/2$ fractional quantum Hall state. Science **320**, 899 (2008)
39. N. Read, D. Green, Paired states of fermions in two dimensions with breaking of parity and time-reversal symmetries and the fractional quantum Hall effect. Phys. Rev. B **61**, 10267 (2000)
40. C. Nayak, F. Wilczek, $2n$-quasihole states realize 2^{n-1}-dimensional spinor braiding statistics in paired quantum Hall states. Nucl. Phys. B **479**, 529 (1996)
41. D.A. Ivanov, Non-Abelian statistics of half-quantum vortices in p-wave superconductors. Phys. Rev. Lett. **86**, 268 (2001)

42. A. Stern, von F. Oppen, E. Mariani, Geometric phases and quantum entanglement as building blocks for non-Abelian quasiparticle statistics. Phys. Rev. B **70**, 205338 (2004)
43. M.H. Freedman, A. Kitaev, Z. Wang, Simulation of topological field theories by quantum computer. Comm. Math. Phys. **227**, 587 (2002)
44. S. Das Sarma, M. Freedman, C. Nayak, Topologically-protected qubits from a possible non-Abelian fractional quantum Hall state. Phys. Rev. Lett. **94**, 166802 (2005)
45. W. Bishara, P. Bonderson, C. Nayak, K. Shtengel, J.K. Slingerland, Interferometric signature of non-Abelian anyons. Phys. Rev. B **80**, 155303 (2009)
46. J.E. Moore, Quasiparticles do the twist. Physics **2**, 82 (2009)
47. M. Greiter, Root configurations and many body interactions for fractionally quantized Hall states. Bull. Am. Phys. Soc. **38**, 137 (1993)
48. S.H. Simon, E.H. Rezayi, N.R. Cooper, I. Berdnikov, Construction of a paired wave function for spinless electrons at filling fraction $\nu = 2/5$. Phys. Rev. B **75**, 075317 (2007)
49. B.A. Bernevig, F.D.M. Haldane, Model fractional quantum Hall states and Jack polynomials. Phys. Rev. Lett. **100**, 246802 (2008)
50. N. Read, E. Rezayi, Beyond paired quantum Hall states: parafermions and incompressible states in the first excited Landau level. Phys. Rev. B **59**, 8084 (1999)
51. D. Yoshioka, B.I. Halperin, P.A. Lee, Ground state of two-dimensional electrons in strong magnetic fields and $\frac{1}{3}$ quantized Hall effect. Phys. Rev. Lett. **50**, 1219 (1983)
52. M. Greiter, R. Thomale, Non-Abelian statistics in a quantum antiferromagnet. Phys. Rev. Lett. **102**, 207203 (2009)
53. M. Greiter, $S=1$ spin liquids: broken discrete symmetries restored. J. Low Temp. Phys. **126**, 1029 (2002)
54. B. Scharfenberger, M. Greiter, manuscript in preparation

Chapter 2
Three Models and a Ground State

2.1 The Laughlin State and Its Parent Hamiltonian

Laughlin's theory [1–6] for a series of fractionally quantized Hall states is first and foremost the key to an explanation for the experimentally observed, fractionally quantized plateaus in the Hall resistivity of a spin-polarized, two-dimensional electron gas realized in semiconductor inversion layers [5, 7–10]. For our purposes here, however, we will view it primarily as an exact model, that is, a ground state which supports fractionally quantized excitations, and a model Hamiltonian for which this ground state is exact.

We will first review the theory in a planar geometry with open boundary conditions, and then turn to the spherical geometry, which will turn out to be the relevant geometry for the mapping of quantized Hall system onto a spin chain. We begin with a review of Landau level quantization in the plane.

2.1.1 Landau Level Quantization in the Planar Geometry

To describe the dynamics of charged particles (e.g. spin-polarized electrons) in a two-dimensional plane subject to a perpendicular magnetic field $\boldsymbol{B} = -B\boldsymbol{e}_z$, it is convenient to introduce complex particles coordinates $z = x + \mathrm{i}y$ and $\bar{z} = x - \mathrm{i}y$ [11, 12]. The associated derivative operators are

$$\frac{\partial}{\partial z} = \frac{1}{2}\left(\frac{\partial}{\partial x} - \mathrm{i}\frac{\partial}{\partial y}\right), \quad \frac{\partial}{\partial \bar{z}} = \frac{1}{2}\left(\frac{\partial}{\partial x} + \mathrm{i}\frac{\partial}{\partial y}\right). \tag{2.1.1}$$

Note that hermitian conjugation yields a $-$ sign,

$$\left(\frac{\partial}{\partial z}\right)^{\dagger} = -\frac{\partial}{\partial \bar{z}}. \tag{2.1.2}$$

M. Greiter, *Mapping of Parent Hamiltonians*, Springer Tracts in Modern Physics 244,
DOI: 10.1007/978-3-642-24384-4_2, © Springer-Verlag Berlin Heidelberg 2011

We further define the complex momentum

$$p \equiv p_\mathrm{x} + \mathrm{i}p_\mathrm{y} = -2\mathrm{i}\hbar\frac{\partial}{\partial\bar{z}}, \quad \bar{p} = p_\mathrm{x} - \mathrm{i}p_\mathrm{y} = -2\mathrm{i}\hbar\frac{\partial}{\partial z}. \tag{2.1.3}$$

The single particle Hamilton operator is obtained by minimally coupling the gauge field to the canonical momentum,

$$H = \frac{1}{2M}\left(\boldsymbol{p} + \frac{e}{c}\boldsymbol{A}\right)^2, \tag{2.1.4}$$

where M is the mass of the particle and $e > 0$. In the symmetric gauge $\boldsymbol{A} = \frac{1}{2}B\boldsymbol{r}\times\boldsymbol{e}_\mathrm{z}$, and with the definition of the magnetic length

$$l = \sqrt{\frac{\hbar c}{eB}}, \tag{2.1.5}$$

we write

$$\begin{aligned} H &= \frac{1}{2M}\left[\left(p_\mathrm{x} + \frac{\hbar}{2l^2}y\right)^2 + \left(p_\mathrm{y} - \frac{\hbar}{2l^2}x\right)^2\right] \\ &= \frac{1}{2M}\left[\Re^2\left(p - \frac{\mathrm{i}\hbar}{2l^2}z\right) + \Im^2\left(p - \frac{\mathrm{i}\hbar}{2l^2}z\right)\right] \\ &= \frac{1}{4M}\left\{p - \frac{\mathrm{i}\hbar}{2l^2}z, \bar{p} + \frac{\mathrm{i}\hbar}{2l^2}\bar{z}\right\}, \\ &= \frac{\hbar^2}{2Ml^2}\{a, a^\dagger\} \end{aligned} \tag{2.1.6}$$

where $\Re$ and $\Im$ denote the real and imaginary part, respectively. In the last line, we have introduced the ladder operators [12–14] [1]

$$a = \frac{l}{\sqrt{2}}\left(2\frac{\partial}{\partial\bar{z}} + \frac{1}{2l^2}z\right), \quad a^\dagger = \frac{l}{\sqrt{2}}\left(-2\frac{\partial}{\partial z} + \frac{1}{2l^2}\bar{z}\right), \tag{2.1.7}$$

which obey

$$\left[a, a^\dagger\right] = 1. \tag{2.1.8}$$

[1] We have not been able to find out who introduced the ladder operators for Landau levels in the plane. The energy eigenfunctions were known since Landau [11]. MacDonald [13] used the ladder operators in 1984, but neither gave nor took credit. Girvin and Jach [14] were aware of two independent ladders a year earlier, but neither spelled out the formalism, nor pointed to references. It appears that the community had been aware of them, but not aware of who introduced them. The clearest and most complete presentation we know of is due to Arovas [12].

With the cyclotron frequency $\omega_c = eB/Mc$ and (2.1.8) we finally obtain

$$H = \hbar\omega_c\left(a^\dagger a + \frac{1}{2}\right). \tag{2.1.9}$$

The kinetic energy of charged particles in a perpendicular magnetic field is hence quantized like a harmonic oscillator. The energy levels are called Landau levels.

It is convenient to write the ladder operators describing the cyclotron variables as

$$a = +\sqrt{2}l\exp\left(-\frac{1}{4l^2}\bar{z}z\right)\frac{\partial}{\partial\bar{z}}\exp\left(+\frac{1}{4l^2}\bar{z}z\right), \tag{2.1.10}$$

$$a^\dagger = -\sqrt{2}l\exp\left(+\frac{1}{4l^2}\bar{z}z\right)\frac{\partial}{\partial z}\exp\left(-\frac{1}{4l^2}\bar{z}z\right), \tag{2.1.11}$$

and introduce a second set of ladder operators for the guiding center variables,

$$b = +\sqrt{2}l\exp\left(-\frac{1}{4l^2}\bar{z}z\right)\frac{\partial}{\partial z}\exp\left(+\frac{1}{4l^2}\bar{z}z\right), \tag{2.1.12}$$

$$b^\dagger = -\sqrt{2}l\exp\left(+\frac{1}{4l^2}\bar{z}z\right)\frac{\partial}{\partial\bar{z}}\exp\left(-\frac{1}{4l^2}\bar{z}z\right). \tag{2.1.13}$$

They likewise obey

$$\left[b, b^\dagger\right] = 1, \tag{2.1.14}$$

and commute with the cyclotron ladder operators:

$$\left[a, b\right] = \left[a, b^\dagger\right] = 0 \tag{2.1.15}$$

A calculation similar to the one presented above for H yields

$$\boldsymbol{L} = \boldsymbol{r} \times \boldsymbol{p} = \hbar\left(b^\dagger b - a^\dagger a\right)\boldsymbol{e}_z \tag{2.1.16}$$

for the *canonical* angular momentum around the origin. [The *kinematical* angular momentum is given by the $a^\dagger a$ term in (2.1.16).]

Since the angular momentum (2.1.16) commutes with the Hamiltonian (2.1.9), we can use it to classify the vastly degenerate states within each Landau level. Specifically, we introduce the basis states

$$|n, m\rangle = \frac{1}{\sqrt{n!}}\frac{1}{\sqrt{m!}}(a^\dagger)^n(b^\dagger)^m|0, 0\rangle, \tag{2.1.17}$$

where the vacuum state is by definition annihilated by both destruction operators,

$$a|0, 0\rangle = b|0, 0\rangle = 0. \tag{2.1.18}$$

Solving (2.1.18) yields the real space representation

$$\phi_0(z) \equiv \phi_0(z, \bar{z}) = \langle \boldsymbol{r}|0, 0\rangle = \frac{1}{\sqrt{2\pi l^2}} \exp\left(-\frac{1}{4l^2}|z|^2\right). \tag{2.1.19}$$

(In the following, we omit $\bar{z}$ from the argument of wave functions as a choice of notation.) The basis states (2.1.17) are trivially eigenstates of both H and L_z,

$$\begin{aligned} H|n, m\rangle &= \hbar\omega_c\left(n + \frac{1}{2}\right) \\ L_z|n, m\rangle &= \hbar(m - n)|n, m\rangle \end{aligned} \tag{2.1.20}$$

The particle coordinate and momentum are given in terms of the ladder operators by

$$z = \sqrt{2}l(a + b^{\dagger}), \quad p = -\frac{i\hbar}{\sqrt{2}l}(a - b^{\dagger}). \tag{2.1.21}$$

This implies that we can write a complete, orthonormal set of basis states in the lowest Landau level ($n = 0$) as

$$\begin{aligned} \phi_m(z) &= \langle \boldsymbol{r}|0, m\rangle \\ &= \frac{1}{\sqrt{m!}}(b^{\dagger})^m \phi_0(z, \bar{z}) \\ &= \frac{1}{\sqrt{2\pi l^2 m!}}(a + b^{\dagger})^m \exp\left(-\frac{1}{4l^2}|z|^2\right) \\ &= \frac{1}{\sqrt{2^{m+1}\pi m!\, l^{m+1}}}\, z^m \exp\left(-\frac{1}{4l^2}|z|^2\right). \end{aligned} \tag{2.1.22}$$

These states is describe narrow rings centered around the origin, with the radius determined by

$$\frac{\partial}{\partial r}\left|\phi_m(r)^2\right|\Bigg|_{r=r_m} \stackrel{!}{=} 0,$$

which yields $r_m = \sqrt{2m}l$. Since there are also m states inside the ring, the areal degeneracy is

$$\frac{\text{number of states}}{\text{area}} = \frac{m}{\pi r_m^2} = \frac{1}{2\pi l^2}, \tag{2.1.23}$$

The magnetic flux required for each state,

$$2\pi l^2 B = \frac{2\pi\hbar c}{e} = \Phi_0,$$

is hence given by the Dirac flux quantum. This implies that in each Landau level, there are as many single particle states in a given area as there are Dirac quanta of magnetic flux going through it. In the following, we set $l = 1$, and no longer keep track of wave function normalizations.

The N particle wave function for a filled lowest Landau level (LLL) on a circular disk is obtained by antisymmetrizing the basis states (2.1.22),

$$\begin{aligned}\psi(z_1, \ldots, z_N) &= \mathcal{A}\left\{z_1^0 z_2^1 \ldots z_N^{N-1}\right\} \cdot \prod_{i=1}^{N} e^{-\frac{1}{4}|z_i|^2} \\ &= \prod_{i<j}^{N} (z_i - z_j) \prod_{i=1}^{N} e^{-\frac{1}{4}|z_i|^2}.\end{aligned} \tag{2.1.24}$$

The most general form for the single particle wave function in the lowest Landau level is

$$\psi(z) = f(z)\, e^{-\frac{1}{4}|z|^2}, \tag{2.1.25}$$

where $f(z)$ is an analytic function of z. Since $\psi(z)$ is annihilated by the destruction operator a, the energy is trivially $\frac{1}{2}\hbar\omega_c$. The most general N particle state in the LLL is given by

$$\psi(z_1, \ldots, z_N) = f(z_1, \ldots, z_N) \prod_{i=1}^{N} e^{-\frac{1}{4}|z_i|^2}, \tag{2.1.26}$$

where $f(z_1, \ldots, z_N)$ is analytic in all the z's, and symmetric or antisymmetric for bosons or fermions, respectively. If we impose periodic boundary conditions [15], we find that $\psi(z_1, z_2, \ldots, z_N)$, when viewed as a function of z_1 while $z_2, \ldots, z_N$ are parameters, has exactly as many zeros as there states in the LLL, i.e. as there are Dirac flux quanta going through the unit cell or principal region. If $\psi(z_1, \ldots, z_N)$ describes fermions and is hence antisymmetric, there will be at least one zero seen by z_1 at each of the other particle positions. The most general wave function is hence

$$\psi(z_1, \ldots, z_N) = P(z_1, \ldots, z_N) \prod_{i<j}^{N} (z_i - z_j) \prod_{i=1}^{N} e^{-\frac{1}{4}|z_i|^2}, \tag{2.1.27}$$

where P is a symmetric polynomial in the z_i's. In the case of a completely filled Landau level, there are only as many zeros as there are particles, which implies that all except one of the zeros in z_1 will be located at the other particle positions $z_2, \ldots, z_N$. This yields (2.1.24) as the unique state for open boundary conditions. For periodic boundary conditions, there is one additional zero as there cannot be a zero seen by z_1 at z_1. The location of this zero, which Haldane and Rezayi [15] refer to as the center-of-mass zero, encodes the information about the boundary phases a test particle acquires as it is taken around one of the meridians of the torus.

To elevate the most general LLL state (2.1.26) into the $(n+1)$-th Landau level, we only have to apply $\left(a^{\dagger}\right)^{n}$ to all the particles in the LLL,

$$\begin{aligned}\psi_n(z_1,\ldots,z_N) &= \prod_{i=1}^{N}\left(a_i^{\dagger}\right)^{n}\psi(z_1,\ldots,z_N)\\ &= \prod_{i=1}^{N} e^{-\frac{1}{4}|z_i|^2}\prod_{i=1}^{N}\left(2\frac{\partial}{\partial z_i}-\bar{z}_i\right)^{n} f(z_1,\ldots,z_N).\end{aligned} \tag{2.1.28}$$

The energy per particle in this state is $\hbar\omega_{\mathrm{c}}\left(n+\frac{1}{2}\right)$.

2.1.2 The Laughlin State

The experimental observation which Laughlin's theory [1] explains is a plateau in the Hall resistivity of a two-dimensional electron gas at a Landau level filling fraction $\nu = 1/3$. The filling fraction denotes the number of particles divided by the number of number of states in each Landau level in the thermodynamic limit, and is defined through

$$\frac{1}{\nu} = \frac{\partial N_{\Phi}}{\partial N}, \tag{2.1.29}$$

where N_{Φ} is the number of Dirac flux quanta through the sample and N is the number of particles. For a wave function at $\nu = 1/3$, we consequently have three times as many zeros seen by z_1 as there are particles, and the polynomial $P(z_1, z_2, \ldots, z_N)$ in (2.1.27) has two zeros per particle. The experimental findings, as well as early numerical work by Yoshioka, Halperin, and Lee [16], are consistent with, if not indicative of, a quantum liquid state at a preferred filling fraction $\nu = 1/3$. Since the kinetic energy is degenerate in each Landau level, such a liquid has to be stabilized by the repulsive Coulomb interactions between the electrons. This implies that the wave function should be highly effective in suppressing configurations in which particles approach each other, as there is a significant potential energy cost associated with it. We may hence ask ourselves whether there is any particular way of efficiently distributing the zeros of $P(z_1, z_2, \ldots, z_N)$ in this regard.

Laughlin's wave function amounts to attaching the additional zeros onto the particles, such that each particle coordinate $z_2, \ldots, z_N$ becomes a triple zero of z_1 when $\psi(z_1, z_2, \ldots, z_N)$ is viewed as a function of z_1 with parameter $z_2, \ldots, z_N$. For filling fraction $\nu = 1/m$, where m is an odd integer if the particles are fermions and an even integer if they are bosons, he proposed the ground state wave function

$$\psi_m(z_1, \ldots, z_N) = \prod_{i<j}^{N} (z_i - z_j)^m \prod_{i=1}^{N} e^{-\frac{1}{4}|z_i|^2}. \tag{2.1.30}$$

There are hence no zeros wasted—all of them contribute in keeping the particles away from each other effectively, as ψ_m vanishes as the mth power of the distance when two particles approach each other. This is the uniquely defining property of Laughlin's state, and also the property which enabled Haldane [3] to identify a parent Hamiltonian, which singles out the state as its unique and exact ground state. We discuss the Hamiltonian in Sect. 2.1.6 below. The wave function (2.1.30) describes an incompressible quantum liquid, as the construction is only possible at filling fractions $\nu = 1/m$.

One of the assumptions of the theory is that we can neglect transitions into higher Landau levels, as the Landau level splitting $\hbar\omega_c$ is much larger then the potential energy per particle, a condition met by the systems amenable to experiment. Formally, the LLL limit requires $\omega_c \to \infty$ while keeping the magnetic length l^2 constant, which is achieved by taking $M \to 0$. The LLL limit is hence a zero mass limit.

Even within this limit, which we assume to hold in the following, the Laughlin state (2.1.30) is not the exact ground state for electrons with (screened) Coulomb interactions at filling fraction $\nu = 1/3$. It is, however, reasonably close in energy and has a significant overlap with the exact ground state for finite systems. The difference between the exact ground state and Laughlin's state is that in the exact ground state, the zeros of $P(z_1, z_2, \ldots, z_N)$ are attached to the particle coordinates, but do not coincide with them [2, 17]. At long distances, the physics described by both states is identical. In particular, the topological quantum numbers of both states, such as the charge and the statistics of the (fractionally) charged excitations, or the degeneracies on closed surfaces of genus one and higher, are identical.

The Laughlin state can be characterized through the notion of "superfermions" [18]. For fermions (bosons), the relative angular momentum is quantized as $\hbar l$, where l is an odd (even) integer, due to the antisymmetry (symmetry) of the wave function under interchange of particles. In the LLL, the relative angular momentum between pairs of fermions can only have components with $l = 1, 3, 5, \ldots$, but no negative values. If we interchange the particles through winding them counterclockwise around each other, these components acquire a phase factor $e^{i\pi l}$. The smallest component hence acquires a phase π, as required by Fermi statistics. For the Laughlin state (2.1.30), the smallest component of relative angular momentum is $l = m$, and the phase this component acquires upon interchange is $m\pi$, while only a phase π is required by Fermi statistics. In this sense, the particles are "superfermions" for m odd, $m > 1$. In the exact ground state for Coulomb interaction, the electrons are "approximate superfermions".

For completeness, we wish to mention that there is a variant of Haldane's parent Hamiltonian [3] for the planar geometry, due to Trugman and Kivelson [19]. They noted that since the Laughlin state (2.1.30) contains a term $(z_i - z_j)^m$ for each pair, it is annihilated by the short range potential interaction

$$V^{(m)} = \sum_{i<j}^{N} \left(\nabla_i^2\right)^{(m-1)/2} \delta^{(2)}(z_i - z_j) \tag{2.1.31}$$

for m odd, and

$$V^{(m)} = \sum_{i<j}^{N} \left(\nabla_i^2\right)^{(m-2)/2} \delta^{(2)}(z_i - z_j) \tag{2.1.32}$$

for m even, as well as by the same terms with any smaller power of the Laplacian. If we combine these terms with the kinetic terms (2.1.9), the resulting Hamiltonian will single out (2.1.30) as the exact and unique ground state.

2.1.3 Fractionally Charged Quasiparticle Excitations

Laughlin [1] created the elementary, charged excitations of the fractionally quantized Hall state (2.1.30) through a *Gedankenexperiment*. If one adiabatically inserts one Dirac quantum of magnetic flux through an infinitely thin solenoid at a position ξ, and then removes this flux quanta via a singular gauge transformation, the final Hamiltonian will be identical to the initial one. The final state will hence be an eigenstate of the initial Hamiltonian as well. The adiabatic insertion of the flux will induce an electric field

$$\oint \boldsymbol{E} \mathrm{d}\boldsymbol{s} = E_\varphi \cdot 2\pi r = -\frac{1}{c}\frac{\partial \phi}{\partial t}, \tag{2.1.33}$$

which in turn will change the canonical angular momentum L_z around ξ by

$$\Delta L_z = \int F_\varphi \cdot r \mathrm{d}t = \frac{e}{2\pi c} \int \frac{\partial \phi}{\partial t} \mathrm{d}t = \frac{e}{2\pi c} \cdot \phi_0 = \hbar. \tag{2.1.34}$$

If we choose a basis of eigenstates of angular momentum around ξ, the basis states evolve according to

$$(z-\xi)^m e^{-\frac{1}{4}|z|^2} \rightarrow (z-\xi)^{m+1} e^{-\frac{1}{4}|z|^2}. \tag{2.1.35}$$

Note that the kinematical angular momentum, which is given by the second term in (2.1.16), has eigenvalue $-\hbar n$, where n labels the Landau level. In this process, it remains zero as the states remain in the lowest Landau level—as there are no states with positive kinematical angular momentum, the insertion of the flux just shifts the states within the LLL.

The Laughlin ground state (2.1.30) evolves in the process into

$$\psi_\xi^{\mathrm{QH}}(z_1, \ldots, z_N) = \prod_{i=1}^{N} (z_i - \xi) \prod_{i<j}^{N} (z_i - z_j)^m \prod_{i=1}^{N} e^{-\frac{1}{4}|z_i|^2}, \tag{2.1.36}$$

which describes a quasihole excitation at ξ. It is easy to see that if the electron charge is $-e$, the charge of the quasihole is $+e/m$. If we were to create m quasiholes at ξ by inserting m Dirac quanta, the final wave function would be

$$\psi_{\xi}^{m\mathrm{QH's}}(z_1,\ldots,z_N)=\prod_{i=1}^{N}(z_i-\xi)^m\prod_{i<j}^{N}(z_i-z_j)^m\prod_{i=1}^{N}e^{-\frac{1}{4}|z_i|^2}, \tag{2.1.37}$$

i.e., we would have created a true hole in the liquid, which is screened as all the other particles. Since the hole has charge $+e$, the quasihole has charge $+e/m$. One may view the quasihole as a zero in the wave function which is not attached to any of the electrons.

The quasielectron, i.e. the antiparticle of the quasihole, has charge $-e/m$ and is created by inserting the flux adiabatically in the opposite direction, thus lowering the angular momentum around some position ξ by $\hbar$, or alternatively, by removing one of the zeros from the wave function. To accomplish this formally, we first rewrite (2.1.36) in terms of ladder operators:

$$\psi_{\xi}^{\mathrm{QH}}(z_1,\ldots,z_N)=\prod_{i=1}^{N}\left(\sqrt{2}b_i^{\dagger}-\xi\right)\prod_{i<j}^{N}(z_i-z_j)^m\prod_{i=1}^{N}e^{-\frac{1}{4}|z_i|^2}. \tag{2.1.38}$$

The insertion of a flux quanta in the opposite direction, or the lowering of angular momentum around ξ, will then correspond to the Hermitian conjugate operation. Laughlin [4] hence proposed for the quasielectron wave function

$$\begin{aligned}\psi_{\xi}^{\mathrm{QE}}(z_1,\ldots,z_N)&=\prod_{i=1}^{N}\left(\sqrt{2}b_i-\bar{\xi}\right)\prod_{i<j}^{N}(z_i-z_j)^m\prod_{i=1}^{N}e^{-\frac{1}{4}|z_i|^2}\\&=\prod_{i=1}^{N}e^{-\frac{1}{4}|z_i|^2}\prod_{i=1}^{N}\left(2\frac{\partial}{\partial z_i}-\bar{\xi}\right)\prod_{i<j}^{N}(z_i-z_j)^m.\end{aligned} \tag{2.1.39}$$

While the quasihole excitation (2.1.36) is still an exact eigenstate of Haldane's parent Hamiltonian, this is not true for the quasielectron (2.1.39). The problem here is that while there is a clean and unique way of introducing an additional zero (we just put it somewhere), there is no such clean way of removing one. One can view the quasielectron as a region, in which n electrons nearby share $2n-1$ zeros attached to the particles. In other words, one zero is missing, but not from any specific electron—rather, the dearth is distributed among all the electrons nearby. The charge of the quasielectron is accordingly not as localized as it is for the quasihole.

The plateau in the observed Hall resistivity occurs because the current in the experiments is carried by edge states, which are sensitive only to the topological quantum numbers of the state. In the vicinity of one of the prefered filling fractions $\nu=1/m$, the excess density of electrons yields to a finite density of quasielectrons or holes, which get pinned by disorder and hence do not contribute to the transport properties.

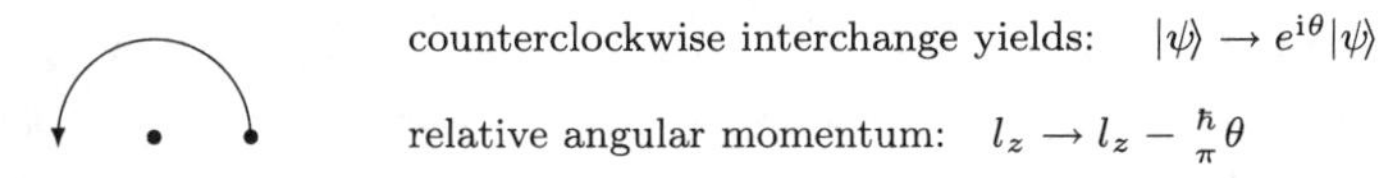

Fig. 2.1 Fractional statistics in two dimensions. The many particle wave function acquires a statistical phase θ whenever we interchange two anyons conterclockwise

2.1.4 Fractional Statistics

Possibly the most interesting property of fractionally quantized Hall states is that the quasiparticle excitations obey fractional statistics [20–22]. The possibility of fractional statistics [23–31] arises in two space dimensions because the space of trajectories for two identical particles consists of an infinite number of topologically distinct sectors, corresponding to the number of times the particles wind around each other. The laws of quantum mechanics allow us to assign distinct phases to paths belonging these sectors, which only need to satisfy the composition principle.

In three or more dimensions, by contrast, there are only two topological distinct sectors, corresponding to interchanging the particles or not interchanging them. The group which classifies all the topologically distinct trajectories is hence the permutation group, and since amplitudes are complex numbers, the possibilities for the quantum statistics are limited to the one-dimensional representations of the permutation group. There are only two such representations, the symmetric and the antisymmetric representation. These correspond to the familiar choices of Bose and Fermi statistics.

In two dimensions, the group is the braid group. The one-dimensional representations are obtained by assigning an arbitrary phase $\tau(T_i) = e^{i\theta}$ for each counterclockwise interchange T_i of the two particles, with statistical parameter $\theta \in]-\pi, \pi]$. Particles interpolating between the familiar choices of bosons ($\theta = 0$) and fermions ($\theta = \pi$) are generically called anyons. We will see in Sect. 2.3.3 that non-Abelian generalizations exist, where successive interchanges of anyons do not commute.

The most direct physical manifestation of the fractional statistics is the quantization of the relative angular momentum of the anyons (see Fig. 2.1). In three dimensions, there are three generators of rotations, and the relative angular momentum is quantized as $\hbar l$, with l an even integer for bosons and an odd integer l odd for fermions. In two dimensions, the wave function may acquire a phase $\exp\left(\frac{i}{\pi}\theta\varphi\right)$ as two anyons wind counterclockwise around each other with winding angle φ, which implies that the relative angular momentum is quantized as

$$L_{\text{rel}} = \hbar\left(-\frac{\theta}{\pi} + 2n\right), \tag{2.1.40}$$

where n is an integer. Note that the possibility of fractional statistics exists only for particles which are strictly two-dimensional, like vortices in an (approximately) two-dimensional quantum fluid.

The only established realization of fractional statistics is provided by the quasiparticles in the fractionally quantized Hall effect [20–22]. When Laughlin introduced the quasiparticles, he introduced them as localized defects or more precisely, vortices in an otherwise uniform quantum liquid. To address the question of their statistics, however, it is propitious to view them as particles, with a Hilbert space spanned by the parent wave function for the electrons. We consider here a Laughlin state with two quasiholes in an eigenstate of relative angular momentum in an "orbit" centered at the origin. Since the quasiholes have charge $e^* = +e/m$, the effective flux quantum seen by them is

$$\Phi_0^* = \frac{2\pi\hbar c}{e^*} = m\Phi_0, \tag{2.1.41}$$

and the effective magnetic length is

$$l^* = \sqrt{\frac{\hbar c}{e^* B}} = \sqrt{m}l. \tag{2.1.42}$$

We expect the single quasihole wave function to describe a particle of charge e^* in the LLL, and hence be of the general form

$$\phi(\bar{\xi}) = f(\bar{\xi})e^{-\frac{1}{4m}|\xi|^2}. \tag{2.1.43}$$

The complex conjugation reflects that the sign of the quasihole charge is reversed relative to the electron charge $-e$.

The electron wave function for the state with two quasiholes in an eigenstate of relative angular momentum is given by

$$\psi(z_1,\ldots,z_N) = \int \mathrm{D}[\xi_1,\xi_2]\phi_{p,m}(\bar{\xi}_1,\bar{\xi}_2)\psi^{\mathrm{QHs}}_{\xi_1,\xi_2}(z_1,\ldots,z_N) \tag{2.1.44}$$

with

$$\phi_{p,m}(\bar{\xi}_1,\bar{\xi}_2) = (\bar{\xi}_1-\bar{\xi}_2)^{p+\frac{1}{m}} \prod_{k=1,2} e^{-\frac{1}{4m}|\xi_k|^2}, \tag{2.1.45}$$

where p is an even integer, and

$$\begin{aligned}\psi^{\mathrm{QHs}}_{\xi_1,\xi_2}(z_1,\ldots,z_N) &= (\xi_1-\xi_2)^{\frac{1}{m}} \prod_{k=1,2} e^{-\frac{1}{4m}|\xi_k|^2} \\ &\quad\cdot \prod_{i=1}^{N}(z_i-\xi_1)(z_i-\xi_2)\prod_{i<j}^{N}(z_i-z_j)^m \prod_{i=1}^{N} e^{-\frac{1}{4}|z_i|^2}.\end{aligned} \tag{2.1.46}$$

The quasihole coordinate integration extends over the complex plane,

$$\int \mathrm{D}[\xi_1,\xi_2] \equiv \int\ldots\int \mathrm{d}x_1\mathrm{d}y_1\mathrm{d}x_2\mathrm{d}y_2,$$

where $\xi_1 = x_1 + \mathrm{i}y_1$ and $\xi_2 = x_2 + \mathrm{i}y_2$.

This needs explanation. We see that both $\phi_{p,m}(\bar{\xi}_1, \bar{\xi}_2)$ and $\psi^{\text{QHs}}_{\xi_1,\xi_2}(z_1, \ldots, z_N)$ contain multiple valued functions of $\bar{\xi}_1 - \bar{\xi}_2$ and $\xi_1 - \xi_2$, respectively, while the product of them is understood to be single valued. The reason for this is that the Hilbert space for the quasiholes at ξ_1 and ξ_2 spanned by $\psi^{\text{QHs}}_{\xi_1,\xi_2}(z_1, \ldots, z_N)$ has to be normalized and is, apart from the exponential, supposed to be analytic in ξ_1 and ξ_2. At the same time, we expect $\phi_{p,m}(\bar{\xi}_1, \bar{\xi}_2)$ to be of the general form (2.1.41), i.e. to be an analytic function of $\bar{\xi}_1, \bar{\xi}_2$ times the exponential.

The form (2.1.45) of the quasihole wave function including its branch cut, is indicative of fractional statistics with statistical parameter $\theta = \pi/m$. This indication, however, is by itself not conclusive, as it is possible to change the representation of the wave function through singular gauge transformations [20, 21, 25], where one removes or adds flux tubes with a fraction of a Dirac flux quanta to the particles, and hence turn an anyonic representation into a bosonic or fermionic one and vice versa. The physically unambivalent quantity is the relative angular momentum of the quasiholes, which for (2.1.45) is given by

$$L_{\text{rel}} = -\hbar\left(p + \frac{1}{m}\right). \tag{2.1.47}$$

Comparing this with (2.1.40) yields $\theta = \pi/m$. This result agrees with the results of Halperin [20, 21] and of Arovas, Schrieffer, and Wilczek [22], who calculated the statistical parameter directly using the adiabatic theorem [32–35].

2.1.5 Landau Level Quantization in the Spherical Geometry

The formalism for Landau level quantization in a spherical geometry, i.e., for the dynamics of a charged particle on the surface of a sphere with radius R, in a magnetic (monopole) field, was pioneered by Haldane for the lowest Landau level [3, 36], and only very recently generalized to higher Landau levels [37]. We will content ourselves here with a review of the formalism for the lowest Landau level.

Following Haldane [3], we assume a radial magnetic field of strength

$$B = \frac{\hbar c s_0}{eR^2} \quad (e > 0). \tag{2.1.48}$$

The number of magnetic Dirac flux quanta through the surface of the sphere is

$$\frac{\Phi_{\text{tot}}}{\Phi_0} = \frac{4\pi R^2 B}{2\pi\hbar c/e} = 2s_0, \tag{2.1.49}$$

which must be integer due to Dirac's monopole quantization condition [38]. In the following, we take $\hbar = c = 1$.

The Hamiltonian is given by

$$H = \frac{\mathbf{\Lambda}^2}{2MR^2} = \frac{\omega_\mathrm{c}}{2s_0}\mathbf{\Lambda}^2, \tag{2.1.50}$$

where $\omega_\mathrm{c} = eB/M$ is the cyclotron frequency,

$$\mathbf{\Lambda} = \boldsymbol{r} \times (-\mathrm{i}\nabla + e\boldsymbol{A}(\boldsymbol{r})) \tag{2.1.51}$$

is the dynamical angular momentum, $\boldsymbol{r} = R\boldsymbol{e}_\mathrm{r}$, and $\nabla \times \boldsymbol{A} = B\boldsymbol{e}_\mathrm{r}$. With (A.4)–(A.6) from Appendix A we obtain

$$\mathbf{\Lambda} = -\mathrm{i}\left(\boldsymbol{e}_\varphi \frac{\partial}{\partial\theta} - \boldsymbol{e}_\theta \frac{1}{\sin\theta}\frac{\partial}{\partial\varphi}\right) + eR(\boldsymbol{e}_\mathrm{r} \times \boldsymbol{A}(\boldsymbol{r})). \tag{2.1.52}$$

Note that

$$\boldsymbol{e}_\mathrm{r}\mathbf{\Lambda} = \mathbf{\Lambda}\boldsymbol{e}_\mathrm{r} = 0, \tag{2.1.53}$$

as one can easily verify with (A.5). The commutators of the Cartesian components of $\mathbf{\Lambda}$ with themselves and with $\boldsymbol{e}_\mathrm{r}$ can easily be evaluated using (2.1.52) and (A.3)–(A.5). This yields

$$\left[\Lambda^i, \Lambda^j\right] = \mathrm{i}\varepsilon^{ijk}(\Lambda^k - s_0\, e_\mathrm{r}^k), \tag{2.1.54}$$

$$\left[\Lambda^i, e_\mathrm{r}^j\right] = \mathrm{i}\epsilon^{ijk} e_\mathrm{r}^k, \tag{2.1.55}$$

where $i, j, k = \mathrm{x}, \mathrm{y}$, or z, and e_r^k is the kth Cartesian coordinate of $\boldsymbol{e}_\mathrm{r}$. From (2.1.53)–(2.1.55), we see that that the operator

$$\boldsymbol{L} = \mathbf{\Lambda} + s_0\boldsymbol{e}_\mathrm{r} \tag{2.1.56}$$

is the generator of rotations around the origin,

$$\left[L^i, X^j\right] = \mathrm{i}\epsilon^{ijk}X^k \quad \text{with} \quad \boldsymbol{X} = \mathbf{\Lambda}, \boldsymbol{e}_\mathrm{r}, \text{ or } \boldsymbol{L}, \tag{2.1.57}$$

and hence the angular momentum. As it satisfies the angular momentum algebra, it can be quantized accordingly. Note that $\boldsymbol{L}$ has a component in the $\boldsymbol{e}_\mathrm{r}$ direction:

$$\boldsymbol{L}\boldsymbol{e}_\mathrm{r} = \boldsymbol{e}_\mathrm{r}\boldsymbol{L} = s_0. \tag{2.1.58}$$

If we take the eigenvalue of $\boldsymbol{L}^2$ to be $s(s+1)$, this implies $s = s_0 + n$, where $n = 0, 1, 2, \ldots$ is a non-negative integer (while s and s_0 can be integer or half integer, according to number of Dirac flux quanta through the sphere).

With (2.1.56) and (2.1.53), we obtain

$$\Lambda^2 = L^2 - s_0^2. \tag{2.1.59}$$

The energy eigenvalues of (2.1.50) are hence

$$\begin{aligned} E_n &= \frac{\omega_c}{2s_0}\left[s(s+1) - s_0^2\right] \\ &= \frac{\omega_c}{2s_0}\left[(2n+1)s_0 + n(n+1)\right] \\ &= \omega_c\left[\left(n+\frac{1}{2}\right) + \frac{n(n+1)}{2s_0}\right]. \end{aligned} \tag{2.1.60}$$

The index n hence labels the Landau levels.

To obtain the eigenstates of (2.1.50), we have to choose a gauge and then explicitly solve the eigenvalue equation. We choose the latitudinal gauge

$$A = -e_\varphi \frac{s_0}{eR}\cot\theta. \tag{2.1.61}$$

The singularities of $B = \nabla \times A$ at the poles are without physical significance. They describe infinitly thin solenoids admitting flux $s_0\Phi_0$ each and reflect our inability to formulate a true magnetic monopole.

The dynamical angular momentum (2.1.52) becomes

$$\Lambda = -\mathrm{i}\left[e_\varphi \frac{\partial}{\partial\theta} - e_\theta \frac{1}{\sin\theta}\left(\frac{\partial}{\partial\varphi} - \mathrm{i}s_0\cos\theta\right)\right]. \tag{2.1.62}$$

With (A.5) we obtain

$$\Lambda^2 = -\frac{1}{\sin\theta}\frac{\partial}{\partial\theta}\left(\sin\theta\frac{\partial}{\partial\theta}\right) - \frac{1}{\sin^2\theta}\left(\frac{\partial}{\partial\varphi} - \mathrm{i}s_0\cos\theta\right)^2. \tag{2.1.63}$$

To formulate the eigenstates, Haldane [3] introduced spinor coordinates for the particle position,

$$u = \cos\frac{\theta}{2}\exp\left(\frac{i\varphi}{2}\right), \; v = \sin\frac{\theta}{2}\exp\left(-\frac{i\varphi}{2}\right), \tag{2.1.64}$$

such that

$$e_r = \Omega(u, v) \equiv (u, v)\sigma\begin{pmatrix}\bar{u}\\ \bar{v}\end{pmatrix}, \tag{2.1.65}$$

where $\sigma = (\sigma_x, \sigma_y, \sigma_z)$ is the vector consisting of the three Pauli matrices

$$\sigma_x = \begin{pmatrix}0 & 1\\ 1 & 0\end{pmatrix}, \quad \sigma_y = \begin{pmatrix}0 & -\mathrm{i}\\ \mathrm{i} & 0\end{pmatrix}, \quad \sigma_z = \begin{pmatrix}1 & 0\\ 0 & -1\end{pmatrix}. \tag{2.1.66}$$

In terms of these, a complete, orthogonal basis of the states spanning the lowest Landau level ($n = 0, s = s_0$) is given by

$$\psi^s_{m,0}(u,v) = u^{s+m}v^{s-m} \tag{2.1.67}$$

with

$$m = -s, s+1, \ldots, s.$$

For these states,

$$\begin{aligned} L^z\psi^s_{m,0} &= m\psi^s_{m,0}, \\ H\psi^s_{m,0} &= \frac{1}{2}\omega_c\psi^s_{m,0}. \end{aligned} \tag{2.1.68}$$

To verify (2.1.68), we consider the action of (2.1.63) on the more general basis states

$$\begin{aligned} \phi^s_{m,p}(u,v) &= \left(\cos\frac{\theta}{2}\right)^{s+m}\left(\sin\frac{\theta}{2}\right)^{s-m} e^{i(m-p)\varphi} \\ &= \begin{cases} \bar{v}^{-p}u^{s+m}v^{s-m+p}, & \text{for } p<0, \\ \bar{u}^{p}u^{s+m-p}v^{s-m}, & \text{for } p\geq 0. \end{cases} \end{aligned} \tag{2.1.69}$$

This yields

$$\begin{aligned} \boldsymbol{\Lambda}^2\phi^s_{m,p} &= \left[s - \left(\frac{s\cos\theta - m}{\sin\theta}\right)^2 + \left(\frac{s_0\cos\theta - m + p}{\sin\theta}\right)^2\right]\phi^s_{m,p} \\ &= \left[s + \frac{2(s\cos\theta - m + p)(p - n\cos\theta) - (p^2 - n^2\cos^2\theta)}{\sin^2\theta}\right]\phi^s_{m,p}, \end{aligned} \tag{2.1.70}$$

For $p=n=0$, this clearly reduces to $\boldsymbol{\Lambda}^2\psi^s_{m,0} = s\psi^s_{m,0}$, and hence (2.1.68). The normalization of (2.1.67) can easily be obtained with the integral

$$\frac{1}{4\pi}\int \mathrm{d}\Omega\, \bar{u}^{S+m'}\bar{v}^{S-m'}u^{s+m}v^{s-m} = \frac{(s+m)!(s-m)!}{(2s+1)!}\delta_{mm'}, \tag{2.1.71}$$

where $\mathrm{d}\Omega = \sin\theta\ \mathrm{d}\theta\ \mathrm{d}\phi$.

To describe particles in the lowest Landau level which are localized at a point $\boldsymbol{\Omega}(\alpha,\beta)$ with spinor coordinates (α,β),

$$\boldsymbol{\Omega}(\alpha,\beta) = (\alpha,\beta)\boldsymbol{\sigma}\begin{pmatrix}\bar{\alpha}\\ \bar{\beta}\end{pmatrix}, \tag{2.1.72}$$

Haldane [3] introduced "coherent states" defined by

$$\{\boldsymbol{\Omega}(\alpha,\beta)\boldsymbol{L}\}\psi^s_{(\alpha,\beta),0}(u,v) = s\psi^s_{(\alpha,\beta),0}(u,v). \tag{2.1.73}$$

In the lowest Landau level, the angular momentum $\boldsymbol{L}$ can be written

$$\boldsymbol{L} = \frac{1}{2}(u, v)\boldsymbol{\sigma} \begin{pmatrix} \dfrac{\partial}{\partial u} \\ \dfrac{\partial}{\partial v} \end{pmatrix}. \tag{2.1.74}$$

Note that u, v may be viewed as Schwinger boson creation, and $\frac{\partial}{\partial u}, \frac{\partial}{\partial v}$ the corresponding annihilation operators (see Sect. 2.4.3). The solutions of (2.1.73) are given by

$$\psi^{s}_{(\alpha,\beta),0}(u, v) = (\bar{\alpha}u + \bar{\beta}v)^{2s}, \tag{2.1.75}$$

as one can verify easily with the identity

$$(\underline{a}\,\boldsymbol{\sigma}\,\underline{b})(\underline{c}\,\boldsymbol{\sigma}\,\underline{d}) = 2(\underline{a}\,\underline{d})(\underline{c}\,\underline{b}) - (\underline{a}\,\underline{b})(\underline{c}\,\underline{d}). \tag{2.1.76}$$

where $\underline{a},\ \underline{b},\ \underline{c},\ \underline{d}$ are two-component spinors.

Haldane [3] further introduced two-particle coherent lowest Landau level states defined by

$$\{\boldsymbol{\Omega}(\alpha, \beta)(\boldsymbol{L}_1 + \boldsymbol{L}_2)\}\psi^{s,j}_{(\alpha,\beta),0}[u, v] = j\psi^{s,j}_{(\alpha,\beta),0}[u, v], \tag{2.1.77}$$

where $[u, v] := (u_1, u_2, v_1, v_2)$ and j is the total angular momentum quantum number,

$$(\boldsymbol{L}_1 + \boldsymbol{L}_2)^2\psi^{s,j}_{(\alpha,\beta),0}[u, v] = j(j+1)\psi^{s,j}_{(\alpha,\beta),0}[u, v]. \tag{2.1.78}$$

The solution of (2.1.77) is given by

$$\psi^{s,j}_{(\alpha,\beta),0}[u, v] = (u_1v_2 - u_2v_1)^{2s-j}\prod_{i=1,2}(\bar{\alpha}u_i + \bar{\beta}v_i)^j. \tag{2.1.79}$$

It describes two particles with relative momentum $2s - j$ precessing about their common center of mass at $\boldsymbol{\Omega}(\alpha, \beta)$.

Since $0 \leq j \leq 2s$, the relative momentum quantum number $l = 2s - j$ has to be a non-negative integer. The restriction to non-negative integers is a consequence of Landau level quantization, and exists in the plane as well, as we discussed in Sect. 2.1.2. For bosons or fermions, l has to be even or odd, respectively. This implies that the projection Π_0 into the lowest Landau level of any rotationally invariant operator $V(\boldsymbol{r}_1 \cdot \boldsymbol{r}_2)$, such as two particle interactions, can be expanded as

$$\Pi_0 V(\boldsymbol{r}_1 \cdot \boldsymbol{r}_2)\Pi_0 = \sum_{l}^{2s} V_l P_{2s-l}(\boldsymbol{L}_1 + \boldsymbol{L}_2), \tag{2.1.80}$$

where the sum over l is restricted to even (odd) integer for bosons (fermions), V_l denotes the so-called pseudopotential coefficients, and $P_j(\boldsymbol{L})$ is the projection operator on states with total momentum $\boldsymbol{L}^2 = j(j+1)$.

As mentioned, this formalism was very recently generalized to include higher Landau levels as well [37]. The key insight permitting this generalization was that there are two mutually commuting SU(2) algebras with spin s, one for the cyclotron variables and one for the guiding center variables. These algebras are analogous to the the two mutually commuting ladder algebras $a, a^\dagger$ and $b, b^\dagger$ in the plane, which we introduced in Sect. 2.1.1.

2.1.6 *The Laughlin State and Its Parent Hamiltonian on the Sphere*

In analogy to (2.1.30), Haldane [3] writes the Laughlin $\nu = 1/m$ state for N particles on a sphere with $2s_0 = m(N-1)$ as

$$\psi_m[u,v] = \prod_{i<j}^{N} (u_i v_j - u_j v_i)^m. \tag{2.1.81}$$

Since the factors $(u_i v_j - u_j v_i)$ commute with the total angular momentum

$$\boldsymbol{L}_{\text{tot}} = \sum_{i=1}^{N} \boldsymbol{L}_i, \tag{2.1.82}$$

(2.1.81) is obviously invariant under spacial rotations around the sphere:

$$\boldsymbol{L}_{\text{tot}} \psi_m = 0. \tag{2.1.83}$$

The Laughlin droplet wave function centered at $\boldsymbol{\Omega}(\alpha, \beta)$ can be recovered by multiplying $\psi_m[u,v]$ by a factor

$$\prod_{i=1}^{N} (\bar{\alpha} u_i + \bar{\beta} v_i)^n,$$

and then taking the limit $n \to \infty$, $R \to \infty$, while $4\pi R^2/n = 2\pi l^2 = \text{const.}$, where l^2 is the magnetic length (2.1.5).

As in the plane, the uniquely specifying property of the Laughlin state (2.1.81) is that the smallest component of relative angular momentum is m, which is even for bosons and odd for fermions. Haldane [3] constructed a model Hamiltonian, which, together with the kinetic Hamiltonian (2.1.50), singles out (2.1.81) as exact and unique zero energy ground state, by assigning a finite energy cost to the components of angular momentum smaller than m. With the most general two-particle interaction Hamiltonian given by

$$H_{\text{int}} = \sum_{i<j}^{N} \left\{ \sum_{l}^{2s} V_l \mathrm{P}_{2s-l}(\boldsymbol{L}_i + \boldsymbol{L}_j) \right\}, \tag{2.1.84}$$

where the values of l are restricted to even (odd) integers for bosons (fermions) and P_{2s-l} is as defined in (2.1.80), Haldane's Hamiltonian amounts to taking

$$V_l = \begin{cases} 1 & \text{for } l < m, \\ 0 & \text{for } l \geq m. \end{cases} \tag{2.1.85}$$

For all practical purposes, we need to rewrite (2.1.84) in terms of boson or fermion creation or annihilation operators,

$$\begin{aligned} H_{\text{int}} = & \sum_{m_1=-s}^{s} \sum_{m_2=-s}^{s} \sum_{m_3=-s}^{s} \sum_{m_4=-s}^{s} a_{m_1}^{\dagger} a_{m_2}^{\dagger} a_{m_3} a_{m_4} \delta_{m_1+m_2,m_3+m_4} \\ & \cdot \sum_{l=0}^{2s} \langle s, m_1; s, m_2 | 2s-l, m_1+m_2 \rangle V_l \langle 2s-l, m_3+m_4 | s, m_3; s, m_4 \rangle, \end{aligned} \tag{2.1.86}$$

where a_m annihilates a boson or fermion in the properly normalized single particle state

$$\psi_{m,0}^{s}(u, v) = \sqrt{\frac{(2s+1)!}{4\pi(s+m)!(s-m)!}}\, u^{s+m} v^{s-m}, \tag{2.1.87}$$

and $\langle s, m_1; s, m_2 | j, m_1+m_2 \rangle$ etc. are Clebsch–Gordan coefficients [39]. Essentially, we take two particles with L_z eigenvalues m_3 and m_4, change the basis into one where $m_3 + m_4$ and the total two particle momentum $2s - l$ are replacing the quantum numbers m_3 and m_4, multiply each amplitude by V_l, and convert the two particles states back into a basis of L_z eigenvalues m_1 and m_2.

The fractionally charged quasihole and quasielectron excitations of the Laughlin state (2.1.81) localized at $\boldsymbol{\Omega}(\alpha, \beta)$ on the sphere are given by

$$\psi_{(\alpha,\beta)}^{\text{QH}}[u, v] = \prod_{i=1}^{N} (\beta u_i - \alpha v_i) \prod_{i<j}^{N} (u_i v_j - u_j v_i)^m \tag{2.1.88}$$

and

$$\psi_{(\alpha,\beta)}^{\text{QE}}[u, v] = \prod_{i=1}^{N} \left(\bar{\beta} \frac{\partial}{\partial u_i} - \bar{\alpha} \frac{\partial}{\partial v_i} \right) \prod_{i<j}^{N} (u_i v_j - u_j v_i)^m, \tag{2.1.89}$$

which increase or decrease the number of flux quanta $2s_0$ through the sphere by one, and decrease or increase $\boldsymbol{\Omega}(\alpha, \beta) \boldsymbol{L}_{\text{tot}}$ by $\frac{1}{2}N$.

Due to the formal simplicity, the sphere is particularly well suited to formulate the hierarchy of quantized Hall states, where all odd-denominator filling fractions can be obtained through successive condensation of quasiparticles into Laughlin-type fluids [3, 20, 21, 40].

2.2 The Haldane–Shastry Model

2.2.1 The $1/r^2$ Model of Haldane and Shastry

The Haldane–Shastry model [41–53] is one of the most important paradigms for a generic spin $\frac{1}{2}$ liquid on a chain. Consider a spin $\frac{1}{2}$ chain with periodic boundary conditions and an even number of sites N on a unit circle embedded in the complex plane:

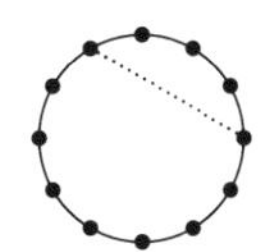

N sites with spin $\frac{1}{2}$ on unit circle:

$$\eta_\alpha = e^{\mathrm{i}\frac{2\pi}{N}\alpha} \quad \text{with } \alpha = 1, \ldots, N$$

The $1/r^2$-Hamiltonian

$$H^{\mathrm{HS}} = \left(\frac{2\pi}{N}\right)^2 \sum_{\alpha<\beta}^{N} \frac{\boldsymbol{S}_\alpha \boldsymbol{S}_\beta}{|\eta_\alpha - \eta_\beta|^2}, \tag{2.2.1}$$

where $|\eta_\alpha - \eta_\beta|$ is the chord distance between the sites α and β, has the exact ground state

$$|\psi_0^{\mathrm{HS}}\rangle = \sum_{\{z_1,\ldots,z_M\}} \psi_0^{\mathrm{HS}}(z_1,\ldots,z_M) S_{z_1}^+ \cdot \ldots \cdot S_{z_M}^+ |\underbrace{\downarrow\downarrow\ldots\ldots\downarrow}_{\text{all } N \text{ spins } \downarrow}\rangle, \tag{2.2.2}$$

where the sum extends over all possible ways to distribute the $M = \frac{N}{2}$ $\uparrow$-spin coordinates z_i on the unit circle and

$$\psi_0^{\mathrm{HS}}(z_1, z_2, \ldots, z_M) = \prod_{i<i}^{M} (z_i - z_j)^2 \prod_{i=1}^{M} z_i. \tag{2.2.3}$$

The ground state has momentum

$$p_0 = -\frac{\pi}{2} N, \tag{2.2.4}$$

where we have adopted a convention according to which the "vacuum" state $|\downarrow\downarrow\ \ldots\ \downarrow\rangle$ has momentum $p = 0$ (and the empty state $|0\rangle$ has $p = \pi(N-1)$) and energy

$$E_0 = -\frac{\pi^2}{24}\left(N + \frac{5}{N}\right). \tag{2.2.5}$$

We will verify (2.2.4) and (2.2.5) in Sects. 2.2.3 and 2.2.4, respectively.

2.2.2 Symmetries and Integrability

The Haldane–Shastry Hamiltonian (2.2.1) is clearly invariant under space translations (rotations of the unit circle), time reversal, parity, and global SU(2) spin rotations generated by

$$\boldsymbol{S}_{\text{tot}} = \sum_{\alpha=1}^{N} \boldsymbol{S}_\alpha, \quad \left[H^{\text{HS}}, \boldsymbol{S}_{\text{tot}}\right] = 0. \tag{2.2.6}$$

The total spin trivially satisfies the standard commutation relations for angular momentum,

$$\left[S^i_{\text{tot}}, S^j_{\text{tot}}\right] = \mathrm{i}\, \varepsilon^{ijk} S^k_{\text{tot}}. \tag{2.2.7}$$

The model possesses an additional symmetry [46, 54] generated by the rapidity operator

$$\boldsymbol{\Lambda} = \frac{\mathrm{i}}{2} \sum_{\substack{\alpha,\beta=1 \\ \alpha\neq\beta}}^{N} \frac{\eta_\alpha + \eta_\beta}{\eta_\alpha - \eta_\beta} \boldsymbol{S}_\alpha \times \boldsymbol{S}_\beta, \quad \left[H^{\text{HS}}, \boldsymbol{\Lambda}\right] = 0, \tag{2.2.8}$$

which measures the spin current. It transforms as a vector under spin rotations,

$$\left[S^i_{\text{tot}}, \Lambda^j\right] = \mathrm{i}\, \varepsilon^{ijk} \Lambda^k. \tag{2.2.9}$$

Note that even though both $\boldsymbol{S}_{\text{tot}}$ and $\boldsymbol{\Lambda}$ commute with the Hamiltonian, they do not commute mutually, but generate an infinite dimensional associative algebra with certain defining relations and consistency conditions, the Yangian $\mathrm{Y}(\mathrm{sl}_2)$ [55, 56]. Since the commutator of the total spin squared with the rapidity operator does not vanish in general,

$$\left[\boldsymbol{S}^2_{\text{tot}}, \Lambda^i\right] = -\mathrm{i}\, \varepsilon^{ijk} \left\{S^j_{\text{tot}}, \Lambda^k\right\}, \tag{2.2.10}$$

elements of the Yangian algebra connect degenerate eigenstates with different total spins. With these elements, it is possible to generate all the eigenstates of the model from all the completely spin polarized eigenstates.

The Yangian symmetry of the model [46, 54] implies significant degeneracies in the spectrum and hence indicates integrability. The model is not integrable in the usual sense, however, as the method of quantum inverse scattering [57] is not applicable to models with long-range interactions. Talstra and Haldane [58] have nonetheless succeeded in constructing an infinite set of mutually commuting integrals of motion for the model by using the determinant rather than the trace of the monodromy matrix. These integrals provide the framework for the model's integrability. The integrability is hence only indirectly related to the Yangian symmetry.

The model is further amenable to exact solution via the asymtotic Bethe Ansatz [44, 47, 54, 59–64], even though the application of this method to models with long-range interactions is likewise heuristic.

2.2.3 Ground State Properties

The ground state (2.2.3) is real (and hence both parity and time-reversal invariant), a spin singlet, and can equivalently be obtained by Gutzwiller projection [65–71], as we will verify now after evaluating the total momentum.

Ground state momentum. To determine the momentum p_0 (in units of inverse lattice spacings $1/a$) we translate the ground state (2.2.3) counterclockwise by one lattice spacing around the unit circle,

$$\boldsymbol{T}\big|\psi_0^{\text{HS}}\big\rangle = e^{\text{i}p_0}\big|\psi_0^{\text{HS}}\big\rangle. \tag{2.2.11}$$

With $\boldsymbol{T}z_i = e^{\text{i}\frac{2\pi}{N}}z_i$, we find

$$p_0 = \frac{2\pi}{N}\left(2\frac{M(M-1)}{2} + M\right) = \pi M,$$

and hence (2.2.4). Note that the sign of p_0 is irrelevant for (2.2.3), as N is always even, and p_0 is *0* or π. The sign will become significant only in Sects. 2.2.6 and 2.2.7 below, when we assign spinons momenta for states with N odd.

Singlet property. Since $S^z_{\text{tot}}\big|\psi_0^{\text{HS}}\big\rangle = 0$, it suffices to show that $\big|\psi_0^{\text{HS}}\big\rangle$ is annihilated by S^-_{tot}:

$$\begin{aligned} S^-_{\text{tot}}\big|\psi_0^{\text{HS}}\big\rangle &= \sum_{\alpha=1}^{N} S^-_\alpha \sum_{\{z_1,\ldots z_M\}} \psi_0^{\text{HS}}(z_1, z_2, \ldots z_M) S^+_{z_1} \ldots S^+_{z_M} |\downarrow\downarrow \ldots \downarrow\rangle \\ &= \sum_{\{z_2,\ldots,z_M\}} \underbrace{\sum_{\alpha=1}^{N} \psi_0^{\text{HS}}(\eta_\alpha, z_2, \ldots, z_M)}_{=0} S^+_{z_2} \ldots S^+_{z_M} |\downarrow\downarrow \ldots \downarrow\rangle, \end{aligned} \tag{2.2.12}$$

since $\psi_0^{\text{HS}}(\eta_\alpha, z_2, \ldots, z_M)$ contains only powers $\eta_\alpha^1, \eta_\alpha^2, \ldots, \eta_\alpha^{N-1}$ and

$$\sum_{\alpha=1}^{N} \eta_\alpha^m = N\delta_{m,0} \mod N. \tag{2.2.13}$$

Parity and time reversal invariance. We begin by showing that ψ_0^{HS} is real. With $\bar{z}_i = 1/z_i$ and hence

$$(z_i - z_j)^2 = -z_i z_j |z_i - z_j|^2, \tag{2.2.14}$$

we write

$$\begin{aligned} \psi_0^{\mathrm{HS}}(z_1, z_2, \ldots, z_M) &= \pm \prod_{i<j}^{M} |z_i - z_j|^2 \prod_{i<j}^{M} z_i z_j \prod_{i=1}^{M} z_i \\ &= \pm \prod_{i<j}^{M} |z_i - z_j|^2 \prod_{i=1}^{M} G(z_i) \end{aligned} \tag{2.2.15}$$

where

$$G(\eta_\alpha) = (\eta_\alpha)^{\frac{N}{2}} = \begin{cases} +1 & \alpha \text{ even} \\ -1 & \alpha \text{ odd.} \end{cases} \tag{2.2.16}$$

The gauge factor $G(z_i)$ effects that the Marshall sign criteria [72] is fulfilled.

Since parity tranforms $\eta_\alpha \rightarrow \eta_{-\alpha} = \bar{\eta}_\alpha$ and hence $z_i \rightarrow \bar{z}_i$, the fact that ψ_0^{HS} is real implies that $\left|\psi_0^{\mathrm{HS}}\right\rangle$ is invariant under parity. Time reversal transforms [73]

$$\mathrm{i} \rightarrow -\mathrm{i}, \quad \boldsymbol{S}_\alpha \rightarrow -\boldsymbol{S}_\alpha, \quad |s, m\rangle \rightarrow \mathrm{i}^{2m} |s, -m\rangle,$$

which implies $z_i \rightarrow \bar{z}_i$, $S_\alpha^+ \rightarrow -S_\alpha^-$, and $|\downarrow\downarrow \ldots \downarrow\rangle \rightarrow (-\mathrm{i})^N |\uparrow\uparrow \ldots \uparrow\rangle$. The basis states in (2.2.2) hence transform according to

$$S_{z_1}^+ \cdot \ldots \cdot S_{z_M}^+ |\downarrow\downarrow \ldots \downarrow\rangle \rightarrow S_{z_1}^- \cdot \ldots \cdot S_{z_M}^- |\uparrow\uparrow \ldots \uparrow\rangle. \tag{2.2.17}$$

Together with the singlet property, this implies that $\left|\psi_0^{\mathrm{HS}}\right\rangle$ is invariant under time reversal.

Generation by Gutzwiller projection. The ground state of the model was first obtained by Gutzwiller projection from a completely filled one-dimensional band which in total contains as many spin $\frac{1}{2}$ fermions as there are lattice sites [65, 68–71]:

$$\left|\psi_0^{\mathrm{HS}}\right\rangle = \mathrm{P_{GW}} \left|\psi_{\mathrm{SD}}^N\right\rangle, \quad \left|\psi_{\mathrm{SD}}^N\right\rangle \equiv \prod_{q \in \mathcal{I}} c_{q\uparrow}^\dagger c_{q\downarrow}^\dagger |0\rangle, \tag{2.2.18}$$

where the Gutzwiller projector

$$\mathrm{P_{GW}} \equiv \prod_{i=1}^{N} \left(1 - c_{i\uparrow}^\dagger c_{i\uparrow} c_{i\downarrow}^\dagger c_{i\downarrow}\right) \tag{2.2.19}$$

eliminates configurations with more than one particle on any site and the interval $\mathcal{I}$ contains $M = \frac{N}{2}$ adjacent momenta. We will now show that (2.2.18) is equivalent to (2.2.3). With lattice constant $a = \frac{2\pi}{N}$, the allowed momenta are given by integers, $q = 0, 1, \ldots, N-1$. With

$$c_q^\dagger = \sum_{\alpha=1}^{N} e^{\mathrm{i}\frac{2\pi}{N}\alpha q} c_\alpha^\dagger = \sum_{\alpha=1}^{N} \eta_\alpha^q c_\alpha^\dagger, \tag{2.2.20}$$

the (unnormalized) single particle momentum eigenstates are given by

$$\phi_q(z) = \langle z|q\rangle = \langle 0|c_z c_q^\dagger|0\rangle = z^q. \tag{2.2.21}$$

The many particle wave function for M fermions with adjacent momenta $q \in \mathcal{I} = [q_1, q_1 + M - 1]$ is hence given by

$$\phi_{\mathcal{I}}(z_1, z_2, \ldots, z_M) = \prod_{i=1}^{M} z_i^{q_1} \cdot \mathcal{A}\{z_1^0 z_2^1 \ldots z_M^{M-1}\} = \prod_{i=1}^{M} z_i^{q_1} \prod_{i<j}^{M} (z_i - z_j). \tag{2.2.22}$$

The Gutzwiller state (2.2.18) is given by

$$\begin{aligned} \left|\psi_0^{\mathrm{HS}}\right\rangle = &\sum_{\{z_1,\ldots,z_M;\, w_1,\ldots,w_M\}} \phi_{\mathcal{I}}(z_1, \ldots, z_M)\phi_{\mathcal{I}}(w_1, \ldots, w_M) \\ &\cdot c_{z_1\uparrow}^\dagger \ldots c_{z_M\uparrow}^\dagger c_{w_1\downarrow}^\dagger \ldots c_{w_M\downarrow}^\dagger |0\rangle, \end{aligned} \tag{2.2.23}$$

where the sum extends over all possible ways to distribute the coordinates z_i and w_k on mutually distinct lattice sites.

Let $\tilde{\mathcal{I}}$ contain all those M momenta not contained in $\mathcal{I}$, and $w_1, \ldots, w_M$ denote the sites which are not occupied by any of the z_i's. Then

$$\begin{aligned} \phi_{\mathcal{I}}(w_1, \ldots, w_M) &= \langle 0|c_{w_M} \ldots c_{w_1} \prod_{q\in\mathcal{I}} c_q^\dagger|0\rangle \\ &= \mathrm{sign}[z; w] \cdot \langle 0| \prod_{q\in\mathcal{I}} c_q \prod_{q\in\tilde{\mathcal{I}}} c_q\, c_{z_1}^\dagger \ldots c_{z_M}^\dagger \prod_{q\in\mathcal{I}} c_q^\dagger|0\rangle \\ &= \mathrm{sign}[z; w] \cdot \langle 0| \prod_{q\in\tilde{\mathcal{I}}} c_q\, c_{z_1}^\dagger \ldots c_{z_M}^\dagger|0\rangle \\ &= \mathrm{sign}[z; w] \cdot \phi_{\tilde{\mathcal{I}}}^{\,*}(z_1, \ldots, z_M) \\ &= \mathrm{sign}[z; w] \cdot \prod_{i=1}^{M} \bar{z}_i^{\,M} \cdot \phi_{\mathcal{I}}^{\,*}(z_1, \ldots, z_M), \end{aligned} \tag{2.2.24}$$

where

$$\mathrm{sign}[z; w] \equiv \langle 0|c_{w_M} \ldots c_{w_1} c_{z_M} \ldots c_{z_1} \prod_{q\in\tilde{\mathcal{I}}} c_q^\dagger \prod_{q\in\mathcal{I}} c_q^\dagger|0\rangle \tag{2.2.25}$$

is an overal sign associated with ordering the z's and w's according to the lattice sites indices α. Since

$$\mathrm{sign}[z; w] \cdot c_{z_1\uparrow}^\dagger \ldots c_{z_M\uparrow}^\dagger c_{w_1\downarrow}^\dagger \ldots c_{w_M\downarrow}^\dagger|0\rangle = S_{z_1}^{+} \cdot \ldots \cdot S_{z_M}^{+}|\downarrow\downarrow \ldots \downarrow\rangle \tag{2.2.26}$$

we may write

$$\left|\psi_0^{\mathrm{HS}}\right\rangle = \sum_{\{z_1,\dots,z_M\}} |\phi_{\mathcal{I}}(z_1,\dots,z_M)|^2 \prod_{i=1}^{M} G(z_i) S_{z_1}^{+} \cdot \ldots \cdot S_{z_M}^{+} |\downarrow\downarrow \ldots \downarrow\rangle. \tag{2.2.27}$$

This is equivalent to (2.2.15).

As an aside, it is very easy to verify the singlet property in the Gutzwiller formulation (2.2.18) of the ground state. To begin with, filling the same single particle states with $\uparrow$ and $\downarrow$ spin fermions obviously yields a singlet,

$$\boldsymbol{S}_{\mathrm{tot}}\left|\psi_{\mathrm{SD}}^{N}\right\rangle = 0. \tag{2.2.28}$$

The Gutzwiller projector (2.2.19), however, commutes with the local spin operators and hence also with the total spin,

$$\left[\mathrm{P_{GW}}, \boldsymbol{S}_{\alpha}\right] = \left[\mathrm{P_{GW}}, \boldsymbol{S}_{\mathrm{tot}}\right] = 0. \tag{2.2.29}$$

Hence

$$\boldsymbol{S}_{\mathrm{tot}}\left|\psi_0^{\mathrm{HS}}\right\rangle = 0. \tag{2.2.30}$$

Norm. The norm of the ground state is [74]

$$\begin{aligned} \sum_{\{z_1,\dots,z_M\}} \prod_{i<j}^{M} |z_i - z_j|^4 &= \left(\frac{N}{2\pi \mathrm{i}}\right)^M \oint \frac{\mathrm{d}z_1}{z_1} \dots \oint \frac{\mathrm{d}z_M}{z_M} \prod_{i\neq j}^{M} \left(1 - \frac{z_i}{z_j}\right)^2 \\ &= \frac{N^M (2M)!}{2^M}. \end{aligned} \tag{2.2.31}$$

Relation to the chiral spin liquid. The Haldane–Shastry ground state may be viewed as the one-dimensional analog of the abelian or $S = \frac{1}{2}$ chiral spin liquid [75–82], which is essentially a Laughlin $m = 2$ quantized Hall state [1] for spin flips on a two dimensional lattice. The spinons in the chiral spin liquid were understood to obey half-Fermi statistics long before this was realized for the Haldane–Shastry model.

2.2.4 Explict Solution

For the explict calculation presented here to be applicable to the one- and two-spinon eigenstates investigated in Sect. 2.2.6 below, we consider wavefunctions of the form [44, 50–52]

$$\psi(z_1,\dots,z_M) = \phi(z_1,\dots,z_M) \cdot \psi_0^{\mathrm{HS}}(z_1,\dots,z_M), \tag{2.2.32}$$

where ψ_0^{HS} is given by (2.2.2) and $\phi[z] \equiv \phi(z_1, \ldots, z_M)$ a polynomial of degree strictly less than $N - 2M + 2$ in each of the z_i's. This implies that the degree of ψ^{HS} is strictly less than $N + 1$. N can be even or odd. This condition enables us to use a Taylor expansion when we calculate the action of the Hamiltonian (2.2.1) on the state. The result is that

$$H^{\text{HS}}|\psi\rangle = \frac{2\pi^2}{N^2}\left(\lambda + \frac{N}{48}(N^2-1) + \frac{M}{6}(4M^2-1) - \frac{N}{2}M^2\right)|\psi\rangle, \tag{2.2.33}$$

provided that ϕ satisfies the eigenvalue equation

$$\sum_{j=1}^{M}\left(\frac{1}{2}z_j^2\frac{\partial^2}{\partial z_j^2} + \sum_{\substack{k=1\\k\neq j}}^{M}\frac{2z_j^2}{z_j - z_k}\frac{\partial}{\partial z_j} - \frac{N-3}{2}z_j\frac{\partial}{\partial z_j}\right)\phi[z] = \lambda\phi[z] \tag{2.2.34}$$

for λ. The derivative operators in (2.2.34) and below are understood to act on the analytic extension of $\phi(z_1, \ldots, z_M)$, in which the z_i's are allowed to take any value in the complex plane. For $\phi[z] = 1$, (2.2.33) shows that $\left|\psi_0^{\text{HS}}\right\rangle$ is an eigenstate of H^{HS} with energy E_0 given by (2.2.5).

Derivation of (2.2.33) *and* (2.2.34). We first use $S^{\pm} = S^{\text{x}} \pm iS^{\text{y}}$ to rewrite (2.2.1) as the sum of a "kinetic" and a "potential" term,

$$H^{\text{HS}} = \frac{2\pi^2}{N^2}\sum_{\alpha\neq\beta}^{N}\frac{1}{|\eta_\alpha - \eta_\beta|^2}\left(S_\alpha^+ S_\beta^- + S_\alpha^{\text{z}} S_\beta^{\text{z}}\right). \tag{2.2.35}$$

We first evaluate the action of the kinetic term on $|\psi\rangle$. Consider first

$$\begin{aligned} S_\alpha^+ S_\beta^-|\psi\rangle &= S_\alpha^+ S_\beta^- \sum_{\{z_2,\ldots,z_M\}} \psi(\eta_\beta, z_2, \ldots, z_M) S_\beta^+ S_{z_2}^+ \cdot \ldots \cdot S_{z_M}^+|\downarrow\downarrow\ldots\downarrow\rangle \\ &= \sum_{\{z_2,\ldots,z_M\}} \psi(\eta_\beta, z_2, \ldots, z_M) S_\alpha^+ S_{z_2}^+ \cdot \ldots \cdot S_{z_M}^+|\downarrow\downarrow\ldots\downarrow\rangle, \end{aligned} \tag{2.2.36}$$

where we have implicitly assumed that each spin configuration in the sum over $\{z_1, z_2, \ldots, z_M\}$ in (2.2.2) appears only once (and not $M!$ times due to permutations of the z_i's). We write this as

$$\left[S_\alpha^+ S_\beta^- \psi\right](\eta_\alpha, z_2, \ldots, z_M) = \psi(\eta_\beta, z_2, \ldots, z_M). \tag{2.2.37}$$

Note in particlular that $\left[S_\alpha^+ S_\beta^- \psi\right](z_1, z_2, \ldots, z_M)$ vanishes unless η_α equals one of the z_i's.

The action of the kinetic term on ψ is given by

$$
\begin{aligned}
T\psi[z] &\equiv \left[\sum_{\alpha\neq\beta}^{N} \frac{S_\alpha^+ S_\beta^-}{|\eta_\alpha - \eta_\beta|^2}\psi\right](z_1,\ldots,z_M) \\
&= \sum_{j=1}^{M} \sum_{\substack{\beta=1 \\ \eta_\beta\neq z_j}}^{N} \frac{\eta_\beta}{|z_j - \eta_\beta|^2} \frac{\psi(z_1,\ldots,z_{j-1},\eta_\beta,z_{j+1},\ldots,z_M)}{\eta_\beta}.
\end{aligned}
\tag{2.2.38}
$$

Since the last fraction is a polynomial of degree strictly less than N in β, we can Taylor expand it around z_j,

$$
\frac{\psi(z_1,\ldots,\eta_\beta,\ldots,z_M)}{\eta_\beta} = \sum_{l=0}^{N-1} \frac{(\eta_\beta - z_j)^l}{l!} \frac{\partial^l}{\partial z_j^l} \frac{\Psi(z_1,\ldots,z_M)}{z_j}.
\tag{2.2.39}
$$

The sum over β yields

$$
\sum_{\substack{\beta=1 \\ \eta_\beta\neq z_j}}^{N} \frac{\eta_\beta(\eta_\beta - z_j)^l}{|z_j - \eta_\beta|^2} = z_j^{l+1} A_l, \quad A_l = -\sum_{\alpha=1}^{N-1} \eta_\alpha^2 (\eta_\alpha - 1)^{l-2},
\tag{2.2.40}
$$

where A_0, A_1, and A_2 are evaluated with (B.14), (B.9), and (B.2) from Appendix B, respectively:

$$
\begin{aligned}
A_0 &= -\sum_{\alpha=1}^{N-1} \frac{\eta_\alpha^2}{(\eta_\alpha - 1)^2} = \frac{(N-1)(N-5)}{12}, \\
A_1 &= -\sum_{\alpha=1}^{N-1} \frac{\eta_\alpha^2}{\eta_\alpha - 1} = -\frac{N-3}{2}, \\
A_2 &= -\sum_{\alpha=1}^{N-1} \eta_\alpha^2 = 1, \\
A_l &= -\sum_{\alpha=1}^{N} \eta_\alpha^2 (\eta_\alpha - 1)^{l-2} = 0 \quad \text{for } 2 < l \leq N-1.
\end{aligned}
$$

In the last line, we have used that $\eta_\alpha^2(\eta_\alpha - 1)^{l-2}$ vanishes for $\eta_\alpha = 1$ and contains only powers $\eta_\alpha^2, \ldots \eta_\alpha^{N-1}$ for $2 < l \leq N-1$. Substituion into (2.2.38) and (2.2.39) yields

$$
\begin{aligned}
T\psi[z] &= \sum_{j=1}^{M}\left(\frac{(N-1)(N-5)}{12}z_j - \frac{N-3}{2}z_j^2\frac{\partial}{\partial z_j} + \frac{1}{2}z_j^3\frac{\partial^2}{\partial z_j^2}\right)\frac{\psi[z]}{z_j} \\
&= \frac{M(N-1)(N-5)}{12}\psi[z] - \frac{N-3}{2}\underbrace{\sum_{j\neq k}^{M}\frac{2z_j}{z_j-z_k}}_{=M(M-1)}\psi[z] \\
&\quad + \sum_{j\neq k}^{M}\frac{z_j^2}{(z_j-z_k)^2}\psi[z] + \underbrace{\sum_{\substack{j,k,m=1\\ j\neq k\neq m\neq j}}^{M}\frac{2z_j^2}{(z_j-z_k)(z_j-z_m)}}_{=2M(M-1)(M-2)/3}\psi[z] \\
&\quad + \sum_{j=1}^{M}\psi_0^{\mathrm{HS}}[z]\left(\frac{1}{2}z_j^2\frac{\partial^2}{\partial z_j^2} + \sum_{k\neq j}^{M}\frac{2z_j^2}{z_j-z_k}\psi_0\frac{\partial}{\partial z_j} - \frac{N-3}{2}z_j\frac{\partial}{\partial z_j}\right)\phi[z],
\end{aligned}
$$

where we have used the algebraic identity (B.7) in the evaluation of the triple sum.

For the action of the potential term we write

$$
S_\alpha^{\mathrm{z}}S_\beta^{\mathrm{z}} = \left(S_\alpha^{\mathrm{z}}+\frac{1}{2}\right)\left(S_\beta^{\mathrm{z}}+\frac{1}{2}\right) - \frac{1}{2}(S_\alpha^{\mathrm{z}}+S_\beta^{\mathrm{z}}) - \frac{1}{4}.
$$

This yields

$$
\begin{aligned}
V\psi[z] &\equiv \left[\sum_{\alpha\neq\beta}^{N}\frac{S_\alpha^{\mathrm{z}}S_\beta^{\mathrm{z}}}{|\eta_\alpha-\eta_\beta|^2}\psi\right](z_1,\ldots,z_M) \\
&= \sum_{j\neq k}^{M}\frac{1}{|z_j-z_k|^2}\psi[z] - \sum_{\alpha\neq\beta}^{N}\frac{S_\alpha^{\mathrm{z}}+\frac{1}{2}}{|\eta_\alpha-\eta_\beta|^2}\psi[z] + \frac{1}{4}\underbrace{\sum_{\alpha\neq\beta}^{N}\frac{1}{|\eta_\alpha-\eta_\beta|^2}}_{=N(N^2-1)/12}\psi[z].
\end{aligned}
\tag{2.2.41}
$$

With

$$
\sum_{j\neq k}^{M}\frac{1}{|z_j-z_k|^2}\psi[z] + \sum_{j\neq k}^{M}\frac{z_j^2}{(z_j-z_k)^2}\psi[z] = \frac{1}{2}M(M-1)\psi[z]
$$

and

$$
\sum_{\alpha\neq\beta}^{N}\frac{S_\alpha^{\mathrm{z}}+\frac{1}{2}}{|\eta_\alpha-\eta_\beta|^2}\psi[z] = \sum_{\alpha=1}^{N}\sum_{\beta=1}^{N-1}\frac{S_\alpha^{\mathrm{z}}+\frac{1}{2}}{|1-\eta_\beta|^2}\psi[z] = M\frac{N^2-1}{12}\psi[z],
$$

where we have substituted $\eta_\beta \to \eta_\beta\eta_\alpha$ and used (B.15), we obtain (2.2.33) and (2.2.34).

2.2.5 Factorization of the Hamiltonian

In Sect. 2.2.4 we have shown that $\left|\psi_0^{\text{HS}}\right\rangle$ is an eigenstate of H^{HS} with energy E_0 given by (2.2.5). To show that $\left|\psi_0^{\text{HS}}\right\rangle$ is the ground state (or at least one of several ground states), we factorize the Haldane–Shastry Hamiltonian [45, 50, 52]. For every site η_α, we define an auxiliary operator $\boldsymbol{D}_\alpha$ by

$$\boldsymbol{D}_\alpha = \frac{1}{2}\sum_{\substack{\beta=1\\ \beta\neq\alpha}}^{N} \frac{\eta_\alpha+\eta_\beta}{\eta_\alpha-\eta_\beta}\Big[\text{i}(\boldsymbol{S}_\alpha\times\boldsymbol{S}_\beta)+\boldsymbol{S}_\beta\Big]. \tag{2.2.42}$$

The rapidity operator (2.2.8) is given in terms of these by

$$\sum_{\alpha=1}^{N}\boldsymbol{D}_\alpha = \boldsymbol{\Lambda}, \tag{2.2.43}$$

as one can easily see with (B.16).

We will show below that H^{HS} can be written as:

$$H^{\text{HS}} = \frac{2\pi^2}{N}\left[\frac{2}{9}\sum_{\alpha=1}^{N}\boldsymbol{D}_\alpha^{\dagger}\boldsymbol{D}_\alpha + \frac{N+1}{12}\boldsymbol{S}_{\text{tot}}^2\right] + E_0, \tag{2.2.44}$$

which consists of two positive semi-definite operators (i.e., operators with only non-negative eigenvalues) and a constant. The lowest energy eigenvalue of H^{HS} is therefore E_0, and $\left|\psi_0^{\text{HS}}\right\rangle$ is a ground state.

Taking the ground state expectation value of (2.2.44) implies with

$$H^{\text{HS}}\left|\psi_0^{\text{HS}}\right\rangle = E_0\left|\psi_0^{\text{HS}}\right\rangle \tag{2.2.45}$$

that

$$\boldsymbol{D}_\alpha\left|\psi_0^{\text{HS}}\right\rangle = 0, \quad \forall\,\alpha = 1,\ldots,N \tag{2.2.46}$$

and $\boldsymbol{S}_{\text{tot}}\left|\psi_0^{\text{HS}}\right\rangle = 0$. This trivially implies

$$\boldsymbol{\Lambda}\left|\psi_0^{\text{HS}}\right\rangle = 0, \tag{2.2.47}$$

i.e., there is no spin current in the ground state. Note that if other ground states were to exist, (2.2.44) shows that they would have to be singlets and likewise be annihilated by $\boldsymbol{D}_\alpha$. It is not very difficult to verify (2.2.46) directly, but since we have verified (2.2.45) in Sect. 2.2.4 and will verify (2.2.44) below, there is no need to do so.

Verification of (2.2.44). For convenience, we define the purely imaginary parameter

$$\theta_{\alpha\beta} \equiv \frac{\eta_\alpha+\eta_\beta}{\eta_\alpha-\eta_\beta}$$

and recall

$$\boldsymbol{D}_\alpha^\dagger = \frac{1}{2}\sum_{\substack{\beta=1\\ \beta\neq\alpha}}^{N} \theta_{\alpha\beta}\big[\mathrm{i}(\boldsymbol{S}_\alpha\times\boldsymbol{S}_\beta) - \boldsymbol{S}_\beta\big],$$

$$\boldsymbol{D}_\alpha = \frac{1}{2}\sum_{\substack{\gamma=1\\ \gamma\neq\alpha}}^{N} \theta_{\alpha\gamma}\big[\mathrm{i}(\boldsymbol{S}_\alpha\times\boldsymbol{S}_\gamma) + \boldsymbol{S}_\gamma\big].$$

For $S=\frac{1}{2}$ and $\alpha\neq\beta,\gamma$, we obtain

$$\begin{aligned}\mathrm{i}(\mathbf{S}_\alpha\times\mathbf{S}_\beta)\mathrm{i}(\mathbf{S}_\alpha\times\mathbf{S}_\gamma) &= \varepsilon^{ijk}\varepsilon^{ilm}S_\beta^j S_\alpha^k S_\alpha^l S_\gamma^m\\ &= \big(\delta^{jl}\delta^{km}-\delta^{jm}\delta^{kl}\big)S_\beta^j\left(\frac{1}{4}\delta^{kl}+\frac{\mathrm{i}}{2}\varepsilon^{kln}S_\alpha^n\right)S_\gamma^m\\ &= -\frac{1}{2}\mathbf{S}_\beta\mathbf{S}_\gamma - \frac{\mathrm{i}}{2}\mathbf{S}_\alpha(\mathbf{S}_\beta\times\mathbf{S}_\gamma),\end{aligned} \tag{2.2.48}$$

and therewith

$$\big[\mathrm{i}(\boldsymbol{S}_\alpha\times\boldsymbol{S}_\beta)-\boldsymbol{S}_\beta\big]\cdot\big[\mathrm{i}(\boldsymbol{S}_\alpha\times\boldsymbol{S}_\gamma)+\boldsymbol{S}_\gamma\big] = -\frac{3}{2}\big[\boldsymbol{S}_\beta\boldsymbol{S}_\gamma - \mathrm{i}\boldsymbol{S}_\alpha(\boldsymbol{S}_\beta\times\boldsymbol{S}_\gamma)\big].$$

This implies

$$\sum_{\alpha=1}^{N}\boldsymbol{D}_\alpha^\dagger\boldsymbol{D}_\alpha = -\frac{3}{8}\sum_{\alpha=1}^{N}\sum_{\substack{\beta=1\\ \beta\neq\alpha}}^{N}\sum_{\substack{\gamma=1\\ \gamma\neq\alpha}}^{N}\theta_{\alpha\beta}\theta_{\alpha\gamma}\left[\boldsymbol{S}_\beta\boldsymbol{S}_\gamma - \mathrm{i}\boldsymbol{S}_\alpha(\boldsymbol{S}_\beta\times\boldsymbol{S}_\gamma)\right].$$

For the terms with $\alpha\neq\beta=\gamma$, we use $\boldsymbol{S}\times\boldsymbol{S}=\mathrm{i}\boldsymbol{S}$ to write

$$\boldsymbol{S}_\beta\boldsymbol{S}_\beta - \mathrm{i}\boldsymbol{S}_\alpha(\boldsymbol{S}_\beta\times\boldsymbol{S}_\beta) = \frac{3}{4}+\boldsymbol{S}_\alpha\boldsymbol{S}_\beta,$$

and observe

$$\theta_{\alpha\beta}^2 = 1-\frac{4}{|\eta_\alpha-\eta_\beta|^2}.$$

For the terms with $\alpha,\beta,$ and γ all distinct, the vector product term vanishes as it changes sign under interchange of the dummy indices β and γ. For these terms we rearrange the sums

$$\sum_{\alpha=1}^{N}\sum_{\substack{\beta=1\\ \beta\neq\alpha}}^{N}\sum_{\substack{\gamma=1\\ \gamma\neq\alpha}}^{N} = \sum_{\beta=1}^{N}\sum_{\gamma=1}^{N}\sum_{\substack{\alpha=1\\ \alpha\neq\beta,\gamma}}^{N}$$

and carry out the summation over α. With

$$\frac{1}{(\eta_\alpha - \eta_\beta)(\eta_\alpha - \eta_\gamma)} = \frac{1}{\eta_\beta - \eta_\gamma}\left(\frac{1}{\eta_\alpha - \eta_\beta} - \frac{1}{\eta_\alpha - \eta_\gamma}\right)$$

and

$$\sum_{\substack{\alpha=1\\ \alpha\neq\beta,\gamma}}^{N} \frac{\eta_\beta}{\eta_\alpha - \eta_\beta} = -\frac{N-1}{2} - \frac{\eta_\beta}{\eta_\gamma - \eta_\beta},$$

which follows directly from (B.12), we obtain

$$\sum_{\substack{\alpha=1\\ \alpha\neq\beta,\gamma}}^{N} \theta_{\alpha\beta}\theta_{\alpha\gamma} = \sum_{\substack{\alpha=1\\ \alpha\neq\beta,\gamma}}^{N} \left(1 + \frac{2\eta_\beta}{\eta_\alpha - \eta_\beta}\right)\left(1 + \frac{2\eta_\gamma}{\eta_\alpha - \eta_\gamma}\right) = N - \frac{8}{|\eta_\beta - \eta_\gamma|^2}.$$

Collecting all the terms yields

$$\begin{aligned}
\frac{8}{3}\sum_{\alpha=1}^{N} \boldsymbol{D}_\alpha^\dagger \boldsymbol{D}_\alpha
&= \sum_{\alpha\neq\beta}^{N} \left(\frac{4}{|\eta_\alpha - \eta_\beta|^2} - 1\right)\left(\frac{3}{4} + \boldsymbol{S}_\alpha \boldsymbol{S}_\beta\right) + \sum_{\beta\neq\gamma}^{N} \left(\frac{8}{|\eta_\beta - \eta_\gamma|^2} - N\right) \boldsymbol{S}_\beta \boldsymbol{S}_\gamma \\
&= 12 \sum_{\alpha\neq\beta}^{N} \frac{\boldsymbol{S}_\alpha \boldsymbol{S}_\beta}{|\eta_\alpha - \eta_\beta|^2} - (N+1)\sum_{\alpha\neq\beta}^{N} \boldsymbol{S}_\alpha \boldsymbol{S}_\beta + \sum_{\alpha\neq\beta}^{N} \left(\frac{3}{|\eta_\alpha - \eta_\beta|^2} - \frac{3}{4}\right).
\end{aligned}$$

With the identities

$$\sum_{\alpha\neq\beta}^{N} \boldsymbol{S}_\alpha \boldsymbol{S}_\beta = \boldsymbol{S}_{\text{tot}}^2 - \frac{3}{4}N$$

and

$$\sum_{\alpha\neq\beta}^{N} \left(\frac{3}{|\eta_\alpha - \eta_\beta|^2} - \frac{3}{4}\right) = \frac{1}{4}N(N^2 - 1) - \frac{3}{4}N(N-1),$$

where we have used (B.15), we obtain

$$\sum_{\alpha\neq\beta}^{N} \frac{\boldsymbol{S}_\alpha \boldsymbol{S}_\beta}{|\eta_\alpha - \eta_\beta|^2} = \frac{2}{9}\sum_{\alpha=1}^{N} \boldsymbol{D}_\alpha^\dagger \boldsymbol{D}_\alpha + \frac{N+1}{12}\boldsymbol{S}_{\text{tot}}^2 - \frac{N(N^2+5)}{48},$$

and hence (2.2.44). □

2.2.6 Spinon Excitations and Fractional Statistics

The elementary excitations for this model are free spinon excitations, which carry spin $\frac{1}{2}$ and no charge. They constitute an instance of fractional quantization, which is both conceptually and mathematically similar to the fractional quantization of charge in the fractional quantum Hall effect [1]. Their fractional quantum number is the spin, which takes the value $\frac{1}{2}$ in a Hilbert space (2.2.2) made out of spin flips S^+, which carry spin 1.

One-spinon states. To write the wave function for a $\downarrow$-spin spinon localized at site η_α, consider a chain with an odd number of sites N and let $M = \frac{N-1}{2}$ be the number of $\uparrow$ or $\downarrow$ spins condensed in the uniform liquid. The spinon wave function is then given by

$$\psi_{\alpha\downarrow}(z_1, z_2, \ldots, z_M) = \prod_{i=1}^{M} (\eta_\alpha - z_i)\psi_0^{\text{HS}}(z_1, z_2, \ldots, z_M), \tag{2.2.49}$$

which we understand substituted into (2.2.2). It is easy to verify $S^{\text{z}}_{\text{tot}}\psi_{\alpha\downarrow} = -\frac{1}{2}\psi_{\alpha\downarrow}$ and $S^{-}_{\text{tot}}\psi_{\alpha\downarrow} = 0$, which shows that the spinon transforms as a spinor under rotations.

The localized spinon (2.2.49) is not an eigenstate of the Hamiltonian (2.2.1). To obtain exact eigenstates, we construct momentum eigenstates according to

$$\psi_{m\downarrow}(z_1, z_2, \ldots, z_M) = \sum_{\alpha=1}^{N} (\bar{\eta}_\alpha)^m \psi_{\alpha\downarrow}(z_1, z_2, \ldots, z_M), \tag{2.2.50}$$

where the integer m corresponds to a momentum quantum number. Since $\psi_{\alpha\downarrow}(z_1, z_2, \ldots, z_M)$ contains only powers $\eta_\alpha^0, \eta_\alpha^1, \ldots, \eta_\alpha^M$ and

$$\sum_{\alpha=1}^{N} \overline{\eta}_\alpha^m \eta_\alpha^n = \delta_{mn} \mod N, \tag{2.2.51}$$

$\psi_{m\downarrow}(z_1, z_2, \ldots, z_M)$ will vanish unless $m = 0, 1, \ldots, M$. There are only roughly half as many spinon orbitals as there are sites. Spinons on neighboring sites hence cannot be orthogonal. With (2.2.33) and (2.2.34), we obtain

$$H^{\text{HS}}|\psi_{m\downarrow}\rangle = \left[-\frac{\pi^2}{24}\left(N - \frac{1}{N}\right) + \frac{2\pi^2}{N^2} m(M - m)\right] |\psi_{m\downarrow}\rangle. \tag{2.2.52}$$

To make a correspondence between m and the spinon momentum p_m, we translate (2.2.50) counterclockwise by one lattice spacing (which we set to unity for present purposes) around the unit circle,

$$\boldsymbol{T}|\psi_{m\downarrow}\rangle = e^{\text{i}(p_0 + p_m)}|\psi_{m\downarrow}\rangle. \tag{2.2.53}$$

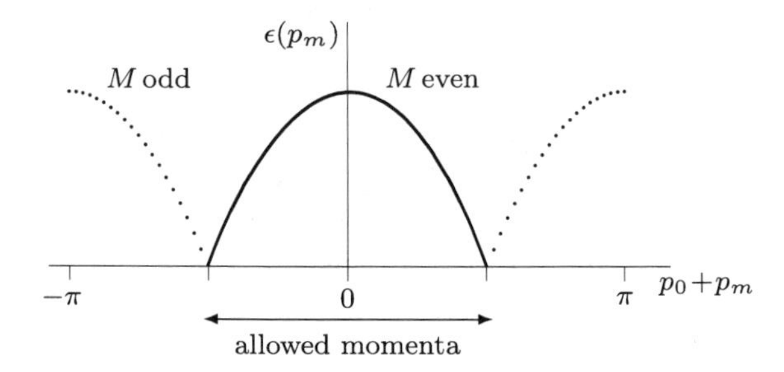

Fig. 2.2 Dispersion of a single spinon in a Haldane–Shastry chain

With ground state momentum $p_0 = -\frac{\pi}{2}N$, we find

$$p_m = \pi - \frac{2\pi}{N}\left(m + \frac{1}{4}\right). \tag{2.2.54}$$

The energy (2.2.52) can be written as $E = E_0 + \epsilon(p_m)$, with the spinon dispersion given by

$$\epsilon(p) = \frac{1}{2}p\,(\pi - p) + \frac{\pi^2}{8N^2}, \tag{2.2.55}$$

as depicted in Fig. 2.2. The interval of allowed spinon momenta spans only half of the Brillouin zone, and alternates with M even vs. M odd.

Two-spinon states. To write the wave function for two $\downarrow$-spin spinons localized at sites η_α and η_β, consider a chain with N even and $M = \frac{N-2}{2}$. The two-spinon state is then given by

$$\psi_{\alpha\beta}(z_1, z_2, \dots, z_M) = \prod_{i=1}^{M}(\eta_\alpha - z_i)(\eta_\beta - z_i)\psi_0^{\text{HS}}(z_1, z_2, \dots, z_M). \tag{2.2.56}$$

A momentum basis for the two-spinon states is given by

$$\psi_{mn}(z_1, z_2, \dots, z_M) = \sum_{\alpha,\beta=1}^{N}(\bar{\eta}_\alpha)^m(\bar{\eta}_\beta)^m\psi_{\alpha\beta}(z_1, z_2, \dots, z_M), \tag{2.2.57}$$

where $M \geq m \geq n \geq 0$. For m or n outside this range, ψ_{mn} vanishes identically, reflecting the overcompleteness of the position space basis. With (2.2.33), (2.2.34), and the algebraic identity

$$\frac{x+y}{x-y}(x^m y^n - x^n y^m) = 2\sum_{l=0}^{m-n} x^{m-l}y^{n+l} - (x^m y^n + x^n y^m), \tag{2.2.58}$$

we obtain [44, 50–52]

$$H^{\mathrm{HS}}|\psi_{mn}\rangle = E_{mn}|\psi_{mn}\rangle + \sum_{l=1}^{l_{\max}} V_l^{mn}|\psi_{m+l,n-l}\rangle \tag{2.2.59}$$

with

$$\begin{aligned} E_{mn} &= -\frac{\pi^2}{24}\left(N - \frac{19}{N} + \frac{24}{N^2}\right) \\ &\quad + \frac{2\pi^2}{N^2}\left[m\left(\frac{N}{2} - 1 - m\right) + n\left(\frac{N}{2} - 1 - n\right) - \frac{m-n}{2}\right], \end{aligned} \tag{2.2.60}$$

$$V_l^{mn} = -\frac{2\pi^2}{N^2}(m - n + 2l), \tag{2.2.61}$$

and $l_{\max} = \min(M - m, n)$. Since the "scattering" of the non-orthogonal basis states $|\psi_{mn}\rangle$ in (2.2.59) only occurs in one direction, increasing $m - n$ while keeping $m + n$ fixed, the eigenstates of H^{HS} have energy eigenvalues E_{mn}, and are of the form

$$|\phi_{mn}\rangle = \sum_{l=0}^{l_M} a_l^{mn}|\psi_{m+l,n-l}\rangle. \tag{2.2.62}$$

A recursion relation for the coefficients a_l^{mn} is readily obtained from (2.2.59).

If we identify the single-spinon momenta for $m \geq n$ according to

$$p_m = \pi - \frac{2\pi}{N}\left(m + \frac{1}{2} + s\right), \quad p_n = \pi - \frac{2\pi}{N}\left(n + \frac{1}{2} - s\right), \tag{2.2.63}$$

with a statistical shift $s = \frac{1}{4}$ [83, 84], we can write the energy

$$E_{mn} = E_0 + \epsilon(p_m) + \epsilon(p_n), \tag{2.2.64}$$

where E_0 is the ground state energy (2.2.5) and $\epsilon(p)$ the spinon dispersion (2.2.55).

Fractional statistics. The mutual half-fermi statistics of the spinons manifests itself in the fractional shift s in the single-spinon momenta (2.2.63), as we will elaborate now [85]. The Ansatz (2.2.57) unambiguously implies that the sum of the two-spinon momenta is given by $q_m + q_n = 2\pi - \frac{2\pi}{N}(m + n + 1)$, and hence (2.2.63). The shift s is determined by demanding that the excitation energy (2.2.64) of the two-spinon state is a sum of single-spinon energies, which in turn is required for the explicit solution here to be consistent with the models solution via the asymptotic Bethe ansatz [54, 83, 86].

The shift decreases the momentum p_m of spinon 1 and increases momentum p_n of spinon 2. This may surprise at first as the basis states (2.2.57) are constructed symmetrically with regard to interchanges of m and n. To understand this asymmetry,

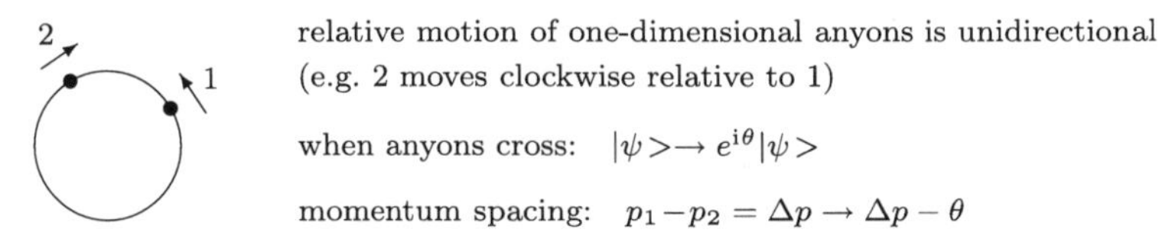

Fig. 2.3 Fractional statistics in one dimension. The crossings of the anyons are unidirectional, and the many particle wave function acquires a statistical phase θ whenever they cross

note that $M \geq m \geq n \geq 0$ implies $0 < p_m < p_n < \pi$. The dispersion (2.2.55) implies that the group velocity of the spinons is given by

$$v_g(p) = \partial_p \epsilon(p) = \frac{\pi}{2} - p, \tag{2.2.65}$$

which in turn implies that $v_g(p_m) > v_g(p_n)$. This means that the *relative motion* of spinon 1 (with q_m) with respect to spinon 2 (with q_n) is *always counterclockwise* on the unit circle (see Fig. 2.3). The shifts in the individual spinon momenta can hence be explained by assuming that the two-spinon state acquires a statistical phase $\theta = 2\pi s$ whenever the spinons pass through each other. This phase implies that q_m is shifted by $-\frac{2\pi}{N}s$ since we have to translate spinon 1 counterclockwise through spinon 2 and hence counterclockwise around the unit circle when obtaining the allowed values for q_m from the PBCs. Similarly, q_n is shifted by $+\frac{2\pi}{N}s$ since we have to translate spinon 2 clockwise through spinon 1 and hence clockwise around the unit circle when obtaining the quantization of q_n.

That the crossing of the spinons occurs only in one direction is a necessary requirement for fractional statistics to exist in one dimension. If the spinons could cross in both directions, the fact that paths interchanging them twice (i.e., once in each direction) are topologically equivalent to paths not interchanging them at all would imply $2\theta = 0 \bmod 2\pi$ for the statistical phase, i.e., only allow for the familiar choices of bosons or fermions. With the scattering occurring in only one direction, arbitrary values for θ are possible. Note that the one-dimensional anyons break neither time-reversal symmetry (T) nor parity (P).

The fractional statistics of the spinons manifests itself further in the fractional exclusion (or generalized Pauli) principle introduced by Haldane [87]. If we consider a state with L spinons, we can easily see from (2.2.50), (2.2.51), and (2.2.57) that the number of orbitals available for further spinons we may wish to create is $M + 1$, where $M = \frac{N-L}{2}$ is the number of $\uparrow$ or $\downarrow$ spins in the remaining uniform liquid. (In this representation, the spinon wave functions are symmetric; two or more spinons can have the same value for m.) In other words, the creation of *two* spinons reduces the number of available single spinon states by *one*. They hence obey half-fermi statistics in the sense of Haldane's exclusion principle. (For fermions, the creation of two particles would decrease the number of available single particle by two, while this number would not change for bosons.)

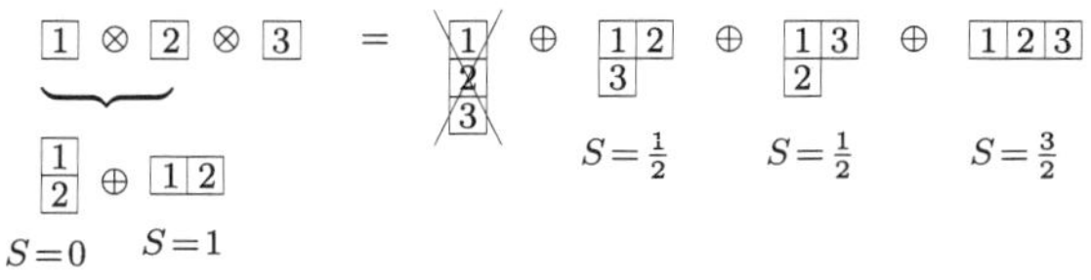

1 ⊗ 2 ⊗ 3 = 1/2/3 (crossed out) ⊕ 12/3 ⊕ 13/2 ⊕ 123

1/2 ⊕ 12

$S=0$ $S=1$ $S=\frac{1}{2}$ $S=\frac{1}{2}$ $S=\frac{3}{2}$

Fig. 2.4 Total spin representations of three $S = \frac{1}{2}$ spins with Young tableaux. For SU(n) with $n > 2$, the tableaux with three boxes on top of each other would exist as well

2.2.7 Young Tableaux and Many Spinon States

The easiest way to obtain the spectrum of the model is through the one-to-one correspondence between the Young tableaux classifying the total spin representations of N spins and the exact eigenstates of the the Haldane–Shastry model for a chain with N sites, which are classified by the total spins and the fractionally spaced single-particle momenta of the spinons [53].

This correspondence yields the allowed sequences of single-spinon momenta $p_1, \ldots, p_L$ as well as the allowed representations for the total spin of the states such that the eigenstates of the Haldane–Shastry model have momenta and energies

$$p = p_0 + \sum_{i=1}^{L} p_i, \quad E = E_0 + \sum_{i=1}^{L} \epsilon(p_i), \tag{2.2.66}$$

where p_0 and E_0 denote the ground state momentum and energy, respectively, and $\epsilon(p)$ is the single-spinon dispersion. The correspondence hence does not only provide the quantum numbers of all the states in the spectrum, but also shows that it is sensible to view the individual spinons as particles, rather than just as solitons or collective excitations in many body condensates. We now proceed by stating these rules without further motivating or even deriving them.

To begin with, the Hilbert space of a system of N identical SU(n) spins can be decomposed into representations of the total spin, which commutes with (2.2.1) and hence can be used to classify the eigenstates. These representations are compatible with the representations of the symmetric group S_N of N elements, which may be expressed in terms of Young tableaux [88, 89]. The general rule for obtaining Young tableaux is illustrated for three $S = \frac{1}{2}$ spins in Fig. 2.4. For each of the N spins, draw a box numbered consecutively from left to right. The representations of SU(n) are constructed by putting the boxes together such that the numbers assigned to them increase in each row from left to right and in each column from top to bottom. Each tableau indicates symmetrization over all boxes in the same row, and antisymmetrization over all boxes in the same column. This implies that we cannot have more than n boxes on top of each other for SU(n) spins. For SU(2), each tableau corresponds to a spin $S = \frac{1}{2}(\lambda_1 - \lambda_2)$ representation, with λ_i the number of boxes in the ith row, and stands for a multiplet $S^z = -S, \ldots, S$.

The one-to-one correspondence between the Young tableaux and the non-interacting many-spinon eigenstates of the Haldane–shastry model is illustrated in

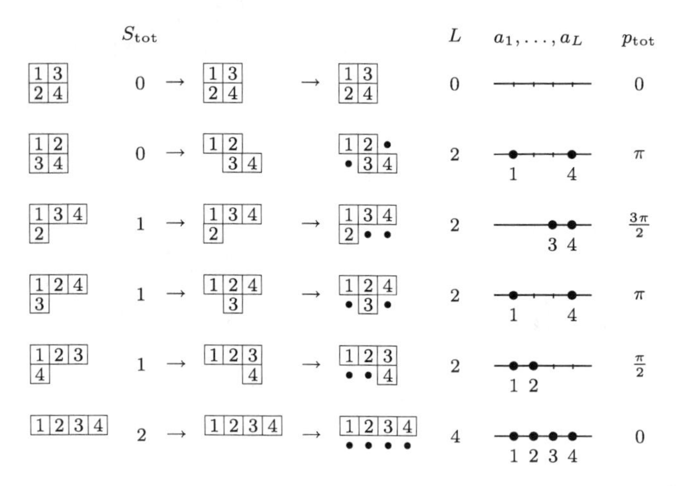

Fig. 2.5 Young tableau decomposition and the corresponding spinon states for an $S = \frac{1}{2}$ spin chain with $N = 4$ sites. The dots represent the spinons. The spinon momentum numbers a_i are given by the numbers in the boxes of the same column. Note that $\sum(2S_{\text{tot}} + 1) = 2^N$

Fig. 2.5 for a chain with $N = 4$ sites. The rule is that in each Young tableau, we shift boxes to the right such that each box is below or in the column to the right of the box with the preceding number. Each missing box in the resulting, extended tableaux represents a spinon. The extended tableaux provide us with the total spin of each multiplet, which is given by the representation specified by the original Young tableau, as well as the number L of spinons present and the individual spinon momentum numbers a_i, which are just the numbers in the boxes above or below the dots representing the spinons. The single-spinon momenta are obtained from those via

$$p_i = \frac{\pi}{N}\left(a_i - \frac{1}{2}\right), \tag{2.2.67}$$

which implies $\delta \leq p_i \leq \pi - \delta$ with $\delta = \frac{\pi}{2N} \to 0$ for $N \to \infty$.

The total momentum and the total energies of the many-spinon states are given by (2.2.66) with

$$p_0 = -\frac{\pi}{2}N, \quad E_0 = -\frac{\pi^2}{24}\left(N + \frac{5}{N}\right), \tag{2.2.68}$$

and the single-spinon dispersion

$$\epsilon(p) = \frac{1}{2}p\,(\pi - p) + \frac{\pi^2}{8N^2}, \tag{2.2.69}$$

where we use a convention according to which the "vacuum" state $|\downarrow\downarrow \ldots \downarrow\rangle$ has momentum $p = 0$ (and the empty state $|0\rangle$ has $p = \pi(N - 1)$).

This correspondence shows that spinons are non-interacting, with momentum spacings appropriate for half-fermions. We may interpret the Haldane–Shastry model as a reparameterization of a Hilbert space spanned by spin flips (2.2.2) into a basis which consists of the Haldane–Shastry ground state plus all possible many spinon states. The reward for such a reparameterization is that a highly non-trivial Hamiltonian in the original basis may be approximately or exactly diagonal in the new basis, as this basis is chosen in accordance with the quantum numbers of the elementary excitations.

2.3 The Moore–Read State and Its Parent Hamiltonian

2.3.1 The Pfaffian State and Its Parent Hamiltonian

The Pfaffian state at even denominator Landau level filling fractions was introduced independently by Moore and Read [90] as an example of a quantized Hall state which supports quasiparticle excitations which obey non-Abelian statistics, and by Wen, Wilczek, and ourselves [91, 92] as a candidate for the observed plateau in Hall resistivity at Landau level filling fraction $\nu = 5/2$, i.e., at $\nu = 1/2$ in the second Landau level [10, 93–96], a proposal which was subsequently strengthened [97–100] and which recently received experimental support through the direct measurement of the quasiparticle charge [101, 102].

The wave function first proposed by Moore and Read [90] is

$$\psi_0(z_1, z_2, \ldots, z_N) = \mathrm{Pf}\left(\frac{1}{z_i - z_j}\right) \prod_{i<j}^{N} (z_i - z_j)^m \prod_{i=1}^{N} e^{-\frac{1}{4}|z_i|^2}, \tag{2.3.1}$$

where the particle number N is even, m is even (odd) for fermions (bosons), and the Pfaffian is is given by the fully antisymmetrized sum over all possible pairings of the N particle coordinates,

$$\mathrm{Pf}\left(\frac{1}{z_i - z_j}\right) \equiv \mathcal{A}\left\{\frac{1}{z_1 - z_2} \cdot \ldots \cdot \frac{1}{z_{N-1} - z_N}\right\}. \tag{2.3.2}$$

The inverse Landau level filling fraction is given by

$$\frac{1}{\nu} = \frac{\partial N_\Phi}{\partial N} = \frac{\partial (m(N-1) - 1)}{\partial N} = m. \tag{2.3.3}$$

The state describes a Laughlin state at $\nu = 1/m$ supplemented by a Pfaffian which implements p-wave pairing correlations. Since the Pfaffian is completely antisymmetric, it reverses the statistics from bosons to fermions or vice versa, but does not change the Landau level filling fraction.

The Pfaffian describes a BCS wave function [103–106] in position space, obtained by projecting on a definite number of particles [107, 108]. To see this, first rewrite the (unnormalized) BCS wave function as

$$\begin{aligned} |\psi_\phi\rangle &= \prod_k \left(1 + e^{i\phi}\frac{v_k}{u_k} c^\dagger_{k\uparrow} c^\dagger_{-k\downarrow}\right)|0\rangle \\ &= \prod_k \exp\left(e^{i\phi}\frac{v_k}{u_k} c^\dagger_{k\uparrow} c^\dagger_{-k\downarrow}\right)|0\rangle \\ &= \exp\left(e^{i\phi}\sum_k \frac{v_k}{u_k} c^\dagger_{k\uparrow} c^\dagger_{-k\downarrow}\right)|0\rangle \\ &= \exp\left(e^{i\phi} b^\dagger\right)|0\rangle, \end{aligned} \tag{2.3.4}$$

where the pair creation operator $b^\dagger$ is given by

$$\begin{aligned} b^\dagger &\equiv \sum_k \frac{v_k}{u_k} c^\dagger_{k\uparrow} c^\dagger_{-k\downarrow} \\ &= \int \mathrm{d}^3\boldsymbol{x}_1 \mathrm{d}^3\boldsymbol{x}_2 \varphi(\boldsymbol{x}_1 - \boldsymbol{x}_2)\psi^\dagger_\uparrow(\boldsymbol{x}_1)\psi^\dagger_\uparrow(\boldsymbol{x}_2)|0\rangle. \end{aligned} \tag{2.3.5}$$

The wave function for each of the individual pairs, which only depends on the relative coordinate, is given by

$$\varphi(\boldsymbol{x}) = \frac{1}{V}\sum_k \frac{v_k}{u_k} e^{ikx}. \tag{2.3.6}$$

If we now project out a state with $N/2$ pairs [107–109], we obtain

$$\begin{aligned} |\psi_N\rangle &= \frac{1}{2\pi}\int_0^{2\pi} \mathrm{d}\phi e^{-iN\phi/2}|\psi_\phi\rangle \\ &= \frac{1}{2\pi}\int_0^{2\pi} \mathrm{d}\phi e^{-iN\phi/2}\exp\left(e^{i\phi}b^\dagger\right)|0\rangle \\ &= \frac{1}{\left(\frac{N}{2}\right)!}\left(b^\dagger\right)^{N/2}|0\rangle, \end{aligned} \tag{2.3.7}$$

which is (up to a normalization) equivalent to

$$\begin{aligned} |\psi_N\rangle = &\int \mathrm{d}^3\boldsymbol{x}_1 \ldots \mathrm{d}^3\boldsymbol{x}_N \varphi(\boldsymbol{x}_1 - \boldsymbol{x}_2)\cdot\ldots\cdot\varphi(\boldsymbol{x}_{N-1} - \boldsymbol{x}_N) \\ &\cdot \psi^\dagger_\uparrow(\boldsymbol{x}_1)\psi^\dagger_\downarrow(\boldsymbol{x}_2)\ldots\psi^\dagger_\uparrow(\boldsymbol{x}_{N-1})\psi^\dagger_\downarrow(\boldsymbol{x}_N)|0\rangle. \end{aligned} \tag{2.3.8}$$

This implies that the many-particle wavefunction is given by a Pfaffian,

$$\psi(\boldsymbol{x}_1 \dots \boldsymbol{x}_N) = \mathrm{Pf}\left(\varphi(\boldsymbol{x}_i - \boldsymbol{x}_j)\right). \tag{2.3.9}$$

This form nicely illustrates that all the pairs have condensed into the same state, which is the essence of superfluidity. For fermion pairings with even relative angular momentum of the pairs, such as s- or d-wave, the wave function $\varphi(\boldsymbol{x}_i - \boldsymbol{x}_j)$ of the pairs is symmetric in real space, and antisymmetric in spin space (i.e., a singlet), while for pairings with odd angular momentum, such as p-wave, $\varphi(\boldsymbol{x}_i - \boldsymbol{x}_j)$ is antisymmetric in real space and symmetric in spin space (i.e., a triplet).

In the quantized Hall state, the requirement of analyticity in the complex coordinates constraints the possible form of the pair wave function decisively. Since the electrons are spin polarized, the only possible choice is the p-wave pairing described by the Pfaffian with $\varphi(z_i - z_j) = 1/(z_i - z_j)$. Note that this pair wave function would not be normalizable if it were not multiplied by at least an $m = 1$ Laughlin state.

One of the most important mathematical properties of the Pfaffian is that its square is equal to the determinant,

$$\mathrm{Pf}\left(\varphi(\boldsymbol{x}_i - \boldsymbol{x}_j)\right)^2 = \det(M_{ij}), \tag{2.3.10}$$

where

$$M_{ij} = \begin{cases} 0 & \text{for } i = j, \\ \varphi(\boldsymbol{x}_i - \boldsymbol{x}_j) & \text{for } i \neq j. \end{cases} \tag{2.3.11}$$

Another important identity, due to Frobenius [110], is given by (2.4.31) in Sect. 2.4.4 below.

The uniquely specifying property of the Pfaffian quantized Hall state (2.3.1) is that the wave function vanishes as the $(3m-1)$th power as *three* particles approach each other. This property simply reflects that there can be at most only one pair among each triplet of particles. This observation has led Wen, Wilczek, and ourselves [91, 92, 111] to propose the parent Hamiltonian

$$V^{(m)} = \sum_{i,j<k}^{N} \left(\nabla_i^2\right)^{(m-1)} \left(\delta^{(2)}(z_i - z_j)\delta^{(2)}(z_i - z_k)\right), \tag{2.3.12}$$

which, when supplemented with the kinetic Hamiltonian (2.1.9) as well as all similar terms with smaller powers of the Laplacian, singles out (2.3.1) as its unique ground state. For all practical purposes, however, it is best to formulate our parent Hamiltonian in terms of three-body pseudopotentials, as we will elaborate in Sect. 2.3.4.

2.3.2 Quasiparticle Excitations and the Internal Hilbert Space

One of the key properties of superconductors is that the magnetic vortices are quantized in units of one half of the Dirac flux quanta $\Phi_0 = 2\pi\hbar c/e$, in accordance to the charge $-2e$ of the Cooper pairs. The paring correlations in the Pfaffian Hall state have a similar effect on the vortices or quasiparticle excitations, which carry one half of the flux and charge they would carry without the pairing, i.e. they carry charge $e^* = e/2m$. The wave function for two flux $\frac{1}{2}$ quasiholes at positions ξ_1 and ξ_2 is easily formulated. We simply replace each factor in the Pfaffian in (2.3.1) by

$$\mathrm{Pf}\left(\frac{1}{z_i - z_j}\right) \rightarrow \mathrm{Pf}\left(\frac{(z_i - \xi_1)(z_j - \xi_2) + (z_i \leftrightarrow z_j)}{z_i - z_j}\right), \tag{2.3.13}$$

such that one member of each electron pair sees the additionally inserted zero at ξ_1 and the other member sees it at ξ_2. If we set $\xi_1 = \xi_2 = \xi$, we will recover a regular quasihole in the Laughlin fluid with charge $e^* = e/m$.

The internal Hilbert space spanned by the quasiparticle excitations only emerges as we consider the wave function for four charge $e^* = e/4$ quasiholes at positions $\xi_1, \ldots, \xi_4$, which is obtained by replacing the Pfaffian in (2.3.1) by

$$\mathrm{Pf}\left(\frac{1}{z_i - z_j}\right) \rightarrow \mathrm{Pf}\left(\frac{(z_i - \xi_1)(z_j - \xi_2)(z_i - \xi_3)(z_j - \xi_4) + (z_i \leftrightarrow z_j)}{z_i - z_j}\right). \tag{2.3.14}$$

We see that ξ_1 and ξ_3 belong to one group in that they constitute additional zeros seen by one member of each electron pair, while ξ_2 and ξ_4 belong to another group as they constitute zeros seen by the other members of each electron pair. The wave function is symmetric (or antisymmetric, depending on the number of electron pairs) under interchange of both groups. The state in the internal Hilbert space spanned by the quasihole affiliations with the two groups will change as we adiabatically interchange two quasiholes belonging to different groups, say ξ_3 and ξ_4. Naively, one might think that the dimension of the internal Hilbert space is given by the number of ways to partition the quasiholes at $\xi_1, \ldots, \xi_{2n}$ into two different groups, i.e., by $(2n-1)!!$ for $2n$ quasiholes. Note that the number of quasiholes has to be even on closed surfaces to satisfy the Dirac flux quantization condition [38]. The true dimension of the internal Hilbert space, however, is only 2^{n-1} [112]. The reason for this is that the internal Hilbert space is spanned by Majorana fermion states in the vortex cores [113], as we will elaborate in the following section.

The statistics is non-Abelian in the sense that the order according to which we interchange quasiholes matters. Let the matrix M_{ij} describe the rotation of the internal Hilbert space state vector which describes the adiabatic interchange two quasiholes at ξ_i and ξ_j:

$$|\psi\rangle \rightarrow M_{ij}|\psi\rangle.$$

The statistics is non-Abelian if the matrices associated with successive interchanges do not commute in general,

$$M_{ij}M_{jk} \neq M_{jk}M_{ij}.$$

Note that the internal state vector is protected in the sense that it is insensitive to local perturbations—it can *only* be manipulated through braiding of the vortices. For a sufficiently large number of vortices, on the other hand, any unitary transformation in this space can be approximated to arbitrary accuracy through successive braiding operations [114]. These properties together render non-Abelions preeminently suited for applications as protected qubits in quantum computation [115–119].

2.3.3 Majorana Fermions and Non-Abelian Statistics

The key to understanding the non-Abelian statistics [119] of the quasiparticle excitations of the Pfaffian state lies in the Majorana fermion modes in the vortices of p-wave superfluids [113, 120–122]. The p-wave pairing symmetry implies that the order parameter for the superfluid acquires a phase of 2π as we go around the Fermi surface,

$$\langle c^{\dagger}_{\boldsymbol{k}} c^{\dagger}_{-\boldsymbol{k}} \rangle = \Delta_0(k) \cdot (k_{\mathrm{x}} + \mathrm{i}k_{\mathrm{y}}), \tag{2.3.15}$$

where $\Delta_0(k)$ can be chosen real. The Hamiltonian for a single vortex at the origin is given by

$$H = \int \mathrm{d}\boldsymbol{r} \left\{ \psi^{\dagger} \left(-\frac{\boldsymbol{\nabla}^2}{2m} - \varepsilon_F \right) \psi + \psi^{\dagger} \left(e^{\mathrm{i}\varphi} \Delta_0(r) * (\partial_x - \mathrm{i}\partial_y) \right) \psi^{\dagger} + \mathrm{h.c.} \right\}, \tag{2.3.16}$$

where $A * B \equiv \frac{1}{2}\{A, B\}$ denotes the symmetrized product, and r and φ are polar coordinates. The order parameter $\Delta_0(r)$ vanishes inside the vortex core. We can obtain the energy eigenstates localized inside the vortex by solving the Bogoliubov–de Gennes equations [105] equations

$$\left[H, \gamma^{\dagger}_n(\boldsymbol{x}) \right] = E_n \gamma^{\dagger}_n(\boldsymbol{x}), \tag{2.3.17}$$

where n labels the modes and

$$\gamma^{\dagger}_n(\boldsymbol{x}) = u_n(\boldsymbol{x})\psi^{\dagger}(\boldsymbol{x}) + v_n(\boldsymbol{x})\psi(\boldsymbol{x}) \tag{2.3.18}$$

are the Bogoliubov quasiparticle operators. The low energy spectrum is given by [113, 120]

$$E_n = n\omega_0, \tag{2.3.19}$$

where n is an integer and $\omega_0 = \Delta^2/\varepsilon_F$ the level spacing. Note that while in an s-wave superfluid, the Bogoliubov operators

$$\gamma_{n\uparrow}(\boldsymbol{x}) = u_{n\uparrow}(\boldsymbol{x})\psi^{\dagger}(\boldsymbol{x}) + v_{n\downarrow}(\boldsymbol{x})\psi(\boldsymbol{x}) \tag{2.3.20}$$

combine ↑-spin electron creation operators with ↓-spin annihilation operators, in the p-wave superfluid, the operators (2.3.18) combine creation and annihilation operators of the *same* spinless (or spin-polarized) fermions. Since the Bogoliubov–de Gennes equations are not able to distinguish between particles and antiparticles, we obtain each physical solution twice: once with positive energy as a solution of the Bogoliubov–de Gennes equation (2.3.17) for the creation operators, and once with negative energy as a solution of the same equation for the annihilation operators,

$$\left[H, \gamma_n(\boldsymbol{x})\right] = -E_n\gamma_n(\boldsymbol{x}), \tag{2.3.21}$$

which is obtained from (2.3.17) by Hermitian conjugation. We resolve this technical artifact by discarding the negative energy solutions as unphysical. For the $n=0$ solution with at $E_0 = 0$, it implies that we get one fermion solution when we overcount by a factor of two. The physical solution at $E = 0$ is hence given by one half of a fermion, or a Majorana fermion, as

$$\gamma_0^{\dagger}(\boldsymbol{x}) = \gamma_0(\boldsymbol{x}). \tag{2.3.22}$$

In general, one fermion ψ, $\psi^{\dagger}$ consists of two Majorana fermions,

$$\psi = \frac{1}{2}(\gamma_1 + \mathrm{i}\gamma_2), \quad \psi^{\dagger} = \frac{1}{2}(\gamma_1 - \mathrm{i}\gamma_2), \tag{2.3.23}$$

which in turn are given by the real and imaginary part of the fermion operators,

$$\gamma_1 = \psi + \psi^{\dagger}, \quad \gamma_2 = -\mathrm{i}(\psi - \psi^{\dagger}). \tag{2.3.24}$$

They obey the anticommutation relations

$$\{\gamma_i, \gamma_j\} = 2\delta_{ij}, \tag{2.3.25}$$

as one may easily verify with (2.3.24). Majorana fermions are their own antiparticles, as $\gamma_i^{\dagger} = \gamma_i$. If we write the basis for a single fermion as $\{|0\rangle, \psi^{\dagger}|0\rangle\}$, we can write the fermion creation and annihilation operators as

$$\psi^{\dagger} = \begin{pmatrix} 0 & 0 \\ 1 & 0 \end{pmatrix}, \quad \psi = \begin{pmatrix} 0 & 1 \\ 0 & 0 \end{pmatrix}. \tag{2.3.26}$$

In this basis, the Majorana fermions are given by the first two Pauli matrices,

$$\gamma_1 = \begin{pmatrix} 0 & 1 \\ 1 & 0 \end{pmatrix} = \sigma_{\mathrm{x}}, \quad \gamma_2 = \begin{pmatrix} 0 & -\mathrm{i} \\ \mathrm{i} & 0 \end{pmatrix} = \sigma_{\mathrm{y}}. \tag{2.3.27}$$

Returning to vortices in a p-wave superfluid, note that the order parameter acquires by definition a phase of 2π as we go around a vortex. This implies that the electron

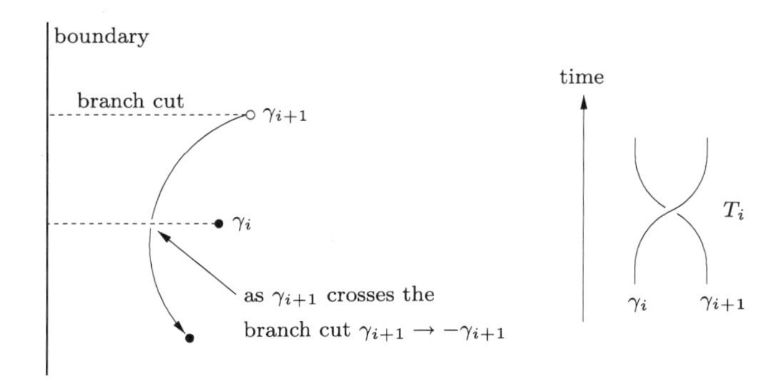

Fig. 2.6 The Majorana fermion γ_{i+1} acquires a $-$ sign as it crosses the branch cut from another vortex

creation and annihilation operators acquire a phase π, or a minus sign, which implies via (2.3.24) that the Majorana fermion states acquire likewise a minus sign,

$$\gamma_i \to -\gamma_i, \tag{2.3.28}$$

as we encircle a vortex. By choice of gauge, we can implement the phase change of 2π in the superconducting order parameter as a branch cut connecting the vortices to the left boundary of the system, and assume a convention according to which the Majorana fermion in each vortex crossing a branch cut acquires a minus sign, as illustrated in Fig. 2.6.

To obtain the non-Abelian statistics, Ivanov [121] considered permutations of $2n$ vortices by braiding, which form the braid group B_{2n} [123]. This group is generated by counterclockwise interchanges T_i of particles i and $i+1$, which are neighbors with regard to the positions of their branch cuts to the boundary. The algebra of the group is given by

$$\begin{aligned} T_i T_j &= T_j T_i \qquad \text{for } |i-j| > 1, \\ T_i T_j T_i &= T_j T_i T_j \qquad \text{for } |i-j| = 1, \end{aligned} \tag{2.3.29}$$

as illustrated in Fig. 2.7. Note that the braid group is different from the permutation group as

$$T_i^{-1} \neq T_i.$$

The convention for the minus signs acquired by the Majorana fermions defined in Fig. 2.6 implies the transformation rule

$$T_i(\gamma_j) = \begin{cases} \gamma_{j+1} & \text{for } i = j, \\ -\gamma_{j-1} & \text{for } i = j-1, \\ \gamma_j & \text{otherwise.} \end{cases} \tag{2.3.30}$$

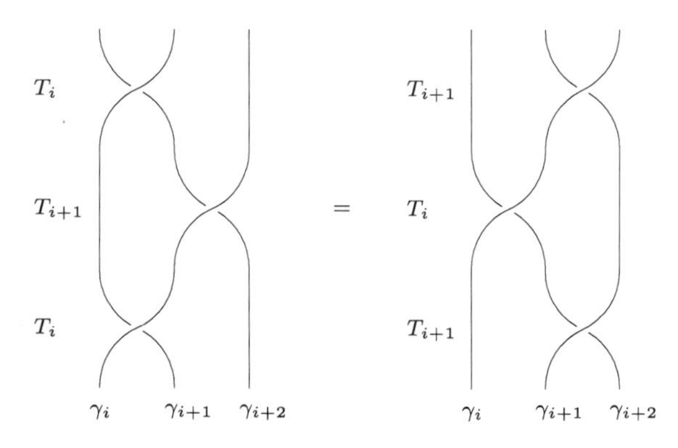

Fig. 2.7 Illustration of the defining algebra of the braid group B_{2n}: $T_i T_{i+1} T_i = T_{i+1} T_i T_{i+1}$

To describe the action of these transformations on the (internal) state vectors, we hence need to find a representation $\tau(T_i)$ of the braid group B_{2n} such that

$$\tau(T_i)\gamma_j \tau(T_i)^{-1} = T_i(\gamma_j) \tag{2.3.31}$$

with $T_i(\gamma_j)$ given by (2.3.30). The solution is [121]

$$\begin{aligned}\tau(T_i) &= \exp\left(\frac{\pi}{4}\gamma_{i+1}\gamma_i\right)\\ &= \cos\left(\frac{\pi}{4}\right) + \gamma_{i+1}\gamma_i \sin\left(\frac{\pi}{4}\right),\\ &= \frac{1}{\sqrt{2}}(1+\gamma_{i+1}\gamma_i),\end{aligned} \tag{2.3.32}$$

as one can easily verify using $(\gamma_{i+1}\gamma_i)^2 = -1$. The inverse transformation is given by

$$\tau(T_i)^{-1} = \frac{1}{\sqrt{2}}(1-\gamma_{i+1}\gamma_i). \tag{2.3.33}$$

A few steps of algebra yield

$$\tau(T_1)\begin{Bmatrix}\gamma_1\\ \gamma_2\end{Bmatrix}\tau(T_1)^{-1} = \begin{Bmatrix}\gamma_2\\ -\gamma_1\end{Bmatrix}.$$

This representation coincides with that of Nayak and Wilczek [112] for the statistics of the quasiholes in the Pfaffian state.

The simplest examples of this representation are the cases of two and four vortices [112, 121, 124], which we will elaborate now. In the case of two vortices, the two Majorana fermions γ_1 and γ_2 can be combined into a single fermion via (2.3.23), and the ground state is hence two-fold degenerate. The braid group B_2 has only one generator T_1 with representation

$$
\begin{aligned}
\tau(T_1) &= \exp\left(\frac{\pi}{4}\gamma_2\gamma_1\right) \\
&= \exp\left(-\mathrm{i}\frac{\pi}{4}(\psi - \psi^\dagger)(\psi + \psi^\dagger)\right) \\
&= \exp\left(\mathrm{i}\frac{\pi}{4}(2\psi^\dagger\psi - 1)\right) \\
&= \exp\left(-\mathrm{i}\frac{\pi}{4}\sigma_z\right),
\end{aligned}
\tag{2.3.34}
$$

where σ_z is the third Pauli matrix (2.1.66) in the basis $\{|0\rangle, \psi^\dagger|0\rangle\}$. The braiding is hence diagonal in this basis, and only gives an overall phase, which depends on whether the fermion state is occupied or not.

The non-Abelian statistics manifests itself only once we consider four vortices. Following Ivanov [121], we combine the four Majorana fermions into two fermions,

$$
\psi_1 = \frac{1}{2}(\gamma_1 + \mathrm{i}\gamma_2), \quad \psi_2 = \frac{1}{2}(\gamma_3 + \mathrm{i}\gamma_4), \tag{2.3.35}
$$

and accordingly for the fermion creation operators $\psi_1^\dagger$, $\psi_2^\dagger$. The braid group B_4 has three generators T_1, T_2, and T_3. Their representations in a basis of fermion occupation numbers

$$
\{|0\rangle, \psi_1^\dagger|0\rangle, \psi_2^\dagger|0\rangle, \psi_1^\dagger\psi_2^\dagger|0\rangle\},
$$

are given by two diagonal operators

$$
\tau(T_1) = \exp\left(\frac{\pi}{4}\gamma_2\gamma_1\right) = \exp\left(-\mathrm{i}\frac{\pi}{4}\sigma_z^{(1)}\right) = \begin{pmatrix} e^{-\mathrm{i}\pi/4} & 0 & 0 & 0 \\ 0 & e^{\mathrm{i}\pi/4} & 0 & 0 \\ 0 & 0 & e^{-\mathrm{i}\pi/4} & 0 \\ 0 & 0 & 0 & e^{\mathrm{i}\pi/4} \end{pmatrix},
$$

$$
\tau(T_3) = \exp\left(\frac{\pi}{4}\gamma_4\gamma_3\right) = \exp\left(-\mathrm{i}\frac{\pi}{4}\sigma_z^{(2)}\right) = \begin{pmatrix} e^{-\mathrm{i}\pi/4} & 0 & 0 & 0 \\ 0 & e^{-\mathrm{i}\pi/4} & 0 & 0 \\ 0 & 0 & e^{\mathrm{i}\pi/4} & 0 \\ 0 & 0 & 0 & e^{\mathrm{i}\pi/4} \end{pmatrix},
$$

and one off-diagonal operator,

$$
\begin{aligned}
\tau(T_2) &= \exp\left(\frac{\pi}{4}\gamma_3\gamma_2\right) \\
&= \frac{1}{\sqrt{2}}\left(1 - \mathrm{i}(\psi_2 + \psi_2^\dagger)(\psi_1 - \psi_1^\dagger)\right) = \begin{pmatrix} 1 & 0 & 0 & -\mathrm{i} \\ 0 & 1 & -\mathrm{i} & 0 \\ 0 & -\mathrm{i} & 1 & 0 \\ -\mathrm{i} & 0 & 0 & 1 \end{pmatrix}.
\end{aligned}
$$

Note that since the representations $\tau(T_i)$ given by (2.3.32) are even in the fermion operators, i.e., change the fermion numbers only by even integers, we may restrict them to only even or odd sectors in the fermion numbers. For the example of four vortices, these sectors are given by $\{|0\rangle, \psi_1^\dagger \psi_2^\dagger |0\rangle\}$ and $\{\psi_1^\dagger |0\rangle, \psi_2^\dagger |0\rangle\}$. Each sector contains 2^{n-1} states, which is the degeneracy found for a Pfaffian state with an even number of electrons [112]. Physically, this reflects that while the number of fermions is not a good quantum number in a superfluid, the number of fermions modulo two, i.e., whether the number is even or odd, is a good quantum number.

Finally, note that the derivation of the non-Abelian statistics depends only on (a) the vortices possessing Majorana fermion modes, and (b) the Majorana fermions changing sign $\gamma_i \to -\gamma_i$ when the order parameter phase changes by 2π, as it does by definition when we go around a vortex.

2.3.4 The Pfaffian State and Its Parent Hamiltonian on the Sphere

The Pfaffian state is readily formulated in the spherical geometry [92]. The wave function for N particles at Landau level filling $\nu = 1/m$ on a sphere with $2s_0 = m(N-1) - 1$ magnetic flux quanta is given by

$$\psi_0[u, v] = \mathrm{Pf}\left(\frac{1}{u_i v_j - u_j v_i}\right) \prod_{i<j}^{N} (u_i v_j - u_j v_i)^m, \tag{2.3.36}$$

where m is even for fermions and odd for bosons. Note that the relation between flux and particle number implies that the states at $\nu = 1/2$ is not its own particle-hole conjugate [125, 126]. The formulation of quasihole excitations generalizes without incident from the planar geometry.

As mentioned in Sect. 2.3.1, the uniquely specifying property of the Pfaffian state (2.3.36) is that it vanishes as the $(3m-1)$-th power of the distance as *three* particles approach each other. For the spherical geometry, the corresponding parent Hamiltonian can be conveniently formulated using three-body pseudopotentials [127]. In analogy to the two-particle interaction Hamiltonian (2.1.84), we write the three-particle interaction Hamiltonian

$$H_{\text{int}}^{(3)} = \sum_{i<j<k}^{N} \left\{ \sum_{l}^{2s} V_l^{(3)} \mathrm{P}_{3s-l}(\boldsymbol{L}_i + \boldsymbol{L}_j + \boldsymbol{L}_k) \right\}. \tag{2.3.37}$$

The three-body parent Hamiltonian proposed by Wen, Wilczek, and ourselves [91, 92] then amounts to taking

$$V_l^{(3)} = \begin{cases} 1 & \text{for } l < 3m-1, \\ 0 & \text{for } l \geq 3m-1. \end{cases} \tag{2.3.38}$$

The form (2.3.37) is not the most general one, as for $l \geq 6$ for bosons ($l \geq 9$ for fermions), the three particle state is no longer uniquely described by the three body angular angular momentum l, and one may assign different pseudopotential coefficients to the different symmetric (antisymmetric), homogeneous, rotationally invariant polynomials of degree l describing the three body states [127]. This, however, should not concern us here as we are only interested in the case $m = 1$ for bosons and $m = 2$ for fermions. Furthermore, as in the case of two-body pseudopotentials, where l had to be even for bosons and odd for fermions, there exists a related restriction for the allowed values of l for three-body pseudopotentials. Specifically, we have no state with $l = 1 (l = 4)$ for bosons (fermions).

For all practical purposes, we once again need to rewrite (2.3.37) in terms of boson or fermion creation or annihilation operators,

$$\begin{aligned} H_{\text{int}}^{(3)} = \sum_{m_1=-s}^{s} \sum_{m_2=-s}^{s} \sum_{m_3=-s}^{s} \sum_{m_4=-s}^{s} \sum_{m_5=-s}^{s} \sum_{m_6=-s}^{s} & a_{m_1}^{\dagger} a_{m_2}^{\dagger} a_{m_3}^{\dagger} a_{m_4} a_{m_5} a_{m_6} \\ & \cdot \delta_{m_1+m_2+m_3, m_4+m_5+m_6} \\ & \cdot \sum_{j=0}^{2s} \sum_{l=3s-(j+s)}^{3s-|j-s|} V_l^{(3)} \langle s, m_1; s, m_2 | j, m_1 + m_2 \rangle \\ & \cdot \langle j, m_1 + m_2; s, m_3 | 3s - l, m_1 + m_2 + m_3 \rangle \\ & \cdot \langle 3s - l, m_4 + m_5 + m_6 | s, m_4; j, m_5 + m_6 \rangle \\ & \cdot \langle j, m_5 + m_6 | s, m_5; s, m_6 \rangle, \end{aligned} \tag{2.3.39}$$

where a_m annihilates a boson or fermion in the properly normalized single particle state

$$\psi_{m,0}^{s}(u, v) = \sqrt{\frac{(2s+1)!}{4\pi (s+m)!(s-m)!}}\, u^{s+m} v^{s-m}, \tag{2.3.40}$$

and $\langle s, m_1; s, m_2 | 2s - l, m_1 + m_2 \rangle$ etc. are Clebsch–Gordan coefficients [39].

2.4 An $S = 1$ Spin Liquid State Described by a Pfaffian

2.4.1 The Ground State

As for the Haldane–Shastry model, we consider a one-dimensional lattice with periodic boundary conditions and an even number of sites N on a unit circle embedded in the complex plane. The only difference is that now the spin on each site is $S = 1$:

N sites with spin 1 on unit circle:

$$\eta_\alpha = e^{i \frac{2\pi}{N} \alpha} \quad \text{with } \alpha = 1, \ldots, N$$

The ground state wave function we consider here [128] is given by a bosonic Pfaffian state in the complex lattice coordinates z_i supplemented by a phase factor,

$$\psi_0^{S=1}(z_1, z_2, \ldots, z_N) = \text{Pf}\left(\frac{1}{z_i - z_j}\right) \prod_{i<j}^{N} (z_i - z_j) \prod_{i=1}^{N} z_i. \tag{2.4.1}$$

The Pfaffian is given by the fully antisymmetrized sum over all possible pairings of the N particle coordinates,

$$\text{Pf}\left(\frac{1}{z_i - z_j}\right) \equiv \mathcal{A}\left\{\frac{1}{z_1 - z_2} \cdot \ldots \cdot \frac{1}{z_{N-1} - z_N}\right\}. \tag{2.4.2}$$

The "particles" z_i represent re-normalized spin flips $\tilde{S}_\alpha^+$ acting on a vacuum with all spins in the $S^z = -1$ state,

$$\left|\psi_0^{S=1}\right\rangle = \sum_{\{z_1, \ldots, z_N\}} \psi_0^{S=1}(z_1, \ldots, z_N)\, \tilde{S}_{z_1}^+ \cdot \ldots \cdot \tilde{S}_{z_N}^+ |-1\rangle_N, \tag{2.4.3}$$

where the sum extends over all possibilities of distributing the N "particles" over the N lattice sites allowing for double occupation,

$$\tilde{S}_\alpha^+ \equiv \frac{S_\alpha^z + 1}{2} S_\alpha^+, \tag{2.4.4}$$

and

$$|-1\rangle_N \equiv \otimes_{\alpha=1}^{N} |1, -1\rangle_\alpha. \tag{2.4.5}$$

This state may be viewed as the one-dimensional analog of the non-Abelian chiral spin liquid [129].

Like the ground state of the Haldane–Shastry model, the $S=1$ state (2.4.1) describes a critical spin liquid in one dimension, with similarly algebraically decaying correlations. It does not, however, serve as a paradigm of the generic $S=1$ spin state, as the generic state possesses a Haldane gap [130–133] in the spin excitation spectrum due to linearly confining forces between the spinons [128, 134–137].

One of the objectives of this work is to identify a parent Hamiltonian for which this state is the exact ground state, and hence accomplish what Haldane and Shastry have accomplished for the spin one-half Gutzwiller wave function.

2.4.2 Symmetries

Translational invariance. As for the Haldane–Shastry model, we obtain the ground state momentum p_0 (in units of inverse lattice spacings $1/a$) by translating the ground state by one lattice spacing around the unit circle,

$$T\left|\psi_0^{S=1}\right\rangle = e^{\mathrm{i}p_0}\left|\psi_0^{S=1}\right\rangle. \tag{2.4.6}$$

With $T z_i = \exp\left(\mathrm{i}\frac{2\pi}{N}\right) z_i$ we find

$$p_0 = \frac{2\pi}{N}\left(-\frac{N}{2} + \frac{N(N-1)}{2} + N\right) = \pi N, \tag{2.4.7}$$

which implies $p_0 = 0$ as N is even.

Invariance under SU(2) spin rotations. The proof of the singlet property is similar to the Haldane–Shastry model, but more instructive as it motivates the re-normalization of the spin-flip operators in (2.4.4).

Since $S^{z}_{\mathrm{tot}}\left|\psi_0^{S=1}\right\rangle = 0$ by construction, it is sufficient to show $S^{-}_{\mathrm{tot}}\left|\psi_0^{S=1}\right\rangle = 0$. Note first that when we substitute (2.4.1) with (2.4.2) into (2.4.3), we may replace the antisymmetrization $\mathcal{A}$ in (2.4.2) by an overall normalization factor, as it is taken care by the commutativity of the bosonic operators $\tilde{S}_\alpha$. Let $\tilde{\psi}_0$ be $\psi_0^{S=1}$ without the antisymmetrization in (2.4.2),

$$\tilde{\psi}_0[z_i] = (N-1)!!\left\{\frac{1}{z_1 - z_2}\cdot\ldots\cdot\frac{1}{z_{N-1} - z_N}\right\}\cdot\prod_{i<j}^{N}(z_i - z_j)\prod_{i=1}^{N} z_i. \tag{2.4.8}$$

Since $\tilde{\psi}_0(z_1, z_2, \ldots, z_N)$ is still symmetric under interchange of pairs, we may assume that a spin flip operator S^{-}_α acting on $|\tilde{\psi}_0\rangle$ will act on the pair (z_1, z_2),

$$\begin{aligned} S^{-}_\alpha\left|\psi_0^{S=1}\right\rangle = \sum_{\{z_3,\ldots,z_N\}}\Bigg\{ &\sum_{z_2(\neq\eta_\alpha)} \tilde{\psi}_0(\eta_\alpha, z_2, z_3, \ldots) S^{-}_\alpha \tilde{S}^{+}_\alpha \tilde{S}^{+}_{z_2} \\ &+ \sum_{z_1(\neq\eta_\alpha)} \tilde{\psi}_0(z_1, \eta_\alpha, z_3, \ldots) S^{-}_\alpha \tilde{S}^{+}_{z_1} \tilde{S}^{+}_\alpha \\ &+ \tilde{\psi}_0(\eta_\alpha, \eta_\alpha, z_3, \ldots) S^{-}_\alpha (\tilde{S}^{+}_\alpha)^2 \Bigg\} \tilde{S}^{+}_{z_3}\ldots\tilde{S}^{+}_{z_N}|-1\rangle_N \\ = \sum_{\{z_3,\ldots,z_N\}}\Bigg\{&\sum_{z_2} 2\tilde{\psi}_0(\eta_\alpha, z_2, z_3, \ldots)\tilde{S}^{+}_{z_2}\Bigg\} \tilde{S}^{+}_{z_3}\ldots\tilde{S}^{+}_{z_N}|-1\rangle_N, \end{aligned} \tag{2.4.9}$$

where we have used

$$S^{-}_\alpha (\tilde{S}^{+}_\alpha)^n |1, -1\rangle_\alpha = n(\tilde{S}^{+}_\alpha)^{n-1}|1, -1\rangle_\alpha, \tag{2.4.10}$$

which follows directly form the definition (2.4.4).

This implies

$$S_{\text{tot}}^{-}\left|\psi_0^{S=1}\right\rangle = \sum_{\alpha=1}^{N} S_\alpha^{-}\left|\psi_0^{S=1}\right\rangle$$

$$= 2 \sum_{\{z_2\ldots,z_N\}} \underbrace{\sum_{\alpha=1}^{N} \tilde{\psi}_0(\eta_\alpha, z_2, \ldots, z_N)}_{=0} \tilde{S}_{z_2}^{+} \ldots \tilde{S}_{z_N}^{+} |-1\rangle_N, \qquad (2.4.11)$$

since $\tilde{\psi}_0(\eta_\alpha, z_2, \ldots, z_N)$ contains only powers $\eta_\alpha^1, \eta_\alpha^2, \ldots, \eta_\alpha^{N-1}$ in η_α, and

$$\sum_{\alpha=1}^{N} \eta_\alpha^m = N\delta_{m,0} \mod N.$$

Parity and time reversal invariance. To show that $\psi_0(z_1, \ldots, z_N)$ is real, and hence that $\left|\psi_0^{S=1}\right\rangle$ is invariant under parity, we calculate its complex conjugate,

$$\begin{aligned}\left(\psi_0^{S=1}[z]\right)^* &= \text{Pf}\left(\frac{1}{\frac{1}{z_i} - \frac{1}{z_j}}\right) \prod_{i<j}^{N} \left(\frac{1}{z_i} - \frac{1}{z_j}\right) \prod_{i=1}^{N} \frac{1}{z_i} \\ &= (-1)^{\frac{N}{2}} \prod_{i=1}^{N} \frac{1}{z_i} \, (-1)^{\frac{N(N-1)}{2}} \prod_{i<j}^{N} \frac{1}{z_i z_j} \prod_{i=1}^{N} \frac{1}{z_i^2} \, \psi_0^{S=1}[z] \\ &= \psi_0^{S=1}[z], \end{aligned} \qquad (2.4.12)$$

as N is even and $z_i^N = 1$ for all i. Time reversal [73] transforms

$$\mathrm{i} \to -\mathrm{i}, \quad z_i \to \bar{z}_i, \quad \boldsymbol{S}_\alpha \to -\boldsymbol{S}_\alpha, \quad |s, m\rangle \to \mathrm{i}^{2m} |s, -m\rangle,$$

which implies that the basis states in (2.4.3) transform according to

$$\tilde{S}_{z_1}^{+} \cdot \ldots \cdot \tilde{S}_{z_N}^{+} |-1\rangle_N \to \tilde{S}_{z_1}^{-} \cdot \ldots \cdot \tilde{S}_{z_N}^{-} |+1\rangle_N, \qquad (2.4.13)$$

where

$$\tilde{S}_\alpha^{-} \equiv \frac{-S_\alpha^{\mathrm{z}} + 1}{2} S_\alpha^{-}, \quad |+1\rangle_N \equiv \otimes_{\alpha=1}^{N} |1, +1\rangle_\alpha. \qquad (2.4.14)$$

Together with the singlet property, this implies that $\left|\psi_0^{S=1}\right\rangle$ is invariant under time reversal.

All the symmetries properties discussed here will emerge almost trivially when we generate the state $\left|\psi_0^{S=1}\right\rangle$ through projection form Gutzwiller (or Haldane–Shastry ground) states in Sect. 2.4.4.

2.4.3 Schwinger Bosons

Schwinger bosons [138, 139] constitute a way to formulate spin-S representations of an SU(2) algebra (which can easily be generalized to SU(n), see e.g. [136]). The spin operators

$$\boldsymbol{S} = \frac{1}{2}\left(a^{\dagger}, b^{\dagger}\right)\boldsymbol{\sigma}\begin{pmatrix} a \\ b \end{pmatrix}, \tag{2.4.15}$$

where $\boldsymbol{\sigma} = (\sigma_{\mathrm{x}}, \sigma_{\mathrm{y}}, \sigma_{\mathrm{z}})$ is the vector consisting of the three Pauli matrices (2.1.66), are given in terms of boson creation and annihilation operators which obey the usual commutation relations

$$\begin{aligned} \left[a, a^{\dagger}\right] &= \left[b, b^{\dagger}\right] = 1, \\ \left[a, b\right] = \left[a, b^{\dagger}\right] &= \left[a^{\dagger}, b\right] = \left[a^{\dagger}, b^{\dagger}\right] = 0. \end{aligned} \tag{2.4.16}$$

It is readily verified with

$$\left[\sigma_i, \sigma_j\right] = 2\mathrm{i}\varepsilon^{ijk}\sigma_k \tag{2.4.17}$$

and (2.4.16), that S^{x}, S^{y}, and S^{z} satisfy the SU(2) algebra

$$\left[S^i, S^j\right] = \mathrm{i}\varepsilon^{ijk}S^k. \tag{2.4.18}$$

Written out in components we have

$$\begin{aligned} S^{\mathrm{x}} + \mathrm{i}S^{\mathrm{y}} &= S^{+} = a^{\dagger}b, \\ S^{\mathrm{x}} - \mathrm{i}S^{\mathrm{y}} &= S^{-} = b^{\dagger}a, \\ S^{\mathrm{z}} &= \frac{1}{2}(a^{\dagger}a - b^{\dagger}b). \end{aligned} \tag{2.4.19}$$

The spin quantum number S is given by half the number of bosons,

$$2S = a^{\dagger}a + b^{\dagger}b, \tag{2.4.20}$$

and the usual spin states (simultaneous eigenstates of $\boldsymbol{S}^2$ and S^{z}) are given by

$$|S, m\rangle = \frac{(a^{\dagger})^{S+m}}{\sqrt{(S+m)!}}\frac{(b^{\dagger})^{S-m}}{\sqrt{(S-m)!}}|0\rangle. \tag{2.4.21}$$

In particular, the spin-$\frac{1}{2}$ states are given by

$$|\uparrow\rangle = c_{\uparrow}^{\dagger}|0\rangle = a^{\dagger}|0\rangle, \quad |\downarrow\rangle = c_{\downarrow}^{\dagger}|0\rangle = b^{\dagger}|0\rangle, \tag{2.4.22}$$

i.e., $a^{\dagger}$ and $b^{\dagger}$ act just like the fermion creation operators $c_{\uparrow}^{\dagger}$ and $c_{\downarrow}^{\dagger}$ in this case. The difference shows up only when two (or more) creation operators act on the same site or orbital. The fermion operators create an antisymmetric or singlet configuration (in accordance with the Pauli principle),

$$|0,0\rangle = c_{\uparrow}^{\dagger} c_{\downarrow}^{\dagger} |0\rangle, \tag{2.4.23}$$

while the Schwinger bosons create a totally symmetric or triplet (or higher spin if we create more than two-bosons) configuration,

$$\begin{aligned} |1,1\rangle &= \frac{1}{\sqrt{2}} (a^{\dagger})^2 |0\rangle, \\ |1,0\rangle &= a^{\dagger} b^{\dagger} |0\rangle, \\ |1,-1\rangle &= \frac{1}{\sqrt{2}} (b^{\dagger})^2 |0\rangle. \end{aligned} \tag{2.4.24}$$

Representations of spin $\frac{1}{2}$ states in terms of Schwinger bosons (rather than fermion creation operators or spin flips) are ideally suited for the construction of higher spin states through projection of $2S$ spin $\frac{1}{2}$'s onto the spin S representations (i.e., the symmetric representation) contained in

$$\underbrace{\mathbf{\frac{1}{2}} \otimes \mathbf{\frac{1}{2}} \otimes \ldots \otimes \mathbf{\frac{1}{2}}}_{2S} = \boldsymbol{S} \oplus (2S-1)\boldsymbol{S-1} \oplus \ldots \tag{2.4.25}$$

Classic examples include the formulation of the Affleck–Kennedy–Lieb–Tasaki (AKLT) model [134, 135] in terms of Schwinger bosons [139, 140] as well as the $S=1$ chirality liquid [128].

2.4.4 Generation by Projection from Gutzwiller States

We will show now that the $S=1$ ground state (2.4.1) can alternatively be generated by considering two (identical) Haldane–Shastry or Gutzwiller states (2.2.3) and projecting onto the triplet or $S=1$ configuration contained in

$$\mathbf{\frac{1}{2}} \otimes \mathbf{\frac{1}{2}} = \mathbf{0} \oplus \mathbf{1} \tag{2.4.26}$$

at each site [128, 129]. To begin with, we rewrite (2.2.2) in terms of Schwinger bosons,

$$\begin{aligned} \left|\psi_0^{\text{HS}}\right\rangle &= \sum_{\{z_1,z_2,\ldots,z_M\}} \psi_0^{\text{HS}}[z]\, S_{z_1}^{+} \cdot \ldots \cdot S_{z_M}^{+} \left|\downarrow\downarrow \ldots \downarrow\right\rangle \\ &= \sum_{\{z_1,\ldots,z_M;w_1,\ldots,w_M\}} \psi_0^{\text{HS}}[z]\, a_{z_1}^{+} \ldots a_{z_M}^{\dagger} b_{w_1}^{+} \ldots b_{w_M}^{\dagger} |0\rangle \\ &\equiv \Psi_0^{\text{HS}}\left[a^{\dagger}, b^{\dagger}\right] |0\rangle, \end{aligned} \tag{2.4.27}$$

where $M = \frac{N}{2}$ and the w_k's are those lattice sites which are not occupied by any of the z_i's. The $S = 1$ state (2.4.1) is then up to an overall normalization factor given by

$$\left|\psi_0^{S=1}\right\rangle = \left(\Psi_0^{\mathrm{HS}}\left[a^\dagger, b^\dagger\right]\right)^2 |0\rangle. \tag{2.4.28}$$

To verify (2.4.28), use the identity

$$\mathcal{S}\left\{\prod_{\substack{i,j=1\\i<j}}^{M}(z_i - z_j)^2 \prod_{\substack{i,j=M+1\\i<j}}^{2M}(z_i - z_j)^2\right\} = \mathrm{Pf}\left(\frac{1}{z_i - z_j}\right)\prod_{i<j}^{2M}(z_i - z_j), \tag{2.4.29}$$

where $\mathcal{S}$ indicates symmetrization over all the variables in the curly brackets, and

$$\frac{1}{\sqrt{2}}(a^\dagger)^n (b^\dagger)^{(2-n)}|0\rangle = (\tilde{S}^+)^n |1, -1\rangle, \tag{2.4.30}$$

which is readily verified with (2.4.19), (2.4.24), and the definition (2.4.4). To proof (2.4.29), use the following identity due to Frobenius [110],

$$\det\left(\frac{1}{z_i - z_{M+j}}\right) = (-1)^{\frac{M(M+1)}{2}} \frac{\prod\limits_{\substack{i,j=1\\i<j}}^{M}(z_i - z_j)\prod\limits_{\substack{i,j=M+1\\i<j}}^{2M}(z_i - z_j)}{\prod\limits_{i=1}^{M}\prod\limits_{j=M+1}^{2M}(z_i - z_j)}. \tag{2.4.31}$$

The projective construction directly reveals several interesting features, which were not nearly as obvious in the previous formulation:

(a) Since the Haldane–Shastry ground state $\left|\psi_0^{\mathrm{HS}}\right\rangle$ is translationally invariant with ground state momentum $p_0 = 0$ or π (depending on whether $\frac{N}{2}$ is even or odd), the $S = 1$ state $\left|\psi_0^{S=1}\right\rangle$ is translationally invariant with $p_0 = 0$.

(b) Since $\left|\psi_0^{\mathrm{HS}}\right\rangle$ is a singlet, and the projection onto spin $S = 1$ on each site commutes with spin rotations, $\left|\psi_0^{S=1}\right\rangle$ has to be a singlet as well.

(c) Since $\psi_0^{\mathrm{HS}}(z_1, \ldots, z_M)$ is real with the sign of each spin configuration given by $\prod_{i=1}^{M} G(z_i)$, the $S = 1$ wave function $\psi_0^{S=1}(z_1, \ldots, z_M)$ is likewise real with the sign given by $\prod_{i=1}^{N} G(z_i)$:

$$\psi_0^{S=1}(z_1, \ldots, z_N) = \left|\mathrm{Pf}\left(\frac{1}{z_i - z_j}\right)\prod_{i<j}^{N}(z_i - z_j)\right|\prod_{i=1}^{N} G(z_i), \tag{2.4.32}$$

with $G(\eta_\alpha) = \pm 1$ depending on whether α even or odd.

(d) Since $\left|\psi_0^{\mathrm{HS}}\right\rangle$ is invariant under parity and and time reversal, $\left|\psi_0^{S=1}\right\rangle$ is invariant as well.

2.4.5 *Topological Degeneracies and Non-Abelian Statistics*

We have seen in Sect. 2.3.3 that $2n$ spatially well separated quasiparticle excitations or vortices carrying half of a Dirac flux quanta each in the non-Abelian quantized Hall state described by the Pfaffian will span an internal or topological Hilbert space of dimensions 2^n (2^{n-1} for either even or odd fermion numbers), in accordance with the existence of one Majorana fermion state at each vortex core. The Majorana fermion states can only be manipulated through braiding of the vortices, with the interchanges being non-commutative or non-Abelian.

The question we wish to address in this section is whether there is any manifestation of this topological space of dimension 2^n, or the $2n$ Majorana fermion states, in the spinon excitation Hilbert space suggested by the $S = 1$ ground state (2.4.1). In Sect. 2.2.6, we have seen that the fractional statistics of the spinons in the Haldane–Shastry model, and presumably in any model supporting one-dimensional anyons, is encoded in the momentum spacings of the excitations. This is not too surprising, as there are no other suitable quantum numbers, like the relative angular momentum for two-dimensional anyons, available. We will propose now that the topological degeneracies, or the occupation numbers of the n fermions consisting of the $2n$ Majorana fermions, are once again encoded in the momentum spacings between single spinon states.

In the Haldane–Shastry model, the spacings between neighboring momenta were always half integer, in accordance with half-fermi statistics, as the difference between consecutive spinon momentum numbers a_i was always an odd integer,

$$a_{i+1} - a_i = \text{odd}. \tag{2.4.33}$$

This follows directly from the construction of the extended Young tableaux illustrated in Fig. 2.5. When two spinons are in neighboring columns, the difference of the a_i is one and hence an odd integer; when we insert complete columns without spinons in between, the number of boxes we insert is always even.

We will now show that for the $S = 1$ chain with the Hilbert space parameterized by the ground state $\left|\psi_0^{S=1}\right\rangle$ and spinon excitations above it, the corresponding rule is

$$\begin{aligned} a_{i+1} - a_i &= \text{even or odd}, \qquad &&\text{for } i \text{ odd},\\ a_{i+1} - a_i &= \text{odd}, &&\text{for } i \text{ even}. \end{aligned} \tag{2.4.34}$$

As $i = 1, 2, \ldots, 2n$, we have a total of n spacings which can be either even or odd, and another n spacings which are always odd. With the single spinon momenta given by

$$p_i = \frac{\pi}{N}\left(a_i - \frac{1}{2}\right), \tag{2.4.35}$$

this yields momentum spacings which can be either an integer or an half-integer times $\frac{2\pi}{N}$ for i odd. This is a topological distinction—for Abelian anyons, one choice

11 ⊗ 22 ⊗ 33 = 1133 / 22•• ⊕ 112• / •233 ⊕ 1123 / •23• ⊕ 11233 / •2•••
$S=1$ $S=0$ $S=1$ $S=2$
= 11 / 22 ⊕ 112 / •2•
⊕ 1122 / •••• ⊕ 1122 / ••33 ⊕ 11223 / •••3• ⊕ 112233 / ••••••
$S=1$ $S=2$ $S=3$

Fig. 2.8 Total spin representations of three $S = 1$ spins in terms of extended Young tableaux

corresponds to bosons or fermions (which are for many purposes equivalent in one dimension), and the other choice to half-fermions. For spinons which are well separated in momentum space, the states spanning this in total 2^n dimensional topological Hilbert space become degenerate as we approach the thermodynamic limit.

To derive (2.4.34), we introduce a second formalism of extended Young tableaux, this time for spin $S = 1$. The general rule we wish to propose for obtaining the tableaux is illustrated in Fig. 2.8 for three spins with $S = 1$. The construction is as follows. For each of the N spins, put a row of two adjacent boxes, which is equivalent to the Young tableau for a single spin without any numbers in the boxes. Put these N small tableaux on a line and number them consecutively from left to right, with the same number in each pair of boxes which represent a single spin. To obtain the product of some extended Young tableau representing spin S_0 on the left with a spin 1 tableau (i.e., a row of two boxes with the same number in it) on the right, we follow the rule

$$\mathbf{S}_0 \otimes \mathbf{1} = \begin{cases} \mathbf{1}, & \text{for } S_0 = 0, \\ \mathbf{S}_0\mathbf{-1} \oplus \mathbf{S}_0 \oplus \mathbf{S}_0\mathbf{+}1, & \text{for } S_0 = 1, 2, \ldots \end{cases} \tag{2.4.36}$$

i.e., we obtain only one new tableau with both boxes from the right added to the top row if the tableau on the left is a singlet, and three new tableaux if it is has spin one or higher. These three tableaux are constructed by adding both boxes to the bottom row (resulting in a representation $\mathbf{S}_0\mathbf{-1}$), by adding the first box to the bottom row and the second box to the top row without stacking them on top of each other (resulting in a representation $\mathbf{S}_0$), and by adding both boxes to the top row (resulting in a representation $\mathbf{S}_0\mathbf{+1}$). In each extended tableau, the boxes must be arranged such that the numbers are strictly increasing in each column from top to bottom, and that they are not decreasing from left to right in that the smallest number in each column cannot be smaller than the largest number in the column to the left of it. In analogy to the Haldane–Shastry model, the empty spaces in between the boxes are filled with dots representing spinons. The spinon momentum number a_i associated with each spinon is given by the number in the box in the same column. A complete table of all the extended Young tableaux for fours $S = 1$ spins is shown in Fig. 2.9. The assignment of physical single spinon momenta to the spinon momentum numbers (2.4.47) is identical to this assignment for the Haldane–Shastry model, as we can obtain the 3^N states of the $S = 1$ Hilbert space by Schwinger boson projection (i.e., by projecting on spin $S = 1$ on each site) from states contained in the $2^N \times 2^N$ dimensional Hilbert space of two $S = \frac{1}{2}$ models, a projection which commutes with

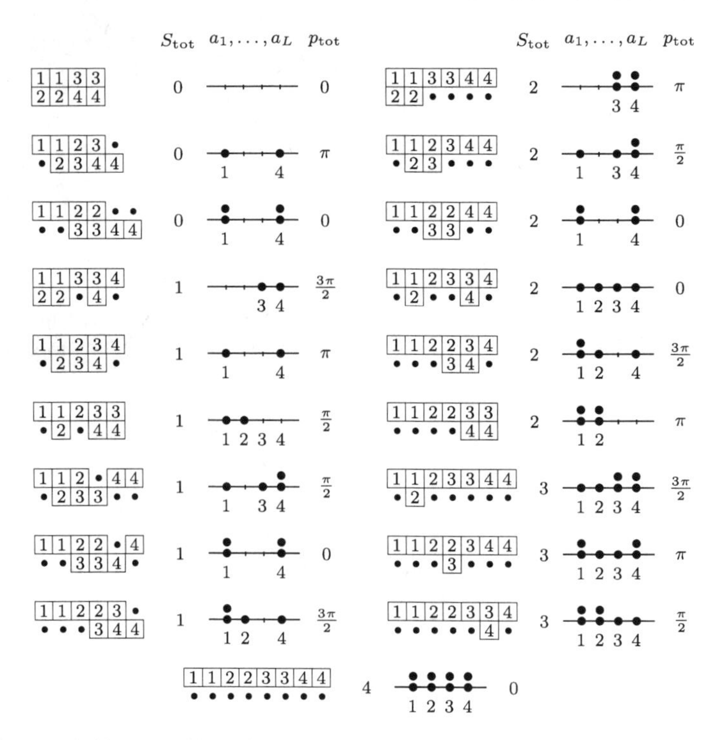

Fig. 2.9 Extended Young tableau decomposition for an $S = 1$ spin chain with $N = 4$ sites. The dots represent the spinons. The spinon momentum numbers a_i are given by the numbers in the boxes of the same column. Note that $\sum(2S_{\text{tot}} + 1) = 3^N$

the total momentum. The correctness of this assignment has further been verified numerically up to $N = 16$ sites [141].

With the tableau structure thus in place, all that is left to show is that the momentum spacings are according to (2.4.34). Looking at any of the tableaux in Fig. 2.9, we note that from left to right, the spinons alternate between being assigned to the first of the two boxes with a given number and being assigned to the second of such two boxes. This follows simply form the fact that the number of boxes in between the columns with the two neighboring spinons must be even. The first spinon momentum number a_1 is always odd, but all the other a_i's can be either even or odd. The rule is therefore that if i is odd, the ith spinon is assigned to the first of the two boxes with number a_i, and the momentum spacing $a_{i+1} - a_i$ can be either even or odd,

3 3 / • • (even) or 3 3 4 / • 4 • (odd) or 3 3 4 5 / • 4 5 • (even) or ...

If i is even, however, the ith spinon is assigned to the second of the two boxes with number a_i, and the momentum spacing $a_{i+1} - a_i$ has to be odd, as we can insert only an even number of columns between the two spinons (recall that we cannot stack two boxes with the same number in it on top of each other):

3 4	or	3 4 4 6	or	3 4 4 6 6 8	or ...
• •		• 5 5 •		• 5 5 7 7 •	
odd		odd		odd	

The spacings between the single spinon momenta are hence as stated in (2.4.34). This result is consistent with the spinon basis proposed by Bouwknegt, Ludwig, and Schoutens [143] for the SU(2) level $k = 2$ Wess-Zumino-Witten model [144, 145].

2.4.6 Generalization to Arbitrary Spin S

The projective generation introduced in Sect. 2.4.4 can be generalized to arbitrary spin $S = s$:

$$\left|\psi_0^S\right\rangle = \left(\Psi_0^{\text{HS}}\left[a^\dagger, b^\dagger\right]\right)^{2s} |0\rangle. \tag{2.4.37}$$

In order to write this state in a form similar to (2.4.1)–(2.4.5),

$$\left|\psi_0^S\right\rangle = \sum_{\{z_1,\ldots,z_{sN}\}} \psi_0^S(z_1,\ldots,z_{sN})\, \tilde{S}_{z_1}^+ \cdot\ldots\cdot \tilde{S}_{z_{sN}}^+ |-s\rangle_N, \tag{2.4.38}$$

where

$$|-s\rangle_N \equiv \otimes_{\alpha=1}^N |s,-s\rangle_\alpha \tag{2.4.39}$$

is the "vacuum" state in which all the spins are maximally polarized in the negative $\hat{z}$-direction, we introduce re-normalized spin flip operators $\tilde{S}^+$ which satisfy

$$\frac{1}{\sqrt{(2s)!}} (a^\dagger)^n (b^\dagger)^{(2s-n)} |0\rangle = (\tilde{S}^+)^n |s,-s\rangle. \tag{2.4.40}$$

If we assume a basis in which S^z is diagonal, we may write

$$\tilde{S}^+ \equiv \frac{1}{b^\dagger b + 1} a^\dagger b = \frac{1}{s - S^z + 1} S^+. \tag{2.4.41}$$

The wave function for the spin S state (2.4.37) is then with $M = \frac{N}{2}$ given by

$$\psi_0^S(z_1,\ldots,z_{sN}) = \prod_{m=1}^{2s} \left(\prod_{\substack{i,j=(m-1)M+1 \\ i<j}}^{mM} (z_i - z_j)^2 \right) \prod_{i=1}^{sN} z_i. \tag{2.4.42}$$

Note that these states are similar to the Read–Rezayi states [142] in the quantized Hall effect.

As for the $S = 1$ state discussed in Sect. 2.4.4, the projective construction (2.4.37) directly implies several symmetries. The state $\left|\psi_0^S\right\rangle$ is translationally invariant with ground state momentum $p_0 = -\pi N s$, a spin singlet, and real:

$$\psi_0^S(z_1, \ldots, z_{sN}) = \left|\psi_0^S(z_1, \ldots, z_{sN})\right| \prod_{i=1}^{sN} G(z_i), \tag{2.4.43}$$

with $G(z_i)$ given by (2.2.16).

2.4.7 *Momentum Spacings and Topological Degeneracies for Arbitrary Spin S*

In Sect. 2.4.5, we have shown that the non-Abelian statistics of the Pfaffian state (2.3.1), and in particular the topological degeneracies associated with the Majorana fermion states in the vortex cores discussed in Sect. 2.3.3, manifests itself in topological choices for the (kinematical) momentum spacings of the spinon excitations above the $S = 1$ ground state (2.4.1). Specifically, we found that if we label the single spinon momenta in ascending order by $p_i < p_{i+1}$, the spacings $p_{i+1} - p_i$ can be either even or odd multiples of $\frac{\pi}{N}$ if i is odd, while it has to be an odd multiple if i is even.

In this Section, we formulate the corresponding restrictions for the general spin S chain with ground state (2.4.37). We will first state the rules and then motivate them. Recall that spinons are represented by dots placed in the empty spaces of extended Young tableaux, and that the momentum number a_i of spinon i is given by the number in the box it shares a column with. For general spin S, the tableau describing the representation on each site is given by

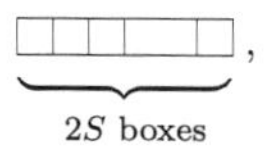

i.e., a horizontal array of $2S$ boxes indicating symmetrization, which all contain the same number.

If this number is n, the spinons we assign to any of these boxes will have momentum number $a_i = n$. Let us denote the number of the box a given spinon i with momentum number a_i is assigned to, by b_i, such that box number $b_i = 1$ corresponds to the first, and box number $b_i = 2S$ to the last box with number n in it:

n	n	n	…	n
•				

$b_i = 1$

n	n	n	…	n
	•			

$b_i = 2$, …

n	n	n	…	n
				•

$b_i = 2S$

We will see below that if a representation of a spin S chain with L spinons is written in terms of an extended Young tableau, the first spinon with momentum number a_1

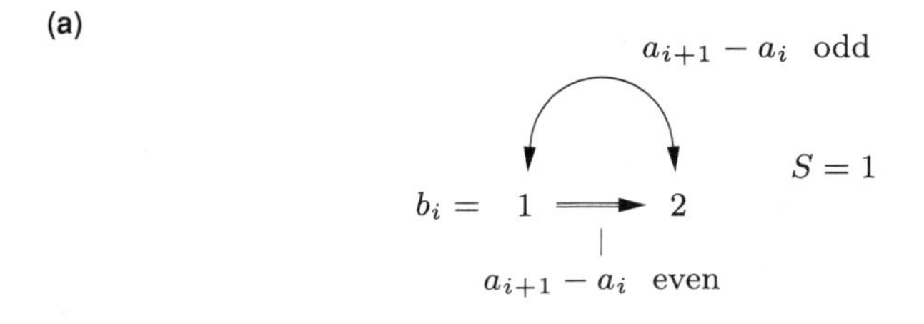

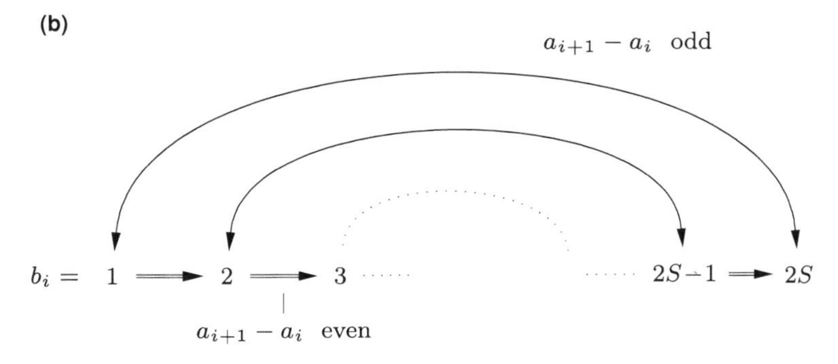

Fig. 2.10 Non-Abelian (SU(2) level $k = 2S$) statistics in one dimension: flow diagram for the (auxiliary) box numbers b_i, which serve to describe the restrictions for the spinon momentum number spacings $a_{i+1} - a_i$ for the critical models of spin chains introduced in Sects. 2.4.1 and 2.4.6 with **a** $S = 1$, and **b** general spin S. The unidirectional, horizontal arrows correspond to even integer momentum number spacings $a_{i+1} - a_i$, while the bidirectional, semicircle arrows correspond to odd integer spacings

will always have box number $b_1 = 1$, and the last spinon with a_L will have $b_L = 2S$. The restrictions corresponding to the non-abelian (SU(2) level $k = 2S$) statistics of the spinons are described by the flow diagram of the numbers b_i shown in Fig. 2.10.

Let us elaborate this diagram first for the case $S = 1$, which we have already studied in Sect. 2.4.5. In this case,

$$b_i = \begin{cases} 1, & \text{for } i \text{ odd,} \\ 2, & \text{for } i \text{ even.} \end{cases} \tag{2.4.44}$$

For i odd, we may move from $b_i = 1$ to $b_{i+1} = 2$ either via the horizontal arrow or via the semicircle in Fig. 2.10a, and $a_{i+1} - a_i$ may hence be either even or odd, respectively. For i even or $b_i = 2$, however, the semicircle is the only available continuation, which implies that the spacing $a_{i+1} - a_i$ must be odd.

For general S, Fig. 2.10b implies that the spacings can be even or odd until $b_i = 2S$ is reached, which is then followed by an odd integer spacing $a_{i+1} - a_i$, as the semicircular arrow is the only possible continuation at this point. Note that for $S \geq 1$, the minimal number of spinons is two (these two spinons then have an odd integer spacing $a_2 - a_1$), and that we cannot have more than $2S$ spinons with the same momentum number $a_i = n$, as $a_{i+1} - a_i = 0$ is even.

We will now motivate this diagram. To begin with, we generalize the formalism of extended Young tableaux to arbitrary spin S. The construction is similar to the one for $S=1$ outlined in Sect. 2.4.5. For each of the N spins, put a row of $2S$ adjacent boxes. Put these N tableaux on a line and number them consecutively from left to right, with the same number in each row of $2S$ boxes representing a single spin. To obtain the product of some extended Young tableau representing spin S_0 on the left with a spin S tableau (i.e., a row of $2S$ boxes with the same number in it) on the right, we first recall

$$\boldsymbol{S}_0 \otimes \boldsymbol{S} = |S_0 - S| \oplus |S_0 - S| + 1 \oplus \ldots \oplus S_0 + S, \tag{2.4.45}$$

which implies that we obtain either $2S_0 + 1$ or $2S + 1$ new tableaux, depending on which number is smaller. In terms of extended Young tableaux, (2.4.44) translates into

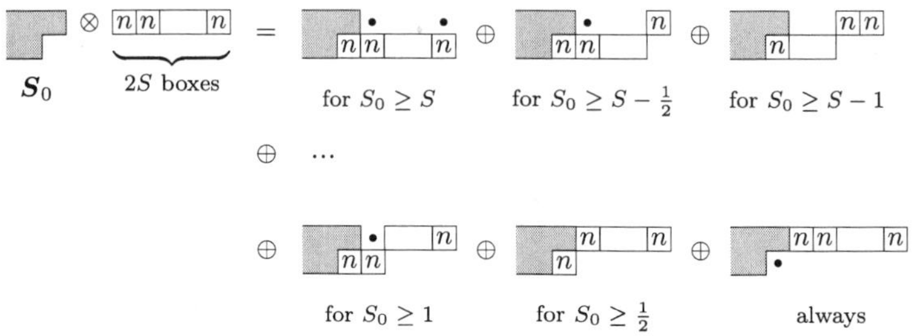

(2.4.46)

The first tableau on the right-hand side of (2.4.46) exists only for $S_0 \geq S$, the second only for $S_0 \geq S - \frac{1}{2}$, and so on. Note that the shape of the right boundary of the extended Young tableaux for $\boldsymbol{S}_0$ does not determine which tableaux are contained in the expansion of $\boldsymbol{S}_0 \otimes \boldsymbol{S}$, as this depends only on the number $S_0 - S$. In the expansion (2.4.46), the $2S$ boxes representing a single spin S always reside in adjacent columns. In an extended tableau, the numbers in the boxes are equal or increasing as we go from left to right, and strictly increasing from top to bottom. The empty spaces we obtain as we build up the tableaux via this method represent the spinons. Note that we cannot take a given tableau and just add a pair of spinons by inserting them somewhere, as the resulting tableau would not occur in the expansion. In Fig. 2.11, we illustrate the principle by writing out a few terms in the expansion for an $S = 2$ chain.

We now turn to the question what this construction implies for the momentum spacings of the spinons. It is very easy to see from Fig. 2.11 that $b_1 = 1$ and a_1 is odd, and that $b_2 = 2S$ and a_L is even (odd) for N even (odd).

Let us assume we have a spinon i with momentum number a_i and box number b_i. If we take $S = 3, a_i = 3,$ and $b_i = 2$, this spinon would be represented by a dot which shares a column with the second box with number 3 in it,

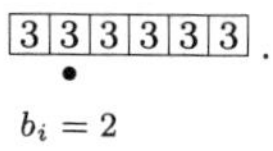

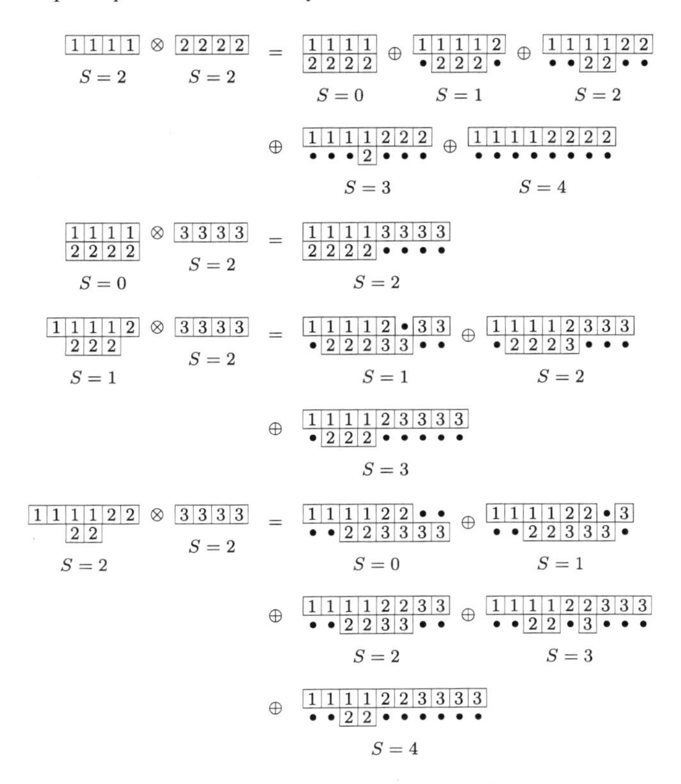

Fig. 2.11 Examples of products of extended tableaux for an $S = 2$ spin chain

For the box number b_{i+1} of the next spinon, there are only two possibilities:

(i) $b_{i+1} = b_i + 1$, which implies that $a_{i+1} - a_i$ is even. The spinons either sit in neighboring columns with $a_{i+1} = a_i$, or contain an even number of spin S representations (with $2S$ boxes each) in between them. For our example, the corresponding tableaux are

3	3	3	3	3	3
	•	•			

$b_i \, b_{i+1}$

$a_{i+1} = a_i$

and

3	3	3	3	3	3	3	4	4	5	5	5	5
•	4	4	4	4	5	5	•					

b_i $\qquad$ b_{i+1}

$a_{i+1} = a_i + 2$

and ...

This possibility produces the unidirectional, horizontal arrows in Fig. 2.10. If $b_i = 2S$, this possibility does not exist, and there are either no further spinons or $a_{i+1} - a_i$ has to be odd.

(ii) $b_{i+1} = 2S - b_i + 1$, which implies that $a_{i+1} - a_i$ is odd. For our example, the tableaux are

$$\begin{array}{c} \boxed{3}\boxed{3}\boxed{3}\boxed{3}\boxed{3}\boxed{3}\boxed{4}\boxed{4} \\ \bullet\,\boxed{4}\boxed{4}\boxed{4}\boxed{4}\,\bullet \\ b_i \qquad\qquad b_{i+1} \\ a_{i+1} = a_i + 1 \end{array} \quad \text{and} \quad \begin{array}{c} \boxed{3}\boxed{3}\boxed{3}\boxed{3}\boxed{3}\boxed{3}\boxed{4}\boxed{4}\boxed{5}\boxed{5}\boxed{5}\boxed{5}\boxed{6}\boxed{6} \\ \bullet\,\boxed{4}\boxed{4}\boxed{4}\boxed{4}\boxed{5}\boxed{5}\boxed{6}\boxed{6}\boxed{6}\boxed{6}\,\bullet \\ b_i \qquad\qquad\qquad\qquad\qquad b_{i+1} \\ a_{i+1} = a_i + 3 \end{array} \quad \text{and} \quad \ldots$$

This possibility produces the bidirectional, semicircle arrows in Fig. 2.10.

This concludes the motivation of the flow diagram in Fig. 2.10b. As in Sects. 2.2.7 and 2.4.5, the single spinon momenta are given by

$$p_i = \frac{\pi}{N}\left(a_i - \frac{1}{2}\right). \tag{2.4.47}$$

This yields momentum spacings $p_{i+1} - p_i$ which can be either an integer or an half-integer times $\frac{2\pi}{N}$.

References

1. R.B. Laughlin, Anomalous quantum Hall effect: an incompressible quantum fluid with fractionally charged excitations. Phys. Rev. Lett. **50**, 1395 (1983)
2. B.I. Halperin, Theory of the quantized Hall conductance. Helv. Phys. Acta **56**, 75 (1983)
3. F.D.M. Haldane, Fractional quantization of the Hall effect: a hierarchy of incompressible quantum fluid states. Phys. Rev. Lett. **51**, 605 (1983)
4. R. Laughlin, Primitive and composite ground states in the fractional quantum Hall effect. Surf. Sci. **142**, 163 (1984)
5. R. Prange, S. Girvin (eds.), *The Quantum Hall Effect*, 2nd edn. (Springer, New York, 1990)
6. T. Chakraborty, P. Pietiläinen, *The Fractional Quantum Hall Effect*, 2nd edn. (Springer, New York, 1995)
7. D.C. Tsui, H.L. Stormer, A.C. Gossard, Two-dimensional magnetotransport in the extreme quantum limit. Phys. Rev. Lett. **48**, 1559 (1982)
8. A.M. Chang, P. Berglund, D.C. Tsui, H.L. Stormer, J.C.M. Hwang, Higherorder states in the multiple-series, fractional, quantum Hall effect. Phys. Rev. Lett. **53**, 997 (1984)
9. R. Clark, R. Nicholas, A. Usher, C. Foxon, J. Harris, Odd and even fractionally quantized states in GaAs–GaAlAs heterojunctions. Surf. Sci. **170**, 141 (1986)
10. R. Willett, J.P. Eisenstein, H.L. Störmer, D.C. Tsui, A.C. Gossard, J.H. English, Observation of an even-denominator quantum number in the fractional quantum Hall effect. Phys. Rev. Lett. **59**, 1776 (1987)
11. L. Landau, Diamagnetismus der Metalle. Z. Phys. **64**, 629 (1930)
12. D.P. Arovas, Fear and loathing in the lowest Landau level, Ph.D. thesis, University of California, Santa Barbara, 1986
13. A.H. MacDonald, Laughlin states in higher Landau levels. Phys. Rev. B **30**, 3550 (1984)
14. S.M. Girvin, T. Jach, Interacting electrons in two-dimensional Landau levels: results for small clusters. Phys. Rev. B **28**, 4506 (1983)
15. F.D.M. Haldane, E.H. Rezayi, Periodic Laughlin–Jastrow wave functions for the fractional quantized Hall effect. Phys. Rev. B **31**, 2529 (1985)
16. D. Yoshioka, B.I. Halperin, P.A. Lee, Ground state of two-dimensional electrons in strong magnetic fields and $\frac{1}{3}$ quantized Hall effect. Phys. Rev. Lett. **50**, 1219 (1983)
17. M. Greiter, Quantum Hall quarks. Phys. E **1**, 1 (1997)
18. M. Greiter, F. Wilczek, Heuristic principle for quantized Hall states. Mod. Phys. Lett. B **4**, 1063 (1990)

19. S.A. Trugman, S. Kivelson, Exact results for the fractional quantum Hall effect with general interactions. Phys. Rev. B **31**, 5280 (1985)
20. B.I. Halperin, Statistics of quasiparticles and the hierarchy of fractional quantized Hall states. Phys. Rev. Lett. **52**, 1583 (1984)
21. B.I. Halperin, Statistics of quasiparticles and the hierarchy of fractional quantized Hall states. Phys. Rev. Lett. **52**, E2390 (1984)
22. D. Arovas, J.R. Schrieffer, F. Wilczek, Fractional statistics and the quantum Hall effect. Phys. Rev. Lett. **53**, 722 (1984)
23. J.M. Leinaas, J. Myrheim, On the theory of identical particles. Nuovo Cimento B **37**, 1 (1977)
24. F. Wilczek, Magnetic flux, angular momentum, and statistics. Phys. Rev. Lett. **48**, 1144 (1982)
25. F. Wilczek, Quantum mechanics of fractional-spin particles. Phys. Rev. Lett. **49**, 957 (1982)
26. Y.-S. Wu, General theory for quantum statistics in two dimensions. Phys. Rev. Lett. **52**, 2103 (1984)
27. D.P. Arovas, R. Schrieffer, F. Wilczek, A. Zee, Statistical mechanics of anyons. Nucl. Phys. B **251**, 117 (1985)
28. J. Fröhlich, P.-A. Marchetti, Quantum field theory of anyons. Lett. Math. Phys. **16**, 347 (1988)
29. A.S. Goldhaber, R. MacKenzie, F. Wilczek, Field corrections to induced statistics. Mod. Phys. Lett. A **4**, 21 (1989)
30. F. Wilczek, *Fractional Statistics and Anyon Superconductivity* (World Scientific, Singapore, 1990)
31. A. Khare, *Fractional Statistics and Quantum Theory* (World Scientific, New Jersey, 2005)
32. M.V. Berry, Quantal phase factors accompanying adiabatic changes. Proc. R. Soc. Lond. A **392**, 4557 (1984)
33. B. Simon, Holonomy, the quantum adiabatic theorem, and Berry's phase. Phys. Rev. Lett. **51**, 2167 (1983)
34. F. Wilczek, A. Zee, Appearance of gauge structure in simple dynamical systems. Phys. Rev. Lett. **52**, 2111 (1984)
35. F. Wilczek, A. Shapere, *Geometric Phases in Physics* (World Scientific, Singapore, 1989)
36. G. Fano, F. Ortolani, E. Colombo, Configuration-interaction calculations on the fractional quantum Hall effect. Phys. Rev. B **34**, 2670 (1986)
37. M. Greiter, Landau level quantization on the sphere. Phys. Rev. B **83**, 115129 (2011)
38. P.A.M. Dirac, Quantised singularities in the electromagnetic field. Proc. Roy. Soc. Lon. Ser. A **133**, 60 (1931)
39. G. Baym, *Lectures on Quantum Mechanics* (Benjamin/Addison Wesley, New York, 1969)
40. M. Greiter, Microscopic formulation of the hierarchy of quantized Hall states. Phys. Lett. B **336**, 48 (1994)
41. Z.N.C. Ha, F.D.M. Haldane, Exact Jastrow-Gutzwiller resonant-valence-bond ground state of the spin- 12 antiferromagnetic Heisenberg chain with $1/r^2$ exchange. Phys. Rev. Lett. **60**, 635 (1988)
42. B.S. Shastry, Exact solution of an $S = \frac{1}{2}$ Heisenberg antiferromagnetic chain with long-ranged interactions. Phys. Rev. Lett. **60**, 639 (1988)
43. V. I. Inozemtsev, On the connection between the one-dimensional $S = \frac{1}{2}$ Heisenberg chain and Haldane–Shastry model. J. Stat. Phys. **59**, 1143 (1990)
44. Z.N.C. Ha, F.D.M. Haldane, "Spinon gas" description of the $S = \frac{1}{2}$ Heisenberg chain with inverse-square exchange: exact spectrum and thermodynamics. Phys. Rev. Lett. **66** 1529 (1991)
45. B.S. Shastry, Taking the square root of the discrete $1/r^2$ model. Phys. Rev. Lett. **69**, 164 (1992)
46. F.D.M. Haldane, Z.N.C. Ha, J.C. Talstra, D. Bernard, V. Pasquier, Yangian symmetry of integrable quantum chains with long-range interactions and a new description of states in conformal field theory. Phys. Rev. Lett. **69**, 2021 (1992)
47. N. Kawakami, Asymptotic Bethe-ansatz solution of multicomponent quantum systems with $1/r^2$ long-range interaction. Phys. Rev. B **46**, 1005 (1992)

48. N. Kawakami, SU(N) generalization of the Gutzwiller–Jastrow wave function and its critical properties in one dimension. Phys. Rev. B **46**, 3191 (1992)
49. J.C. Talstra, Integrability and applications of the exactly-solvable Haldane–Shastry one-dimensional quantum spin chain, Ph.D. thesis, Department of Physics, Princeton University, 1995
50. R.B. Laughlin, D. Giuliano, R. Caracciolo, O.L. White, Quantum number fractionalization in antiferromagnets. In: G. Morandi, P. Sodano, A. Tagliacozzo, V. Tognetti (eds) *Field Theories for Low-Dimensional Condensed Matter Systems* (Springer, Berlin, 2000)
51. B.A. Bernevig, D. Giuliano, R.B. Laughlin, Spinon attraction in spin-1/2 antiferromagnetic chains. Phys. Rev. Lett. **86**, 3392 (2001)
52. B.A. Bernevig, D. Giuliano, R.B. Laughlin, Coordinate representation of the two-spinon wave function and spinon interaction in the Haldane–Shastry model. Phys. Rev. B **64**, 024425 (2001)
53. M. Greiter, D. Schuricht, Many-spinon states and the secret significance of Young tableaux. Phys. Rev. Lett. **98**, 237202 (2007)
54. Z.N.C. Ha, F.D.M. Haldane, Squeezed strings and Yangian symmetry of the Heisenberg chain with long-range interaction. Phys. Rev. B **47**, 12459 (1993)
55. V.G. Drinfel'd, Hopf algebras and the quantum Yang–Baxter equation. Sov. Math. Dokl. **31**, 254 (1985)
56. V. Chari, A. Pressley, *A Guide to Quantum Groups* (Cambridge University Press, Cambridge, 1998)
57. V.E. Korepin, N.M. Bogoliubov, A.G. Izergin, *Quantum Inverse Scattering Method and Correlation Functions* (Cambridge University Press, Cambridge, 1997)
58. J.C. Talstra, F.D.M. Haldane, Integrals of motion of the Haldane–Shastry model. J. Phys. A Math. Gen. **28**, 2369 (1995)
59. B. Sutherland, Quantum many-body problem in one dimension: ground state. J. Math. Phys. **12**, 246 (1971)
60. B. Sutherland, Quantum many-body problem in one dimension: thermodynamics. J. Math. Phys. **12**, 251 (1971)
61. B. Sutherland, Exact result for a quantum many-body problem in one dimension. Phys. Rev. A **4**, 2019 (1971)
62. B. Sutherland, Exact result for a quantum many-body problem in one dimension. II, Phys. Rev. A **5**, 1372 (1972)
63. Z.N.C. Ha, F.D.M. Haldane, Elementary excitations of one-dimensional *t–J* model with inverse-square exchange. Phys. Rev. Lett. **73**, 2887 (1994)
64. Z.N.C. Ha, F.D.M. Haldane, Elementary excitations of one-dimensional *t–J* model with inverse-square exchange. Phys. Rev. Lett. **74**, E3501 (1995)
65. M.C. Gutzwiller, Effect of correlation on the ferromagnetism of transition metals. Phys. Rev. Lett. **10**, 159 (1963)
66. M. Gaudin, Gaz coulombien discretà une dimension. J. Phys. (Paris) **34**, 511 (1973)
67. M.L. Mehta, G.C. Mehta, Discrete Coulomb gas in one dimension: correlation functions. J. Math. Phys. **16**, 1256 (1975)
68. T.A. Kaplan, P. Horsch, P. Fulde, Close relation between localized-electron magnetism and the paramagnetic wave function of completely itinerant electrons. Phys. Rev. Lett. **49**, 889 (1982)
69. C. Gros, R. Joynt, T.M. Rice, Antiferromagnetic correlations in almostlocalized fermi liquids. Phys. Rev. B **36**, 381 (1987)
70. W. Metzner, D. Vollhardt, Ground-state properties of correlated fermions: exact analytic results for the Gutzwiller wave function. Phys. Rev. Lett. **59**, 121 (1987)
71. F. Gebhard, D. Vollhardt, Correlation functions for Hubbard-type models: the exact results for the Gutzwiller wave function in one dimension. Phys. Rev. Lett. **59**, 1472 (1987)
72. W. Marshall, Antiferromagnetism. Proc. R. Soc. (London), Ser. A **232**, 48 (1955)
73. K. Gottfried, *Quantum mechanics*, vol. I, Fundamentals (Benjamin/Addison Wesley, New York, 1966)

74. K.G. Wilson, Proof of a conjecture by Dyson. J. Math. Phys. **3**, 1040 (1962)
75. V. Kalmeyer, R.B. Laughlin, Equivalence of the resonating-valence-bond and fractional quantum Hall states. Phys. Rev. Lett. **59**, 2095 (1987)
76. S.A. Kivelson, D.S. Rokhsar, Quasiparticle statistics in time-reversal invariant states. Phys. Rev. Lett. **61**, 2630 (1988)
77. Z. Zou, B. Doucot, B.S. Shastry, Equivalence of fractional Hall and resonatingvalence-bond states on a square lattice. Phys. Rev. B **39**, 11424 (1989)
78. X.G. Wen, F. Wilczek, A. Zee, Chiral spin states and superconductivity. Phys. Rev. B **39**, 11413 (1989)
79. V. Kalmeyer, R.B. Laughlin, Theory of the spin liquid state of the heisenberg antiferromagnet. Phys. Rev. B **39**, 11879 (1989)
80. R.B. Laughlin, Z. Zou, Properties of the chiral-spin-liquid state. Phys. Rev. B **41**, 664 (1990)
81. D.F. Schroeter, E. Kapit, R. Thomale, M. Greiter, Spin Hamiltonian for which the chiral spin liquid is the exact ground state. Phys. Rev. Lett. **99**, 097202 (2007)
82. R. Thomale, E. Kapit, D.F. Schroeter, M. Greiter, Parent Hamiltonian for the chiral spin liquid. Phys. Rev. B **80**, 104406 (2009)
83. M. Greiter, D. Schuricht, No attraction between spinons in the Haldane–Shastry model. Phys. Rev. B **71**, 224424 (2005)
84. M. Greiter, D. Schuricht, Comment on "Spinon Attraction in Spin-1/2 Antiferromagnetic Chains". Phys. Rev. Lett. **96**, 059701 (2006)
85. M. Greiter, Statistical phases and momentum spacings for one-dimesional anyons. Phys. Rev. B **79**, 064409 (2009)
86. F.H.L. Eßler, A note on dressed S-matrices in models with long-range interactions. Phys. Rev. B **51**, 13357 (1995)
87. Z.N.C. Ha, F.D.M. Haldane, "Fractional statistics" in arbitrary dimensions: a generalization of the Pauli principle. Phys. Rev. Lett. **67**, 937 (1991)
88. M. Hamermesh, *Group Theory and its Application to Physical Problems* (Addison-Wesley, Reading, 1962)
89. T. Inui, Y. Tanabe, Y. Onodera, *Group Theory and Its Applications in Physics* (Springer, Berlin, 1996)
90. G. Moore, N. Read, Nonabelions in the fractional quantum Hall effect. Nucl. Phys. B **360**, 362 (1991)
91. M. Greiter, X.G. Wen, F. Wilczek, Paired Hall state at half filling. Phys. Rev. Lett. **66**, 3205 (1991)
92. M. Greiter, X.G. Wen, F. Wilczek, Paired Hall states. Nucl. Phys. B **374**, 567 (1992)
93. W. Pan, J.-S. Xia, V. Shvarts, D.E. Adams, H.L. Stormer, D.C. Tsui, L.N. Pfeiffer, K.W. Baldwin, K.W. West, Exact quantization of the even-denominator fractional quantum Hall state at $\nu = 5/2$ Landau level filling factor. Phys. Rev. Lett. **83**, 3530 (1999)
94. J.S. Xia, W. Pan, C.L. Vicente, E.D. Adams, N.S. Sullivan, H.L. Stormer, D.C. Tsui, L.N. Pfeiffer, K.W. Baldwin, K.W. West, Electron correlation in the second Landau level: a competition between many nearly degenerate quantum phases. Phys. Rev. Lett. **93**, 176809 (2004)
95. W. Pan, J.S. Xia, H.L. Stormer, D.C. Tsui, C. Vicente, E.D. Adams, N.S. Sullivan, L.N. Pfeiffer, K.W. Baldwin, K.W. West, Experimental studies of the fractional quantum Hall effect in the first excited Landau level. Phys. Rev. B **77**, 075307 (2008)
96. C. Zhang, T. Knuuttila, Y. Dai, R.R. Du, L.N. Pfeiffer, K.W. West, $\nu = 5/2$ fractional quantum Hall effect at 10 T: implications for the Pfaffian state. Phys. Rev. Lett. **104**, 166801 (2010)
97. R.H. Morf, Transition from quantum Hall to compressible states in the second Landau level: new light on the $\nu = 5/2$ enigma. Phys. Rev. Lett. **80**, 1505 (1998)
98. G. Möller, S.H. Simon, Paired composite-fermion wave functions. Phys. Rev. B **77**, 075319 (2008)
99. M. Storni, R.H. Morf, S.D. Sarma Fractional quantum Hall state at $\nu = 5/2$ and the Moore-Read Pfaffian. Phys. Rev. Lett. **104**, 076803 (2010)

100. R. Thomale, A. Sterdyniak, N. Regnault, B.A. Bernevig, Entanglement gap and a new principle of adiabatic continuity. Phys. Rev. Lett. **104**, 180502 (2010)
101. M. Dolev, M. Heiblum, V. Umansky, A. Stern, D. Mahalu, Observation of a quarter of an electron charge at the $\nu = 5/2$ quantum Hall state. Nature **452**, 829 (2008)
102. I.P. Radu, J.B. Miller, C.M. Marcus, M.A. Kastner, L.N. Pfeiffer, K.W. West, Quasi-particle properties from tunneling in the $\nu = 5/2$ fractional quantum Hall state. Science **320**, 899 (2008)
103. J. Bardeen, L.N. Cooper, J.R. Schrieffer, Microscopic theory of superconductivity. Phys. Rev. **106**, 162 (1957)
104. J.R. Schrieffer, *Theory of Superconductivity* (Benjamin/Addison Wesley, New York, 1964)
105. de P.G. Gennes, *Superconductivity of Metals and Alloys* (Benjamin/Addison Wesley, New York, 1966)
106. M. Tinkham, *Introduction to Superconductivity* (McGraw Hill, New York, 1996)
107. F. Dyson, quoted in [104], page 42.
108. M. Greiter, Is electromagnetic gauge invariance spontaneously violated in superconductors? Ann. Phys. **319**, 217 (2005)
109. P. Anderson, Considerations on the flow of superfluid helium. Rev. Mod. Phys. **38**, 298 (1966)
110. G. Frobenius, Über die elliptischen Funktionen zweiter Art. J. Reine Angew. Math. **93**, 53 (1882)
111. M. Greiter, F. Wilczek, Exact solutions and the adiabatic heuristic for quantum Hall states. Nucl. Phys. B **370**, 577 (1992)
112. C. Nayak, F. Wilczek, $2n$-quasihole states realize $2n-1$-dimensional spinor braiding statistics in paired quantum Hall states. Nucl. Phys. B **479**, 529 (1996)
113. N. Read, D. Green, Paired states of fermions in two dimensions with breaking of parity and time-reversal symmetries and the fractional quantum Hall effect. Phys. Rev. B **61**, 10267 (2000)
114. M.H. Freedman, A. Kitaev, Z. Wang, Simulation of topological field theories by quantum computer. Comm. Math. Phys. **227**, 587 (2002)
115. S.D. Sarma, M. Freedman, C. Nayak, Topologically-protected qubits from a possible non-Abelian fractional quantum Hall state. Phys. Rev. Lett. **94**, 166802 (2005)
116. C. Nayak, S.H. Simon, A. Stern, M. Freedman, S.D. Sarma, Rev. Mod. Phys. **80**, 1083 (2008)
117. W. Bishara, P. Bonderson, C. Nayak, K. Shtengel, J.K. Slingerland, Interferometric signature of non-Abelian anyons. Phys. Rev. B **80**, 155303 (2009)
118. J.E. Moore, Quasiparticles do the twist. Physics **2**, 82 (2009)
119. A. Stern, Non-Abelian states of matter. Nature **464**, 187 (2010)
120. N.B. Kopnin, M. M. Salomaa, Mutual friction in superfluid ^{3}he: effects of bound states in the vortex core. Phys. Rev. B **44**, 9667 (1991)
121. D.A. Ivanov, Non-Abelian statistics of half-quantum vortices in p-wave superconductors. Phys. Rev. Lett. **86**, 268 (2001)
122. A. Stern, von F. Oppen, E. Mariani, Geometric phases and quantum entanglement as building blocks for non-Abelian quasiparticle statistics. Phys. Rev. B **70**, 205338 (2004)
123. L.H. Kauffman, *Knots and Physics* (World Scientific, Singapore, 1993)
124. E. Fradkin, C. Nayak, A. Tsvelik, F. Wilczek, A chern-simons effective field theory for the Pfaffian quantum Hall state. Nucl. Phys. B **516**, 704 (1998)
125. M. Levin, B.I. Halperin, B. Rosenow, Particle-hole symmetry and the Pfaffian state. Phys. Rev. Lett. **99**, 236806 (2007)
126. S.-S. Lee, S. Ryu, C. Nayak, M.P.A. Fisher, Particle-hole symmetry and the $\nu = 5/2$ quantum Hall state. Phys. Rev. Lett. **99**, 236807 (2007)
127. S.H. Simon, E.H. Rezayi, N.R. Cooper, Pseudopotentials for multiparticle interactions in the quantum Hall regime. Phys. Rev. B **75**, 195306 (2007)
128. M. Greiter, S=1 spin liquids: broken discrete symmetries restored. J. Low Temp. Phys. **126**, 1029 (2002)

129. M. Greiter, R. Thomale, Non-Abelian statistics in a quantum antiferromagnet. Phys. Rev. Lett. **102**, 207203 (2009)
130. F.D.M. Haldane, Contiuum dynamics of the 1-D Heisenberg antiferromagnet: identification with the O(3) nonlinear sigma model. Phys. Lett. **93 A**, 464 (1983)
131. F.D.M. Haldane, Nonlinear field theory of large-spin Heisenberg antiferromagnets: semiclassically quantized solitons of the one-dimensional easy-axis Néel state. Phys. Rev. Lett. **50**, 1153 (1983)
132. I. Affleck, Field theory methods and quantum critical phenomena, in *Fields Strings and Critical Phenomena*, vol. XLIX, Les Houches Lectures, ed. by E. Brézin, J. Zinn-Justin (Elsevier, Amsterdam, 1990)
133. E. Fradkin, *Field Theories of Condensed Matter Systems number 82 in Frontiers in Physics* (Addison Wesley, Redwood City, 1991)
134. I. Affleck, T. Kennedy, E.H. Lieb, H. Tasaki, Rigorous results on valence-bond ground states in antiferromagnets. Phys. Rev. Lett. **59**, 799 (1987)
135. I. Affleck, T. Kennedy, E.H. Lieb, H. Tasaki, Valence bond ground states in isotropic quantum antiferromagnets. Commun. Math. Phys. **115**, 477 (1988)
136. M. Greiter, S. Rachel, Valence bond solids for SU(n) spin chains: exact models, spinon confinement, and the Haldane gap. Phys. Rev. B **75**, 184441 (2007)
137. M. Greiter, Quantum many-body physics: confinement in a quantum magnet. Nat. Phys. **6**, 5 (2010)
138. J. Schwinger, in *Quantum Theory of Angular Momentum*, ed. by L. Biedenharn, H. van Dam (Academic Press, New York, 1965)
139. A. Auerbach, *Interacting Electrons and Quantum Magnetism* (Springer, New York, 1994)
140. D.P. Arovas, A. Auerbach, F.D.M. Haldane, Extended Heisenberg models of antiferromagnetism: analogies to the fractional quantum Hall effect. Phys. Rev. Lett. **60**, 531 (1988)
141. B. Scharfenberger, M. Greiter, manuscript in preparation
142. N. Read, E. Rezayi, Beyond paired quantum Hall states: parafermions and incompressible states in the first excited Landau level. Phys. Rev. B **59**, 8084 (1999)
143. P. Bouwknegt, A.W.W. Ludwig, K. Schoutens, Spinon basis for higher level SU(2) WZW models. Phys. Lett. B **359**, 304 (1995)
144. J. Wess, B. Zumino, Consequences of anomalous ward identities. Phys. Lett. **37**, 95 (1971)
145. E. Witten, Non-Abelian bosonization in two dimensions. Commun. Math. Phys. **92**, 455 (1984)

Chapter 3
From a Laughlin State to the Haldane–Shastry Model

3.1 General Considerations

In this section, we wish to derive, or maybe better obtain, the Haldane–Shastry model (see Sect. 2.2) from the bosonic $m=2$ Laughlin state and its parent Hamiltonian (see Sect. 2.1.2). At first sight, this does not appear to be a sensible endeavor. Let us briefly recall both models.

3.1.1 Comparison of the Models

The Haldane–Shastry model describes a spin $\frac{1}{2}$ chain with periodic boundary conditions. The Hamiltonian is

$$H^{\mathrm{HS}} = \left(\frac{2\pi}{N}\right)^2 \sum_{\alpha<\beta}^{N} \frac{\boldsymbol{S}_\alpha \boldsymbol{S}_\beta}{\left|\eta_\alpha - \eta_\beta\right|^2}, \tag{3.1.1}$$

where $\eta_\alpha = e^{\mathrm{i}\frac{2\pi}{N}\alpha}$ with $\alpha = 1, \dots, N$ are sites on a unit circle embedded in the complex plane. Written as a wave function for the position of the $M = \frac{N}{2}$ $\uparrow$-spin coordinates z_i, the ground state is given by

$$\psi_0^{\mathrm{HS}}(z_1, \dots, z_M) = \prod_{i<i}^{M} (z_i - z_j)^2 \prod_{i=1}^{M} z_i \,. \tag{3.1.2}$$

The bosonic $m = 2$ Laughlin state for M particles,

$$\psi_0(z_1, \dots, z_M) = \prod_{i<j}^{M} (z_i - z_j)^2 \prod_{i=1}^{M} e^{-\frac{1}{4}|z_i|^2}, \tag{3.1.3}$$

M. Greiter, *Mapping of Parent Hamiltonians*, Springer Tracts in Modern Physics 244,
DOI: 10.1007/978-3-642-24384-4_3, © Springer-Verlag Berlin Heidelberg 2011

is the exact ground state of the δ-function potential interaction Hamiltonian

$$V = \sum_{i<j}^{M} \delta^{(2)}(z_i - z_j) \tag{3.1.4}$$

in the lowest Landau level. Obviously, both models share the factor

$$\prod_{i<j}^{M} (z_i - z_j)^2 \tag{3.1.5}$$

in their ground state wave function, a connection which was exploited recently by Thomale et al. [1] in their study of the entanglement spectrum of spin chains, but this seems to be about it. The Gutzwiller or Haldane–Shastry ground state is invariant under P and T, under translations along the chain, and under global SU(2) spin rotations (see Sect. 2.2.3). The model further possesses a Yangian symmetry and is integrable (see Sect. 2.2.2). The Laughlin ground state is up to a gauge transformation invariant under rotations around the origin. The geometries of both models differ.

Let us proceed by clearing some obvious hurdles to our endeavor of connecting the models. To begin with, the circular droplet described by the Laughlin wave function (3.1.3) has a boundary, while the Haldane–Shastry ground state describes a spin liquid on a compact surface. This problem, however, is easily circumvented by formulating the quantum Hall model on the sphere (see Sect. 2.1.6). Then the bosonic $m = 2$ Laughlin state for M particles on a sphere with $2s = 2M - 2$ flux quanta is given by

$$\psi_0[u, v] = \prod_{i<j}^{M} (u_i v_j - u_j v_i)^2. \tag{3.1.6}$$

Within the lowest Landau level, it is the exact and unique zero-energy ground state of the interaction Hamiltonian

$$\begin{aligned} V^{\text{qh}} = &\sum_{m_1=-s}^{s} \sum_{m_2=-s}^{s} \sum_{m_3=-s}^{s} \sum_{m_4=-s}^{s} a_{m_1}^{\dagger} a_{m_2}^{\dagger} a_{m_3} a_{m_4}\, \delta_{m_1+m_2,\, m_3+m_4} \\ &\cdot \langle s, m_1; s, m_2 | 2s, m_1 + m_2 \rangle \langle 2s, m_3 + m_4 | s m_3, s m_4 \rangle, \end{aligned} \tag{3.1.7}$$

where a_m annihilates a boson in the properly normalized single particle state

$$\psi_{m,0}^{s}(u, v) = \langle u, v | a_m^{\dagger} | 0 \rangle = \sqrt{\frac{(2s+1)!}{4\pi\, (s+m)!\, (s-m)!}}\, u^{s+m} v^{s-m}, \tag{3.1.8}$$

and $\langle s, m_1; s, m_2 | j, m_1 + m_2 \rangle$ etc. are Clebsch–Gordan coefficients [2]. The Hamiltonian (3.1.7) assigns a finite energy cost whenever the relative angular momentum of a pair of particles is zero. The expansion coefficients of the polynomial (3.1.6) are still identical to those of (3.1.5).

3.1.2 A Hole at a Pole

The Haldane–Shastry ground state wave function (3.1.2), however, contains an additional factor $\prod_i z_i$. This is related to another problem. The dimension of the single particle Hilbert space for the bosons on the sphere is $2s+1 = 2M-1$, while the dimension of the single particle Hilbert space for the spin flips on the unit circle is equal to the number of sites, $N=2M$. The Hilbert space dimensions of both models hence do not match. We can adapt the quantum Hall state by insertion of a quasihole at the south pole $(\alpha, \beta) = (0, 1)$ of the sphere. This leads to the wave function

$$\psi_0^{\mathrm{qH}}[u, v] = \prod_{i<j}^{M} (u_i v_j - u_j v_i)^2 \prod_{i=1}^{M} u_i, \tag{3.1.9}$$

on a sphere with $2s+1 = 2M$ single particle states. It is the exact and unique ground state of

$$H^{\mathrm{qH}} = V^{\mathrm{qH}} + U^{\mathrm{qH}} \tag{3.1.10}$$

with

$$U^{\mathrm{qH}} = U_0\, a_{-s}^{\dagger} a_{-s} \tag{3.1.11}$$

for $U_0 > 0$ if we restrict our Hilbert space again to the lowest Landau level. In (3.1.10), we have added a local repulsive potential U_0 for the single particle state with $m = -s$, i.e., the state at the south pole, to the interaction Hamiltonian (3.1.7). Note that both V^{qH} and U^{qH} annihilate the ground state (3.1.9) individually. The single particle Hilbert space dimensions match now, $2s+1 = 2M = N$, and the expansion coefficients $C_{q_1,\ldots,q_M}$ for the polynomials

$$\psi_0^{\mathrm{HS}}[z] = \sum_{\{q_1,\ldots,q_M\}} C_{q_1,\ldots,q_M}\, z_1^{q_1} \cdots z_M^{q_M} \tag{3.1.12}$$

and

$$\psi_0^{\mathrm{qH}}[u, v] = \sum_{\{q_1,\ldots,q_M\}} C_{q_1,\ldots,q_M}\, u_1^{q_1} v_1^{2s-q_1} \ldots u_M^{q_M} v_M^{2s-q_M} \tag{3.1.13}$$

are identical. Note that in both states, the amplitudes are non-zero only for $1 \le q_i \le N$, i.e., the $q=0$ state is never occupied. For the Haldane–Shastry ground state, this means that we never flip a spin S_q^+ with momentum $q=0$, which is a necessary requirement for the singlet property (see Sect. 2.2.3). For the quantized Hall state, it means no particle occupies the $m = -s$ state at the south pole of the sphere. Note further that the Hamiltonians for the sphere (3.1.10) and for the spin chain (3.1.1) are formulated in different spaces. The Hamiltonian (3.1.10) with (3.1.7) on the sphere scatters bosons in a basis of (angular) momentum eigenstates m, while the Haldane–Shastry Hamiltonian (3.1.1) scatters bosonic spin-flips in a position space basis of sites η_α.

3.2 Hilbert Space Renormalization

There is yet another significant difference between both models. We noted above that the coefficients in the polynomial expansions of the ground states (3.1.12) and (3.1.13) are identical. The expansions of both states in terms of single particle states, however, are different, due to different normalizations of the polynomials. In the Haldane–Shastry model, the wave function acts on a Hilbert space constructed out of spin flips at positions z_i,

$$\left|\psi_0^{\mathrm{HS}}\right\rangle = \sum_{\{z_1,\ldots,z_M\}} \psi_0^{\mathrm{HS}}(z_1,\ldots,z_M)\, S_{z_1}^+ \cdot \ldots \cdot S_{z_M}^+ \,\big|\underbrace{\downarrow\downarrow \cdots\cdots \downarrow}_{\text{all } N \text{ spins } \downarrow}\big\rangle. \tag{3.2.1}$$

The polynomial $\frac{1}{\sqrt{N}} z^q$ describes the normalized single particle state

$$\frac{1}{\sqrt{N}} \sum_{\{z\}} z^q S_z^+ |\downarrow\downarrow \cdots \downarrow\rangle = \frac{1}{\sqrt{N}} \sum_{\alpha} \eta_\alpha^q S_\alpha^+ |\downarrow\downarrow \cdots \downarrow\rangle \equiv \hat{S}_q^+ |\downarrow\downarrow \cdots \downarrow\rangle, \tag{3.2.2}$$

and we can rewrite the state vector (3.2.1) in terms of (3.1.12) as

$$\left|\psi_0^{\mathrm{HS}}\right\rangle = \sum_{\{q_1,\ldots,q_M\}} c_{q_1,\ldots,q_M}\, \hat{S}_{q_1}^+ \cdot \ldots \cdot \hat{S}_{q_M}^+ |\downarrow\downarrow \cdots\cdots \downarrow\rangle. \tag{3.2.3}$$

The polynomials $u^{s+m} v^{s-m}$, by contrast, describe the unnormalized single particle states

$$g_m \left|\psi_{m,0}^s\right\rangle = g_m a_m^\dagger |0\rangle, \tag{3.2.4}$$

where

$$g_m = \sqrt{\frac{4\pi\,(s+m)!\,(s-m)!}{(2s+1)!}} \tag{3.2.5}$$

is the normalization factor from (3.1.8) and $a_m^\dagger$ is the associated, properly normalized creation operator. The state vector for the quantum Hall state is hence given by

$$\left|\psi_0^{\mathrm{qH}}\right\rangle = \sum_{\{m_1,\ldots,m_M\}} c_{m_1+s,\ldots,m_M+s}\, g_{m_1} \cdots g_{m_M}\, a_{m_1}^\dagger \cdots a_{m_M}^\dagger |0\rangle \tag{3.2.6}$$

This means that not only the Hamiltonians, but also the coefficients in the ground state vectors, are different. In particular, we we diagonalize the Haldane–Shastry Hamiltonian (3.1.1) for a finite chain, we obtain a ground state vector which is quite different from the ground state vector of (3.1.10) with (3.1.7). If two models have different symmetries, different Hamiltonian and different ground states, it is not clear what the connection should be.

If we think about the problem from a scholarly perspective, the conclusion would probably be to abandon our undertaking. The scholarly approach, however, is not always the most fruitful one. Haldane [3] invented the parent Hamiltonian (2.1.85) because he was looking for an economical way to write out the coefficients of a Laughlin state for a significant number of particles without expanding the polynomial, which he could then compare to the ground state for Coulomb interactions. Along these lines, note that since the single particle normalizations g_m are known, it is easy to obtain the coefficients in (3.2.3) from the coefficients in (3.2.6) and vice versa. So regardless of how different the two states are from a scholarly point of view, there may be practical benefit in exploring the common features.

In fact, even though the quantum Hall Hamiltonian (3.1.10) with (3.1.7) cannot be used directly to obtain the Haldane–Shastry ground state (3.2.3), we can construct a parent Hamiltonian for (3.2.3) from (3.1.10). To do so, consider first the following theorem.

Theorem 3.1 *Let* $|\psi_0\rangle$ *be the exact and non-degenerate zero-energy ground state of* H,

$$\begin{aligned} H|\psi_0\rangle = 0, \quad \langle\psi H|\psi\rangle &\geq 0 \quad \forall\, |\psi\rangle, \\ &= 0 \quad \textit{if and only if} \quad |\psi\rangle = |\psi_0\rangle, \end{aligned}$$

and let G *be an invertible matrix,* $G^{-1}G = 1$. *Then* $G^{-1}|\psi_0\rangle$ *is the exact and non-degenerate zero-energy ground state of* $G^{\dagger}HG$.

Proof Trivially, $G^{\dagger}HG\,G^{-1}|\psi_0\rangle = 0$. With $|\psi'\rangle \equiv G|\psi\rangle$, we have

$$\begin{aligned} \langle\psi G^{\dagger}HG|\psi\rangle = \langle\psi' H|\psi'\rangle &\geq 0 \quad \forall\, |\psi'\rangle \quad \text{and hence } \forall\, |\psi\rangle, \\ &= 0 \quad \text{if and only if} \quad |\psi'\rangle = |\psi_0\rangle, \\ &\qquad \text{i.e., } |\psi\rangle = G^{-1}|\psi_0\rangle. \end{aligned}$$

□

Note that this transformation is not just a rotation of the basis. It completely changes the Hamiltonian, but has the benefit instructing us how to obtain the zero energy ground state of the new Hamiltonian from the original one.

While this theorem points in the right direction, we are not aware of any way of arriving at a convenient parent Hamiltonian by employing it directly. On the positive side, if we choose

$$G = \prod_{m=-s}^{s} (g_m)^{a_m^{\dagger} a_m}, \quad G^{-1} = \prod_{m=-s}^{s} \left(\frac{1}{g_m}\right)^{a_m^{\dagger} a_m}, \tag{3.2.7}$$

we obtain

$$G^{-1}\left|\psi_0^{\mathrm{qH}}\right\rangle = \sum_{\{m_1,\ldots,m_M\}} c_{m_1+s,\ldots,m_M+s}\, a_{m_1}^{\dagger} \cdots a_{m_M}^{\dagger} |0\rangle, \tag{3.2.8}$$

which is identical to the Haldane–Shastry ground state (3.2.3) if we were to substitute[1] $a_m^\dagger \to \hat{S}_{s+m}^+$. On the negative side, the Hamiltonian $G^\dagger H^{\text{qH}} G$ is unnecessarily complicated. To obtain a convenient parent Hamiltonian for (3.2.8), we avail ourselves of another theorem.

Theorem 3.2 *Let $|\psi_0\rangle$ be a zero-energy eigenstate of the interaction Hamiltonian*

$$H = \sum_{\{m_1,m_2,m_3,m_4\}} a_{m_1}^\dagger a_{m_2}^\dagger \, V_{m_1,m_2,m_3,m_4} \, a_{m_3} a_{m_4},$$

and let G be an invertible matrix, $G^{-1}G = 1$. Then $G^{-1}|\psi_0\rangle$ is a zero-energy eigenstate of

$$H' = \sum_{\{m_1,m_2,m_3,m_4\}} G^\dagger a_{m_1}^\dagger a_{m_2}^\dagger G^{\dagger -1} \, V_{m_1,m_2,m_3,m_4} \, G^{-1} a_{m_3} a_{m_4} G.$$

Proof The property $H|\psi_0\rangle = 0$ implies

$$\sum_{\{m_3,m_4\}} V_{m_1,m_2,m_3,m_4} \, a_{m_3} a_{m_4} \, |\psi_0\rangle = 0 \quad \forall\, m_1,\, m_2,$$

and hence

$$\sum_{\{m_3,m_4\}} V_{m_1,m_2,m_3,m_4} \, G^{-1} \, a_{m_3} a_{m_4} \, G \, G^{-1} |\psi_0\rangle = 0 \quad \forall\, m_1,\, m_2,$$

which in turn implies $H' G^{-1} |\psi_0\rangle = 0$. □

Remark The theorem holds for n-body interactions as well.

The choice (3.2.7) implies $G^\dagger = G$ and

$$G^{-1} \, a_m \, G = g_m a_m, \qquad G \, a_m \, G^{-1} = \frac{1}{g_m} a_m, \tag{3.2.9}$$

$$G^{-1} \, a_m^\dagger \, G = \frac{1}{g_m} a_m^\dagger, \qquad G \, a_m^\dagger \, G^{-1} = g_m a_m^\dagger. \tag{3.2.10}$$

Theorem 3.2 implies that the "renormalized" quantum Hall state (3.2.8) is a zero-energy eigenstate of

---

[1] It is not clear whether such a substitution is sensible, since the operators $a_m^\dagger$ and $\hat{S}_{s+m}^+$ obey different commutation relations. For this reason, we do not implement it, but merely mention the possibility. We will see below that a similar transition from the Fourier transforms of $a_m^\dagger$ to local spin flips S_α^+ can be implemented sensibly.

$$V = \sum_{m_1=-s}^{s} \sum_{m_2=-s}^{s} \sum_{m_3=-s}^{s} \sum_{m_4=-s}^{s} a^{\dagger}_{m_1} a^{\dagger}_{m_2} a_{m_3} a_{m_4}\, \delta_{m_1+m_2,\, m_3+m_4}$$
$$\cdot\, g_{m_1} g_{m_2} \langle s, m_1; s, m_2 | 2s, m_1+m_2 \rangle \langle 2s, m_3+m_4 | s m_3, s m_4 \rangle g_{m_3} g_{m_4}. \qquad (3.2.11)$$

Since (3.2.8) is likewise annihilated by (3.1.11), it is also a zero energy state of

$$H = V + U^{\mathrm{qH}}. \qquad (3.2.12)$$

We will see in Sect. 3.3.2 that (3.2.8) is a ground state of (3.2.12), but we have not been able to deduce this from the considerations presented so far. For our purposes, however, it is sufficient to know that (3.2.12) annihilates the state (3.2.8).

With (3.2.5) and the explicit formula

$$\langle s, m_1; s, m_2 | 2s, m_1+m_2 \rangle = \frac{\sqrt{(2s-m_1+m_2)!\,(2s+m_1+m_2)!}}{\sqrt{(s-m_1)!\,(s+m_1)!\,(s-m_2)!\,(s+m_2)!}} \cdot \frac{\sqrt{s}\cdot(2s-1)!}{\sqrt{(4s-1)!}} \qquad (3.2.13)$$

for the Clebsch–Gordan coefficients [2], we obtain

$$g_{m_1} g_{m_2} \langle s, m_1; s, m_2 | 2s, m_1+m_2 \rangle = \sqrt{(2s-m_1+m_2)!\,(2s+m_1+m_2)!} \cdot \frac{2\pi}{(2s+1)\sqrt{s\,(4s-1)!}}. \qquad (3.2.14)$$

The second factor in (3.2.14) does not depend on any m_i and can hence be absorbed by rescaling V accordingly. This yields

$$V = \sum_{m_1=-s}^{s} \sum_{m_2=-s}^{s} \sum_{m_3=-s}^{s} \sum_{m_4=-s}^{s} a^{\dagger}_{m_1} a^{\dagger}_{m_2} a_{m_3} a_{m_4}\, V_{m_1, m_2, m_3, m_4} \qquad (3.2.15)$$

with

$$V_{m_1, m_2, m_3, m_4} = V_{m_1+m_2} \cdot \delta_{m_1+m_2,\, m_3+m_4}, \qquad (3.2.16)$$

$$V_m = (2s-m)!\,(2s+m)!. \qquad (3.2.17)$$

The essential simplification we have encountered so far is that the scattering matrix elements V_{m_1, m_2, m_3, m_4} in (3.2.15) depend only on the conserved total value of L^z, $m_1 + m_2 = m_3 + m_4$, and not on the (angular) momentum transfer.

Even though the Hamiltonian (3.2.12) with (3.2.15) and (3.1.11) annihilates the Haldane–Shastry ground state (3.2.8), we a still very far from having derived the Haldane–Shastry Hamiltonian (3.1.1). First, (3.2.15) scatters single particle states in momentum space, since $m = q - s$ is effectively a momentum quantum number. Second, (3.2.12) is not likely to share the symmetries of (3.1.1). Third, we do not even know whether (3.2.8) is the (non-degenerate) ground state of (3.2.12).

3.3 Fourier Transformation

3.3.1 Particle Creation and Annihilation Operators

We proceed by transforming the interaction Hamiltonian (3.2.15) into Fourier space. To this end, we define the transformations

$$a_m = \frac{1}{\sqrt{N}} \sum_{\alpha=1}^{N} (\bar{\eta}_\alpha)^{s+m} a_\alpha, \quad a_m^\dagger = \frac{1}{\sqrt{N}} \sum_{\alpha=1}^{N} (\eta_\alpha)^{s+m} a_\alpha^\dagger, \tag{3.3.1}$$

where $N = 2s + 1$, $\eta_\alpha = e^{i\frac{2\pi}{N}\alpha}$, and $\bar{\eta}_\alpha = e^{-i\frac{2\pi}{N}\alpha}$. We may interpret α as site indices of a periodic chain with N sites, and η_α as the positions of these sites when the periodic chain is embedded as a unit circle in the complex plane.

The Fourier transformation yields

$$V = \frac{1}{N^2} \sum_{\{\alpha_1, \alpha_2, \alpha_3, \alpha_4\}} a_{\alpha_4}^\dagger a_{\alpha_3}^\dagger a_{\alpha_2} a_{\alpha_1} V_{\alpha_1, \alpha_2, \alpha_3, \alpha_4} \tag{3.3.2}$$

with

$$\begin{aligned} V_{\alpha_1, \alpha_2, \alpha_3, \alpha_4} = & \sum_{m_1=-s}^{s} \sum_{m_2=-s}^{s} \sum_{m_3=-s}^{s} \sum_{m_4=-s}^{s} V_{m_1+m_2} \delta_{m_1+m_2, m_3+m_4} \\ & \cdot (\eta_{\alpha_4})^{s+m_4} (\eta_{\alpha_3})^{s+m_3} (\bar{\eta}_{\alpha_2})^{s+m_2} (\bar{\eta}_{\alpha_1})^{s+m_1} \end{aligned} \tag{3.3.3}$$

for the interaction Hamiltonian (3.2.15) and

$$\begin{aligned} |\psi_0\rangle &= G^{-1} |\psi_0^{\text{qH}}\rangle \\ &= \sum_{\{\alpha_1,\ldots,\alpha_M\}} \frac{1}{\sqrt{N}^M} \sum_{\{m_1,\ldots,m_M\}} C_{m_1+s,\ldots,m_M+s} (\eta_{\alpha_1})^{s+m_1} \ldots (\eta_{\alpha_M})^{s+m_M} a_{\alpha_1}^\dagger \ldots a_{\alpha_M}^\dagger |0\rangle \\ &= \sum_{\{\alpha_1,\ldots,\alpha_M\}} \psi_0^{\text{HS}}(\eta_{\alpha_1}, \ldots, \eta_{\alpha_M}) a_{\alpha_1}^\dagger \ldots a_{\alpha_M}^\dagger |0\rangle \end{aligned} \tag{3.3.4}$$

for the ground state it annihilates. In (3.3.4), we have used the definition of the coefficients $C_{m_1+s,\ldots,m_M+s}$ from (3.2.1) to (3.2.3). Since $\psi_0^{\text{HS}}(\eta_{\alpha_1}, \ldots, \eta_{\alpha_M})$ vanishes identically whenever two coordinates η_α coincide, we are allowed to discard configurations with multiply occupied sites. This yields a reduced Hilbert space in which the boson creation and annihilation operators $a^\dagger$ and a obey the same commutation relations as the spin flip operators S^+ and S^-. We may hence substitute one for the each other.

If we substitute $a_{\alpha_i}^\dagger \to S_{z_i}^+$, $a_{\alpha_i} \to S_{z_i}^-$, in (3.3.2) and (3.3.4), we find that the Haldane–Shastry ground state (3.2.1) with (3.1.2) is annihilated by the interaction Hamiltonian

$$V = \frac{1}{N^2} \sum_{\{\alpha_1, \alpha_2, \alpha_3, \alpha_4\}} S^+_{\alpha_4} S^+_{\alpha_3} S^-_{\alpha_2} S^-_{\alpha_1} \, V_{\alpha_1, \alpha_2, \alpha_3, \alpha_4} \tag{3.3.5}$$

with the matrix elements (3.3.3). For the on-site potential term (3.1.11), Fourier transformation and subsequent substitution yields

$$U^{\mathrm{qH}} = \frac{1}{N} U_0 \, S^+_{\mathrm{tot}} S^-_{\mathrm{tot}}, \tag{3.3.6}$$

where $\boldsymbol{S}_{\mathrm{tot}}$ is defined in (2.2.6). This term annihilates any singlet state, and in particular the Haldane–Shastry ground state (3.2.1) with (3.1.2). It will not be helpful in constructing a parent Hamiltonian, but it might be useful to keep in mind that this term was required to single out the ground state wave function on the quantum Hall sphere.

These observations, and in particular (3.3.5) with (3.3.3) and (3.2.17), are the results of the considerations presented so far, and the starting point for the analysis below.

3.3.2 Renormalized Matrix Elements

In this section, we wish to obtain an explicit expression for the scattering matrix elements (3.3.3) of (3.3.5) for general V_m by direct evaluation. For convenience, we assume $\alpha_1 \neq \alpha_2$ and $\alpha_3 \neq \alpha_4$, as enforced by the spin flips in (3.3.5).

This transformation may look trivial at first, but it is not. When we perform a conventional Fourier transform from real space into momentum space or vice versa, both spaces are periodic. In particular, if we scatter a momentum across the boundary at one end of the Brillouin zone, it will just reappear at the other boundary. The distinguishing feature of the L^z angular momentum quantum number m is that it is not subject to periodic, but to hard wall boundary conditions if we attempt to scatter m to values smaller than $-s$ or larger than s. This does not preclude a Fourier transformation, but it does lead to phase space restrictions we have to take into account.

The δ-function in (3.3.3) allows us to eliminate the two summations over m_3 and m_4 in favor of a single summation,

$$V_{\alpha_1, \alpha_2, \alpha_3, \alpha_4} = \sum_{m_1=-s}^{s} \sum_{m_2=-s}^{s} V_{m_1+m_2} \sum_q{}' \, (\eta_{\alpha_4})^{s+m_2-q} (\eta_{\alpha_3})^{s+m_1+q} (\bar{\eta}_{\alpha_2})^{s+m_2} (\bar{\eta}_{\alpha_1})^{s+m_1}, \tag{3.3.7}$$

where

$$\sum_q{}' \equiv \sum_{q=\max\{-s-m_1,\, -s+m_2\}}^{\min\{s-m_1,\, s+m_2\}} .$$

With $m = m_2 + m_1,\ p = m_2 - m_1$, we write

$$\sum_q{}' = \sum_{q=-s+\frac{1}{2}\max\{p-m,\,p+m\}}^{s+\frac{1}{2}\min\{p-m,\,p+m\}} = \sum_{q=-s+\frac{p}{2}+\frac{|m|}{2}}^{s+\frac{p}{2}-\frac{|m|}{2}}.$$

With

$$\sum_{q=a}^{b} x^q = \frac{x^{b+1} - x^a}{x-1} = \frac{x^{b+\frac{1}{2}} - x^{a-\frac{1}{2}}}{x^{\frac{1}{2}} - x^{-\frac{1}{2}}}, \qquad b \geq a, \tag{3.3.8}$$

we obtain

$$\sum_q{}' (\eta_{34})^q = \frac{(\eta_{34})^{s+\frac{p}{2}-\frac{|m|}{2}+\frac{1}{2}} - (\eta_{34})^{-s+\frac{p}{2}+\frac{|m|}{2}-\frac{1}{2}}}{(\eta_{34})^{\frac{1}{2}} - (\eta_{34})^{-\frac{1}{2}}},$$

where $\eta_{34} \equiv \eta_{\alpha_3-\alpha_4} = \eta_{\alpha_3}\bar{\eta}_{\alpha_4}$. Note that $\eta_{34} \neq 1$ as $\alpha_3 \neq \alpha_4$. Using the periodicity in Fourier space,

$$(\eta_\alpha)^{-s} = (\eta_\alpha)^{s+1}, \tag{3.3.9}$$

we can rewrite the second term in the numerator, and obtain

$$\sum_q{}' (\eta_{34})^q = -(\eta_{34})^{s+\frac{p}{2}+\frac{1}{2}} J(|m|, \alpha_3 - \alpha_4),$$

where

$$J(|m|, \alpha) \equiv \frac{(\eta_\alpha)^{\frac{|m|}{2}} - (\eta_\alpha)^{-\frac{|m|}{2}}}{(\eta_\alpha)^{\frac{1}{2}} - (\eta_\alpha)^{-\frac{1}{2}}}. \tag{3.3.10}$$

Substitution into (3.3.7) yields

$$\begin{aligned} V_{\alpha_1,\alpha_2,\alpha_3,\alpha_4} = \sum_{m_1=-s}^{s} \sum_{m_2=-s}^{s} & V_m \cdot (\eta_{42})^{s+m_2} (\eta_{31})^{s+m_1} \\ & \cdot (-1) \cdot (\eta_{34})^{s+\frac{p}{2}+\frac{1}{2}} J(|m|, \alpha_3 - \alpha_4). \end{aligned} \tag{3.3.11}$$

With $m_1 = \frac{m-p}{2}$, $m_2 = \frac{m+p}{2}$, we can rewrite the sums as

$$\sum_{m_1=-s}^{s} \sum_{m_2=-s}^{s} = \sum_{m=-2s}^{2s} \sum_{\substack{p=-2s+|m| \\ \text{even or odd}}}^{2s-|m|},$$

where the last sum extends only over even (odd) values of p for m odd (even). (Since $N = 2s + 1$ is even, $2s$ is odd.) This yields

$$V_{\alpha_1,\alpha_2,\alpha_3,\alpha_4} = \sum_{m=-2s}^{2s} V_m \cdot (-1) \cdot J(|m|, \alpha_3 - \alpha_4) \cdot \sum_{\substack{p=-2s+|m| \\ \text{even or odd}}}^{2s-|m|} (\eta_{42})^{s+\frac{m}{2}+\frac{p}{2}} (\eta_{31})^{s+\frac{m}{2}-\frac{p}{2}} (\eta_{34})^{s+\frac{p}{2}+\frac{1}{2}}. \tag{3.3.12}$$

We proceed by evaluating the sum over the terms which depend on p,

$$\begin{aligned}\sum_{\substack{p=-2s+|m| \\ \text{even or odd}}}^{2s-|m|} (\eta_{42})^{\frac{p}{2}} (\eta_{31})^{-\frac{p}{2}} (\eta_{34})^{\frac{p}{2}} &= \sum_{k=-s+\frac{|m|}{2}}^{s-\frac{|m|}{2}} (\eta_{12})^k \\ &= \frac{(\eta_{12})^{s-\frac{|m|}{2}+\frac{1}{2}} - (\eta_{12})^{-s+\frac{|m|}{2}-\frac{1}{2}}}{(\eta_{12})^{\frac{1}{2}} - (\eta_{12})^{-\frac{1}{2}}} \\ &= -(\eta_{12})^{s+\frac{1}{2}} J(|m|, \alpha_1 - \alpha_2),\end{aligned}$$

where we have used $(\eta_{12})^{-s} = (\eta_{12})^{s+1}$ and $\eta_{12} \neq 1$. Substitution into (3.3.12) yields

$$V_{\alpha_1,\alpha_2,\alpha_3,\alpha_4} = \sum_{m=-2s}^{2s} V_m \cdot J(|m|, \alpha_3 - \alpha_4)\, J(|m|, \alpha_1 - \alpha_2) \cdot (\eta_{42}\eta_{31})^{s+\frac{m}{2}} (\eta_{34}\eta_{12})^{s+\frac{1}{2}}. \tag{3.3.13}$$

Writing out the factors in the second line yields

$$(\eta_{\alpha_4})^{\frac{m}{2}-\frac{1}{2}} (\eta_{\alpha_3})^{2s+\frac{m}{2}+\frac{1}{2}} (\eta_{\alpha_2})^{-2s-\frac{m}{2}-\frac{1}{2}} (\eta_{\alpha_1})^{-\frac{m}{2}+\frac{1}{2}} = \left(\frac{\eta_{\alpha_4}\eta_{\alpha_3}}{\eta_{\alpha_2}\eta_{\alpha_1}}\right)^{\frac{m}{2}-\frac{1}{2}}.$$

With the definition (3.3.10) we obtain

$$V_{\alpha_1,\alpha_2,\alpha_3,\alpha_4} = \sum_{m=-2s}^{2s} V_m \cdot \frac{(\eta_{34})^{\frac{|m|}{2}} - (\eta_{34})^{-\frac{|m|}{2}}}{(\eta_{34})^{\frac{1}{2}} - (\eta_{34})^{-\frac{1}{2}}} \cdot \frac{(\eta_{12})^{\frac{|m|}{2}} - (\eta_{12})^{-\frac{|m|}{2}}}{(\eta_{12})^{\frac{1}{2}} - (\eta_{12})^{-\frac{1}{2}}} \cdot \left(\frac{\eta_{\alpha_4}\eta_{\alpha_3}}{\eta_{\alpha_2}\eta_{\alpha_1}}\right)^{\frac{m}{2}-\frac{1}{2}}. \tag{3.3.14}$$

Note that we may omit the absolute value signs from m, as both fractions in (3.3.14) change their sign with m. This yields

$$V_{\alpha_1,\alpha_2,\alpha_3,\alpha_4} = \sum_{m=-2s}^{2s} V_m \cdot \frac{\eta_{\alpha_4}^m - \eta_{\alpha_3}^m}{\eta_{\alpha_4} - \eta_{\alpha_3}} \cdot \frac{\bar{\eta}_{\alpha_2}^m - \bar{\eta}_{\alpha_1}^m}{\bar{\eta}_{\alpha_2} - \bar{\eta}_{\alpha_1}}. \tag{3.3.15}$$

3.3.3 An Alternative Derivation

Inspired by the result (3.3.15), we realize that there is an alternative derivation, which will lend itself to generalization to the case $S=1$. To begin with, note that the matrix elements (3.3.3) may be written

$$V_{\alpha_1,\alpha_2,\alpha_3,\alpha_4} = \sum_{m=-2s}^{2s} V_m \cdot \bar{A}_{m;\,\alpha_4,\alpha_3}\, A_{m;\,\alpha_2,\alpha_1}, \tag{3.3.16}$$

where we have defined the sums

$$\begin{aligned} A_{m;\,\alpha_1,\alpha_2} &= \sum_{m_1=-s}^{s}\sum_{m_2=-s}^{s} (\bar{\eta}_{\alpha_2})^{s+m_2}(\bar{\eta}_{\alpha_1})^{s+m_1}\,\delta_{m,\,m_1+m_2},\\ \bar{A}_{m;\,\alpha_3,\alpha_4} &= \sum_{m_3=-s}^{s}\sum_{m_4=-s}^{s} (\eta_{\alpha_4})^{s+m_4}(\eta_{\alpha_3})^{s+m_3}\,\delta_{m,\,m_3+m_4}. \end{aligned} \tag{3.3.17}$$

As these sums are complex conjugates to each other, it is sufficient to evaluate $A_{m;\,\alpha_1,\alpha_2}$. With $m_2 = m - m_1$ and the restriction $-s \le m_2 \le s$, we find $-s+m \le m_1 \le s+m$. This yields

$$A_{m;\,\alpha_1,\alpha_2} = (\bar{\eta}_{\alpha_2})^{2s+m} \sum_{m_1=\max\{-s,\,-s+m\}}^{\min\{s,\,s+m\}} (\bar{\eta}_{12})^{s+m_1}, \tag{3.3.18}$$

where $\bar{\eta}_{12} = \bar{\eta}_{\alpha_1-\alpha_2}$. With (3.3.8), the sum gives for $0 \le m \le 2s$

$$\sum_{m_1=-s+m}^{s} (\bar{\eta}_{12})^{s+m_1} = \frac{(\bar{\eta}_{12})^{2s+1} - (\bar{\eta}_{12})^m}{\bar{\eta}_{12} - 1},$$

and for $-2s \le m < 0$

$$\sum_{m_1=-s}^{s+m} (\bar{\eta}_{12})^{s+m_1} = \frac{(\bar{\eta}_{12})^{2s+1+m} - 1}{\bar{\eta}_{12} - 1}.$$

With $(\bar{\eta}_\alpha)^{2s+1} = 1$, we obtain

$$\begin{aligned} A_{m;\,\alpha_1,\alpha_2} &= -\mathrm{sign}(m)\cdot(\bar{\eta}_{\alpha_2})^{m-1}\,\frac{(\bar{\eta}_{\alpha_1}\eta_{\alpha_2})^m - 1}{\bar{\eta}_{\alpha_1}\eta_{\alpha_2} - 1}\\ &= -\mathrm{sign}(m)\cdot\frac{\bar{\eta}_{\alpha_1}^m - \bar{\eta}_{\alpha_2}^m}{\bar{\eta}_{\alpha_1} - \bar{\eta}_{\alpha_2}}, \end{aligned} \tag{3.3.19}$$

where we have defined

$$\mathrm{sign}(m) \equiv \begin{cases} 1, & m > 0,\\ 0, & m = 0,\\ -1, & m < 0. \end{cases} \tag{3.3.20}$$

Since the signs cancels in the sum (3.3.16), we obtain (3.3.15).

3.4 The Defining Condition for the Gutzwiller State

3.4.1 Annihilation Operators

So far, we have shown that the Gutzwiller or Haldane–Shastry ground state $\left|\psi_0^{\mathrm{HS}}\right\rangle$ given by (3.2.1) with (3.1.2) above is annihilated by the interaction Hamiltonian

$$V = \frac{1}{N^2} \sum_{\{\alpha_1,\alpha_2,\alpha_3,\alpha_4\}} S_{\alpha_4}^{+} S_{\alpha_3}^{+} S_{\alpha_2}^{-} S_{\alpha_1}^{-} V_{\alpha_1,\alpha_2,\alpha_3,\alpha_4} \tag{3.4.1}$$

with the matrix elements (3.3.15) and

$$V_m = (2s-m)!\,(2s+m)!.$$

If we now define an operator

$$A_m \equiv \frac{1}{N} \sum_{\alpha_1 \neq \alpha_2}^{N} \frac{\bar{\eta}_{\alpha_2}^m - \bar{\eta}_{\alpha_1}^m}{\bar{\eta}_{\alpha_2} - \bar{\eta}_{\alpha_1}} S_{\alpha_2}^{-} S_{\alpha_1}^{-} = \frac{2}{N} \sum_{\alpha_1 \neq \alpha_2}^{N} \frac{\bar{\eta}_{\alpha_1}^m}{\bar{\eta}_{\alpha_1} - \bar{\eta}_{\alpha_2}} S_{\alpha_2}^{-} S_{\alpha_1}^{-}, \tag{3.4.2}$$

we may rewrite (3.4.1) as

$$V = \sum_{m=-2s}^{2s} V_m A_m^{\dagger} A_m. \tag{3.4.3}$$

The fact that V annihilates the Gutzwiller state $\left|\psi_0^{\mathrm{HS}}\right\rangle$ implies

$$\begin{aligned} \left\langle\psi_0^{\mathrm{HS}}\right| V \left|\psi_0^{\mathrm{HS}}\right\rangle &= \sum_{m=-2s}^{2s} V_m \left\langle\psi_0^{\mathrm{HS}}\right| A_m^{\dagger} A_m \left|\psi_0^{\mathrm{HS}}\right\rangle \\ &= \sum_{m=-2s}^{2s} V_m \left\| A_m \left|\psi_0^{\mathrm{HS}}\right\rangle \right\|^2 = 0. \end{aligned} \tag{3.4.4}$$

Since all the values V_m for $-2s \leq m \leq 2s$ are positive, and the norms of the vectors by definition non-negative, (3.4.4) implies that the vectors $A_m\left|\psi_0^{\mathrm{HS}}\right\rangle$ must vanish for all values of $m \in [-2s, 2s]$. Since A_m is further periodic under $m \to m+N$ and $N \leq 4s+1$, we have

$$A_m \left|\psi_0^{\mathrm{HS}}\right\rangle = 0 \quad \forall\, m. \tag{3.4.5}$$

This a much stronger condition than we could have hoped to obtain. As an aside, the form (3.4.3) implies that the spectrum of V is positive semi-definite, i.e., all the eigenvalues are non-negative, and hence that $\left|\psi_0^{\mathrm{HS}}\right\rangle$ is a ground state. Of course, we do not know whether it is the only ground state.

Since the Gutzwiller or Haldane–Shastry state $\left|\psi_0^{\text{HS}}\right\rangle$ is real or invariant under parity, i.e., under $\eta_\alpha \to \bar{\eta}_\alpha$, as shown in Sect. 2.2.3, it is also annihilated by the complex conjugates $\bar{A}_m$ of A_m for all m.

The state $\left|\psi_0^{\text{HS}}\right\rangle$ is further annihilated by the operators

$$\begin{aligned}\Omega_\alpha^{\text{HS}} &\equiv \frac{1}{2}\sum_{m=0}^{N} \bar{\eta}_\alpha^m \bar{A}_m \\ &= \sum_{\substack{\beta=1\\ \beta\neq\alpha}}^{N} \frac{1}{\eta_\alpha - \eta_\beta} S_\alpha^- S_\beta^-, \quad \Omega_\alpha^{\text{HS}}\left|\psi_0^{\text{HS}}\right\rangle = 0 \quad \forall\,\alpha,\end{aligned} \tag{3.4.6}$$

which are obtained from the complex conjugate of (3.4.2) by Fourier transformation, as well as their complex conjugates:

$$\bar{\Omega}_\alpha^{\text{HS}} = \sum_{\substack{\beta=1\\ \beta\neq\alpha}}^{N} \frac{1}{\bar{\eta}_\alpha - \bar{\eta}_\beta} S_\alpha^- S_\beta^-, \quad \bar{\Omega}_\alpha^{\text{HS}}\left|\psi_0^{\text{HS}}\right\rangle = 0 \quad \forall\,\alpha. \tag{3.4.7}$$

Note that we would not need to exclude configurations with $\beta = \alpha$, as the spin operators exclude these automatically.

In Sect. 3.6, we will use the operators Ω_α to construct a parent Hamiltonian, which is translationally invariant, invariant under P and T, and invariant under SU(2) spin rotations, for the Gutzwiller state $\left|\psi_0^{\text{HS}}\right\rangle$. Not surprisingly, this Hamiltonian will turn out to be the Haldane–Shastry Hamiltonian (3.1.1) plus a constant to account for the ground state energy (2.2.5).

This implies that the Haldane–Shastry Hamiltonian is completely specified by the condition (3.4.6) plus the symmetries mentioned in the previous paragraph. Therefore, we will refer to (3.4.6) as the *defining condition* of the Gutzwiller or Haldane–Shastry ground state. The universality of this condition is such that both the parent Hamiltonian of the bosonic Laughlin state and the Haldane–Shastry Hamiltonian secretly use (3.4.6) or (3.4.7) to single out the Jastrow polynomial (3.1.5) as their ground state.

3.4.2 Direct Verification

Before proceeding, however, we wish to verify the defining condition (3.4.6) directly for the Haldane–Shastry ground state (3.1.2). This only takes a few lines, and is reassuring after the acrobatics we performed to derive it. We have

$$\Omega_\alpha^{\text{HS}}|\psi_0^{\text{HS}}\rangle = \sum_{\substack{\beta=1\\ \beta\neq\alpha}}^{N} \frac{1}{\eta_\alpha - \eta_\beta} S_\alpha^- S_\beta^- \sum_{\{z_1,\ldots,z_M\}} \psi_0^{\text{HS}}(z_1, z_2, \ldots, z_M)\, S_{z_1}^+ \cdots S_{z_M}^+ |\downarrow \cdots \downarrow\rangle$$
$$= \sum_{\{z_1,\ldots,z_M\}} \underbrace{\sum_{\substack{\beta=1\\ \beta\neq\alpha}}^{N} \frac{\psi_0^{\text{HS}}(\eta_\alpha, \eta_\beta, z_3, \ldots, z_M)}{\eta_\alpha - \eta_\beta}}_{=0} S_{z_3}^+ \cdots S_{z_M}^+ |\downarrow \cdots \downarrow\rangle, \tag{3.4.8}$$

since

$$\frac{\psi_0^{\text{HS}}(\eta_\alpha, \eta_\beta, z_3, \ldots, z_M)}{\eta_\alpha - \eta_\beta} = (\eta_\alpha - \eta_\beta)\eta_\alpha\eta_\beta \prod_{i=3}^{M} (\eta_\alpha - z_i)^2 (\eta_\beta - z_i)^2 z_i \prod_{3\leq i<j}^{M} (z_i - z_j)^2$$

vanishes for $\beta = \alpha$ and contains only powers $\eta_\beta^1, \eta_\beta^2, \ldots, \eta_\beta^{N-2}$. Note that the calculation for $\bar{\Omega}_\alpha^{\text{HS}}$ is almost identical, since

$$\frac{\psi_0^{\text{HS}}(\eta_\alpha, \eta_\beta, z_3, \ldots, z_M)}{\bar{\eta}_\alpha - \bar{\eta}_\beta} = -\eta_\alpha\eta_\beta \frac{\psi_0^{\text{HS}}(\eta_\alpha, \eta_\beta, z_3, \ldots, z_M)}{\eta_\alpha - \eta_\beta}$$

vanishes also for $\beta = \alpha$ and contains only powers $\eta_\beta^2, \eta_\beta^3, \ldots, \eta_\beta^{N-1}$.

3.4.3 The Role of the Hole

In Sect. 3.1.2, we introduced a quasihole at the south pole of the quantum Hall sphere, such that the quantum Hall and the Haldane–Shastry ground state wave functions would resemble each more closely and the Hilbert space dimensions of both models would match. We introduced an additional term (3.1.11) for the quantum Hall Hamiltonian, which morphed into the total spin term (3.3.6) under Fourier transformation, and has played no role since.

The attentive reader will have noticed that the creation of the quasihole has played no role in our analysis up to (3.4.5) whatsoever. In other words, if we had not created it, instead of (3.4.5) we would have found that the state

$$|\psi_0^{N=2M-1}\rangle = \sum_{\{z_1,\ldots,z_M\}} \psi_0^{N=2M-1}(z_1, \ldots, z_M)\, S_{z_1}^+ \cdot \ldots \cdot S_{z_M}^+ |\underbrace{\downarrow\downarrow \cdots\cdots \uparrow}_{\text{all } N \text{ spins } \downarrow}\rangle \tag{3.4.9}$$

with

$$\psi_0^{N=2M-1}(z_1, \ldots, z_M) = \prod_{i<i}^{M} (z_i - z_j)^2 \tag{3.4.10}$$

on a unit circle with $N = 2M - 1$ sites is annihilated by A_m as defined in (3.4.2),

$$\bar{A}_m \left| \psi_0^{N=2M-1} \right\rangle = 0 \quad \forall m. \tag{3.4.11}$$

The state (3.4.9) is likewise annihilated by $\bar{\Omega}_\alpha^{\mathrm{HS}} \forall \alpha$, which can easily be verified directly along the lines of (3.4.8), as

$$\frac{\psi_0^{N=2M-1}(\eta_\alpha, \eta_\beta, z_3, \ldots, z_M)}{\bar{\eta}_\alpha - \bar{\eta}_\beta}$$
$$= -\eta_\alpha \eta_\beta (\eta_\alpha - \eta_\beta) \prod_{i=3}^{M} (\eta_\alpha - z_i)^2 (\eta_\beta - z_i)^2 \prod_{3 \le i < j}^{M} (z_i - z_j)^2$$

vanishes for $\beta = \alpha$ and contains only powers $\eta_\beta^1, \eta_\beta^2, \ldots, \eta_\beta^{N-1}$ for $N = 2M - 1$.

The state (3.4.9) with (3.4.10), however, is neither real (and hence not invariant under P and T) nor a spin singlet. It is not annihilated by $\Omega_\alpha^{\mathrm{HS}}$ for any α. It is not a sensible spin liquid, and we have no symmetries to construct a Hamiltonian. We conclude that while the quasihole is not essential to the mapping of the model itself, it is essential to obtaining a sensible spin model via this mapping.

3.5 Rotations and Spherical Tensor Operators

As mentioned in Sect. 3.4.1 above, we intend to use the defining condition (3.4.6) to formulate a parent Hamiltonian for the Gutzwiller ground state (3.1.2). We wish the Hamiltonian to be invariant under translations, parity and time reversal transformations, and SU(2) spin rotations. This last invariance states that the Hamiltonian must transform as a scalar under spin rotations, while $\Omega_\alpha^{\mathrm{HS}}$ transforms as a tensor of 2nd order. When we construct the Hamiltonian, we will project out certain tensor components (like the scalar or vector component) from operators which do not have simple transformation properties (i.e., which consist of tensor components of different orders). For example, the operator $S_\alpha^+ S_\beta^- + S_\alpha^- S_\beta^+$ consists of both a scalar component and a second order tensor component. In Chap. 4, we will have to analyze the tensor content of more complicated operators, like $S_\alpha^{z\,2}(S_\beta^+ S_\gamma^- + S_\beta^- S_\gamma^+)$.

In this section, we review the rotation properties of tensor operators [4, 2] including the use of Clebsch–Gordan coefficients for projections onto certain tensor components.

3.5.1 Representations of Rotations

The angular momentum operator $\boldsymbol{J}$ is the generator of SU(2) rotations. Specifically, the operator

$$R_\omega = e^{-\mathrm{i}\boldsymbol{J}\omega}, \tag{3.5.1}$$

rotates a state vector by an angle $|\boldsymbol{\omega}|$ around the axis $\boldsymbol{\omega}$. Let $|j, m\rangle$ be an eigenstate of $\boldsymbol{J}^2$ and J^z with eigenvalues $j(j+1)$ and m, respectively. Since (3.5.1) commutes with the total angular momentum, the action of R_{ω} on this state can only change m, i.e.,

$$R_{\omega}|j, m\rangle = \sum_{m'=-j}^{j} |j, m'\rangle d^{(j)}_{m'm}(\boldsymbol{\omega}). \tag{3.5.2}$$

Since the states $|j, m\rangle$ form a complete basis set which does not contain any subgroup of states which only transform under themselves, the matrices

$$d^{(j)}_{m'm}(\boldsymbol{\omega}) = \langle j, m'|e^{-\mathrm{i}\boldsymbol{J}\boldsymbol{\omega}}|j, m\rangle \tag{3.5.3}$$

describe an irreducible, $2j+1$ dimensional representation of the group SU(2).[2]

3.5.2 Tensor Operators

We can further use the operators (3.5.1) to rotate operators,

$$A \to R_{\omega} A R_{\omega}^{-1}, \tag{3.5.4}$$

such that the expectation value of an operator A in a state $|\psi\rangle$ is equal to the expectation value of the rotated operator $R_{\omega} A R_{\omega}^{-1}$ in the rotated state $R_{\omega}|\psi\rangle$. Certain operators transform as scalars under rotations, which means that they commute with $\boldsymbol{J}$ and remain unchanged under (3.5.4). Other operators, like the position vector $\boldsymbol{r}$ or the angular momentum operator $\boldsymbol{J}$, transform as vectors. In general, an irreducible tensor operator $T^{(j)}$ of order j has $2j+1$ components $T^{(j)^m}$, $m = -j, \ldots, j$, which transform among themselves under rotations according to

$$R_{\omega}\, T^{(j)^m} R_{\omega}^{-1} = \sum_{m'=-j}^{j} T^{(j)^{m'}} d^{(j)}_{m'm}(\boldsymbol{\omega}), \tag{3.5.5}$$

where the coefficients $d^{(j)}_{m'm}(\boldsymbol{\omega})$ are given by (3.5.3). Clearly, a scalar is an irreducible tensor of order $j=0$, and a vector is an irreducible tensor of order $j=1$.

If we write out (3.5.5) for infinitesimal rotations

$$R_{\epsilon} = e^{-\mathrm{i}\boldsymbol{J}\boldsymbol{\epsilon}} \approx 1 - \mathrm{i}\boldsymbol{J}\boldsymbol{\epsilon}, \tag{3.5.6}$$

and compare coefficients to first order in $\boldsymbol{\epsilon}$, we obtain

---

[2] For half integer j, these matrices constitute double valued representation of the rotation group O(3), and a single valued representation of the larger group SU(2). For integer j, they are single valued representations of both groups.

$$\left[\boldsymbol{J}, T^{(j)^m} \right] = \sum_{m'=-j}^{j} T^{(j)^{m'}} \langle j, m' | \boldsymbol{J} | j, m \rangle. \tag{3.5.7}$$

With (C.6), this implies

$$\left[J^z, T^{(j)^m} \right] = m\, T^{(j)^m}, \tag{3.5.8}$$

$$\left[J^{\pm}, T^{(j)^m} \right] = \sqrt{j(j+1) - m(m \pm 1)}\; T^{(j)^{m\pm 1}}, \tag{3.5.9}$$

where $J^{\pm} \equiv J^x \pm i J^y$. Equations (3.5.8) and (3.5.9) are fully equivalent to (3.5.5), but much more convenient to use in practise.

Since a vector operator $\boldsymbol{V}$ obeys the commutation relations

$$\left[J^i, V^j \right] = \mathrm{i}\epsilon^{ijk} V^k,$$

(3.5.8) and (3.5.9) imply that the tensor components are (up to an overall normalization factor) given by

$$V^{m=1} = -\frac{V^x + \mathrm{i}V^y}{\sqrt{2}}, \quad V^{m=0} = V^z, \quad V^{m=-1} = \frac{V^x - \mathrm{i}V^y}{\sqrt{2}}. \tag{3.5.10}$$

Note that the J^z eigenvalue of

$$T^{(j)^m} | j', m' \rangle$$

is $m + m'$, as one can easily verify by either considering a rotation (3.5.5) around the z-axis, or directly with (3.5.8),

$$\begin{aligned} J^z\, T^{(j)^m} | j', m' \rangle &= \left[J^z, T^{(j)^m} \right] | j', m' \rangle + T^{(j)^m} J^z | j', m' \rangle \\ &= (m + m')\, T^{(j)^m} | j', m' \rangle. \end{aligned} \tag{3.5.11}$$

The tensor operator $T^{(j)^m}$ hence increases the eigenvalue of J^z by m.

3.5.3 Products of Tensor Operators

Similarly, the J^z quantum number m of a product of two tensors

$$T^{(j_1)^{m_1}} T^{(j_2)^{m_2}} \tag{3.5.12}$$

is simply the sum of the J^z quantum numbers of the individual tensors, $m = m_1 + m_2$. We can again verify this by considering a rotation (3.5.5) around the z-axis, or directly with (3.5.8). The product (3.5.12), however, is not an irreducible tensor, but in general rather a sum of irreducible tensors of orders $|j_1 - j_2|, \ldots, j_1 + j_2$.

We can combine two tensors using Clebsch–Gordan coefficients, however, to obtain a tensor of well-defined order j. Specifically, we can write

$$T^{(j)^m} = \sum_{m_1=-j_1}^{j_1} \sum_{m_2=-j_2}^{j_2} T^{(j_1)^{m_1}} T^{(j_2)^{m_2}} \langle j_1, m_1;\ j_2,\ m_2 | j, m\rangle, \tag{3.5.13}$$

where $\langle j_1, m_1;\ j_2, m_2 | j, m\rangle$ are Clebsch–Gordan coefficients. To verify that the left-hand side of (3.5.13) is an irreducible tensor of order j, consider its transformation properties under a rotation (3.5.5) with coefficient matrices (3.5.3):

$$\begin{aligned}
R_\omega T^{(j)^m} R_\omega^{-1} &= \sum_{m_1, m_2} R_\omega T^{(j_1)^{m_1}} R_\omega^{-1} R_\omega T^{(j_2)^{m_2}} R_\omega^{-1} \langle j_1,\ m_1;\ j_2,\ m_2 | j, m\rangle \\
&= \sum_{m_1', m_2'} T^{(j_1)^{m_1'}} T^{(j_2)^{m_2'}} \\
&\quad \cdot \sum_{m_1, m_2} \langle j_1, m_1';\ j_2, m_2' | e^{-\mathrm{i}\boldsymbol{J}\boldsymbol{\omega}} | j_1,\ m_1;\ j_2,\ m_2\rangle \langle j_1,\ m_1;\ j_2,\ m_2 | j, m\rangle \\
&= \sum_{m_1', m_2'} T^{(j_1)^{m_1'}} T^{(j_2)^{m_2'}} \sum_{j', m'} \langle j_1,\ m_1';\ j_2,\ m_2' | j', m'\rangle \langle j', m' | e^{-\mathrm{i}\boldsymbol{J}\boldsymbol{\omega}} | j, m\rangle \\
&= \sum_{m'} T^{(j)^{m'}} d^{(j)}_{m'm}(\boldsymbol{\omega}).
\end{aligned} \tag{3.5.14}$$

Here we have used the completeness relations

$$\sum_{m_1=-j_1}^{j_1} \sum_{m_2=-j_2}^{j_2} | j_1, m_1;\ j_2, m_2\rangle \langle j_1, m_1;\ j_2, m_2 | = 1, \tag{3.5.15}$$

$$\sum_{j=|j_1-j_2|}^{j_1+j_2} \sum_{m=-j}^{j} | j, m\rangle \langle j, m | = 1, \tag{3.5.16}$$

of the Clebsch–Gordan algebra, which are understood to be valid in a Hilbert space with fixed j_1 and j_2.

We can use the relations (3.5.15) and (3.5.16) further to invert (3.5.13). This yields

$$T^{(j_1)^{m_1}} T^{(j_2)^{m_2}} = \sum_{j=|j_1-j_2|}^{j_1+j_2} \sum_{m=-j}^{j} T^{(j)^m} \langle j, m | j_1, m_1;\ j_2, m_2\rangle. \tag{3.5.17}$$

Let us denote the projection of a tensor A onto its jth order component tensor by $\{A\}_j$. Then (3.5.17) implies

$$\left\{T^{(j_1)^{m_1}} T^{(j_2)^{m_2}}\right\}_j = T^{(j)^{m_1+m_2}} \langle j, m_1 + m_2 | j_1, m_1; j_2, m_2 \rangle, \tag{3.5.18}$$

where $T^{(j)^{m_1+m_2}}$ is given by (3.5.13), i.e.,

$$T^{(j)^m} = \sum_{m_1=\max\{-j_1,-j_2+m\}}^{\min\{j_1,j_2+m\}} T^{(j_1)^{m_1}} T^{(j_2)^{m-m_1}} \langle j_1, m_1; j_2, m - m_1 | j, m \rangle. \tag{3.5.19}$$

For $m = m_1 = m_2 = 0$, we obtain

$$\left\{T^{(j_1)^0} T^{(j_2)^0}\right\}_j = \langle j, 0 | j_1, 0; j_2, 0 \rangle \sum_{m=-\min\{j_1,j_2\}}^{\min\{j_1,j_2\}} T^{(j_1)^m} T^{(j_2)^{-m'}} \langle j_1, m; j_2, -m | 0, 0 \rangle. \tag{3.5.20}$$

We will use this formula repeatedly below.

The tensors we can form out of up to three spin operators, and the tensor decomposition of expressions like $S_1^+ S_2^-$ or $S_1^z S_2^+ S_3^-$, are given in Appendix D.

3.6 Construction of a Parent Hamiltonian for the Gutzwiller State

We now turn to the construction of a parent Hamiltonian for the Gutzwiller state (3.2.1) with (3.1.2) using the annihilation operator (3.4.6), i.e.,

$$\Omega_\alpha^{\text{HS}} \left| \psi_0^{\text{HS}} \right\rangle = 0 \quad \forall \alpha, \quad \text{where} \quad \Omega_\alpha^{\text{HS}} = \sum_{\substack{\beta=1 \\ \beta \neq \alpha}}^{N} \frac{1}{\eta_\alpha - \eta_\beta} S_\alpha^- S_\beta^-. \tag{3.6.1}$$

The Hamiltonian has to be Hermitian, and we wish it to be invariant under translations, time reversal (T), parity (P), and SU(2) spin rotations.

3.6.1 *Translational, Time Reversal, and Parity Symmetry*

The operator $\Omega_\alpha^{\text{HS}\dagger} \Omega_\alpha^{\text{HS}}$ is Hermitian and positive semi-definite, meaning that all the eigenvalues are non-negative. A translationally invariant operator is given by

$$H_0 = \sum_{\alpha=1}^{N} \Omega_\alpha^{\mathrm{HS}\dagger} \Omega_\alpha^{\mathrm{HS}} = \sum_{\substack{\alpha,\beta,\gamma \\ \alpha\neq\beta,\gamma}} \frac{1}{\bar{\eta}_\alpha - \bar{\eta}_\beta} \frac{1}{\eta_\alpha - \eta_\gamma} S_\alpha^+ S_\alpha^- S_\beta^+ S_\gamma^-$$
$$= \sum_{\substack{\alpha,\beta,\gamma \\ \alpha\neq\beta,\gamma}} \omega_{\alpha\beta\gamma} \left(S_\alpha^z + \frac{1}{2} \right) S_\beta^+ S_\gamma^-, \tag{3.6.2}$$

where we have defined

$$\omega_{\alpha\beta\gamma} \equiv \frac{1}{\bar{\eta}_\alpha - \bar{\eta}_\beta} \frac{1}{\eta_\alpha - \eta_\gamma}. \tag{3.6.3}$$

The transformation properties of the individual entities in (3.6.2) under time reversal (T) are [4]

$$\text{T:} \quad \eta_\alpha \rightarrow \Theta \eta_\alpha \Theta = \bar{\eta}_\alpha, \quad \boldsymbol{S} \rightarrow \Theta \boldsymbol{S} \Theta = -\boldsymbol{S}, \tag{3.6.4}$$

and hence

$$\omega_{\alpha\beta\gamma} \rightarrow \omega_{\alpha\beta\gamma}, \quad S^+ \rightarrow -S^-, \quad S^- \rightarrow -S^+, \quad S^z \rightarrow -S^z. \tag{3.6.5}$$

The operator (3.6.2) transforms into

$$\Theta H_0 \Theta = \sum_{\substack{\alpha,\beta,\gamma \\ \alpha\neq\beta,\gamma}} \omega_{\alpha\beta\gamma} \left(-S_\alpha^z + \frac{1}{2} \right) S_\gamma^- S_\beta^+. \tag{3.6.6}$$

We proceed with the T invariant operator

$$H_0^{\mathrm{T}} = \frac{1}{2} (H_0 + \Theta H_0 \Theta) = H_0^{\mathrm{T}=} + H_0^{\mathrm{T}\neq}, \tag{3.6.7}$$

where

$$H_0^{\mathrm{T}=} = \sum_{\substack{\alpha,\beta \\ \alpha\neq\beta}} \omega_{\alpha\beta\beta} \left(\frac{1}{2} S_\alpha^z [S_\beta^+, S_\beta^-] + \frac{1}{4} \{S_\beta^+, S_\beta^-\} \right)$$
$$= \sum_{\substack{\alpha,\beta \\ \alpha\neq\beta}} \omega_{\alpha\beta\beta} \left(S_\alpha^z S_\beta^z + \frac{1}{4} \right), \tag{3.6.8}$$

$$H_0^{\mathrm{T}\neq} = \frac{1}{2} \sum_{\substack{\alpha,\beta,\gamma \\ \alpha\neq\beta\neq\gamma\neq\alpha}} \omega_{\alpha\beta\gamma} S_\beta^+ S_\gamma^-. \tag{3.6.9}$$

The transformation properties of the individual operators under parity (P) are [4]

$$\text{P:}\quad \eta_\alpha \to \Pi\eta_\alpha\Pi = \bar{\eta}_\alpha, \quad \boldsymbol{S} \to \Theta\boldsymbol{S}\Theta = \boldsymbol{S}, \tag{3.6.10}$$

and hence $\omega_{\alpha\beta\gamma} \to \omega_{\alpha\gamma\beta}$. We proceed with the P and T invariant operator

$$H_0^{\text{PT}} = \frac{1}{2}\left(H_0^{\text{T}} + \Pi H_0^{\text{T}}\Pi\right) = H_0^{\text{PT}=} + H_0^{\text{PT}\neq}, \tag{3.6.11}$$

where

$$H_0^{\text{PT}=} = H_0^{\text{T}=},\; H_0^{\text{PT}\neq} = \frac{1}{4}\sum_{\substack{\alpha,\beta,\gamma\\ \alpha\neq\beta\neq\gamma\neq\alpha}} \omega_{\alpha\beta\gamma}\left(S_\beta^+S_\gamma^- + S_\beta^-S_\gamma^+\right). \tag{3.6.12}$$

Since the operator $S_\beta^+S_\gamma^- + S_\beta^-S_\gamma^+$ is symmetric under interchange of β and γ, we can use (B.20) from Appendix B to obtain

$$H_0^{\text{PT}\neq} = \frac{1}{2}\sum_{\alpha\neq\beta}\omega_{\alpha\beta\beta}\left(S_\alpha^+S_\beta^- + S_\alpha^-S_\beta^+\right) - \frac{1}{8}\sum_{\alpha\neq\beta}\left(S_\alpha^+S_\beta^- + S_\alpha^-S_\beta^+\right). \tag{3.6.13}$$

Adding (3.6.8) and (3.6.13) together, we obtain with (B.15)

$$H_0^{\text{PT}} = \sum_{\alpha\neq\beta}\frac{\boldsymbol{S}_\alpha\boldsymbol{S}_\beta}{|\eta_\alpha-\eta_\beta|^2} + \frac{N(N^2-1)}{48} - \frac{1}{8}\sum_{\alpha\neq\beta}\left(S_\alpha^+S_\beta^- + S_\alpha^-S_\beta^+\right). \tag{3.6.14}$$

3.6.2 Spin Rotation Symmetry

The Haldane–Shastry ground state $\left|\psi_0^{\text{HS}}\right\rangle$ is annihilated by (3.6.14), and is also a spin singlet. Since the different tensor components of (3.6.14) yield states which transform according to different representations under SU(2) spin rotations when we act with them on $\left|\psi_0^{\text{HS}}\right\rangle$, each tensor component must annihilate $\left|\psi_0^{\text{HS}}\right\rangle$ individually.

With the exception of the last term, (3.6.14) transforms like a scalar under spin rotations. With (D.2.4), we find that the scalar component of the last term of (3.6.14) is given by

$$-\frac{1}{6}\sum_{\alpha\neq\beta}\boldsymbol{S}_\alpha\boldsymbol{S}_\beta = -\frac{1}{6}\boldsymbol{S}_{\text{tot}}^2 + \frac{1}{6}\sum_\alpha\boldsymbol{S}_\alpha^2 = -\frac{1}{6}\boldsymbol{S}_{\text{tot}}^2 + \frac{N}{8}. \tag{3.6.15}$$

The scalar component of (3.6.14) is therefore given by

$$\left\{H_0^{\text{PT}}\right\}_0 = \sum_{\alpha\neq\beta}\frac{\boldsymbol{S}_\alpha\boldsymbol{S}_\beta}{|\eta_\alpha-\eta_\beta|^2} + \frac{N(N^2+5)}{48} - \frac{\boldsymbol{S}_{\text{tot}}^2}{6}. \tag{3.6.16}$$

We have hence derived that $\left|\psi_0^{\mathrm{HS}}\right\rangle$ is an eigenstate of

$$H^{\mathrm{HS}} = \frac{2\pi^2}{N^2} \sum_{\alpha\neq\beta} \frac{\boldsymbol{S}_\alpha \boldsymbol{S}_\beta}{|\eta_\alpha - \eta_\beta|^2} \tag{3.6.17}$$

with energy eigenvalue

$$E_0^{\mathrm{HS}} = -\frac{2\pi^2}{N^2} \frac{N(N^2+5)}{48}. \tag{3.6.18}$$

In other words, we have derived the Haldane–Shastry model.

This derivation by (conceptually) straightforward projection onto the scalar component is instructive as we will employ this method for the $S=1$ spin chain in Sect. 4.5. It has the disadvantage, however, that the information regarding the semi-positive definiteness has been lost. There are two ways to restore this information. The first is via an alternative derivation of the model without projection from (3.6.14), we will explain now. The second way is to derive first a vector annihilation operator for $\left|\psi_0^{\mathrm{HS}}\right\rangle$, and then construct the Hamiltonian from there, as explained in Sect. 3.7.

3.6.3 An Alternative Derivation

The operators H_0, H_0^{T}, and H_0^{PT} constructed in Sect. 3.6.1 are all sums of terms of the form $A^\dagger A$, and are hence all positive semi-definite, i.e., have only non-negative eigenvalues. Since $\left|\psi_0^{\mathrm{HS}}\right\rangle$ is an eigenstate with eigenvalue zero, it is also a ground state of these operators when we view them as Hamiltonians.

We now wish to employ (3.6.14) to derive that $\left|\psi_0^{\mathrm{HS}}\right\rangle$ is not only an eigenstate of (3.6.17) with energy (3.6.18), but also a ground state. For this purpose, we rewrite (3.6.14) as

$$\begin{aligned} H_0^{\mathrm{PT}} &+ \frac{1}{8}\left(S_{\mathrm{tot}}^+ S_{\mathrm{tot}}^- + S_{\mathrm{tot}}^- S_{\mathrm{tot}}^+\right) \\ &= \sum_{\alpha\neq\beta} \frac{\boldsymbol{S}_\alpha \boldsymbol{S}_\beta}{|\eta_\alpha - \eta_\beta|^2} + \frac{N(N^2-1)}{48} + \frac{1}{8}\sum_\alpha \left(S_\alpha^+ S_\alpha^- + S_\alpha^- S_\alpha^+\right) \\ &= \sum_{\alpha\neq\beta} \frac{\boldsymbol{S}_\alpha \boldsymbol{S}_\beta}{|\eta_\alpha - \eta_\beta|^2} + \frac{N(N^2+5)}{48}, \end{aligned} \tag{3.6.19}$$

where we have used $S_\alpha^+ S_\alpha^- + S_\alpha^- S_\alpha^+ = 1$ for spin $\frac{1}{2}$. Since the left-hand side of (3.6.19) is a sum of positive semi-definite operators which annihilate $\left|\psi_0^{\mathrm{HS}}\right\rangle$, $\left|\psi_0^{\mathrm{HS}}\right\rangle$ has to be a zero energy ground state of the right-hand side as well, i.e., a ground state of (3.6.17) with energy (3.6.18).

3.7 The Rapidity Operator and More

3.7.1 Annihilation Operators Which Transform Even Under T

We can use the defining condition (3.6.1) further to construct a vector annihilation operator. First note that since

$$\Omega_\alpha^{\mathrm{HS}} \left|\psi_0^{\mathrm{HS}}\right\rangle = 0 \quad \forall \alpha,$$

$\left|\psi_0^{\mathrm{HS}}\right\rangle$ is also annihilated by the Hermitian operator

$$H_\alpha = \Omega_\alpha^{\mathrm{HS}\dagger}\Omega_\alpha^{\mathrm{HS}} = \sum_{\substack{\beta,\gamma\\ \beta,\gamma\neq\alpha}} \omega_{\beta\gamma}\left(S_\alpha^z + \frac{1}{2}\right) S_\beta^+ S_\gamma^-, \tag{3.7.1}$$

which is just the operator (3.6.2) without the sum over α. Constructing an operator which is even under T,

$$H_\alpha^{\mathrm{T}} = \frac{1}{2}\left(H_\alpha + \Theta H_\alpha \Theta\right) = H_\alpha^{\mathrm{T}=} + H_\alpha^{\mathrm{T}\neq}, \tag{3.7.2}$$

with

$$H_\alpha^{\mathrm{T}=} = \sum_{\substack{\beta\\ \beta\neq\alpha}} \omega_{\alpha\beta\beta}\left(S_\alpha^z S_\beta^z + \frac{1}{4}\right), \quad H_\alpha^{\mathrm{T}\neq} = \frac{1}{2}\sum_{\substack{\beta\neq\gamma\\ \beta,\gamma\neq\alpha}} \omega_{\alpha\beta\gamma} S_\beta^+ S_\gamma^-, \tag{3.7.3}$$

and odd under P, we obtain

$$H_\alpha^{\bar{\mathrm{P}}\mathrm{T}} = \frac{1}{2}\left(H_\alpha^{\mathrm{T}} - \Pi H_\alpha^{\mathrm{T}} \Pi\right) = H_\alpha^{\bar{\mathrm{P}}\mathrm{T}=} + H_\alpha^{\bar{\mathrm{P}}\mathrm{T}\neq}, \tag{3.7.4}$$

where

$$H_\alpha^{\bar{\mathrm{P}}\mathrm{T}=} = 0, \quad H_\alpha^{\bar{\mathrm{P}}\mathrm{T}\neq} = \frac{1}{4}\sum_{\substack{\beta\neq\gamma\\ \beta,\gamma\neq\alpha}} \omega_{\alpha\beta\gamma}\left(S_\beta^+ S_\gamma^- - S_\beta^- S_\gamma^+\right). \tag{3.7.5}$$

With

$$\begin{aligned}
\omega_{\alpha\beta\gamma} - \omega_{\alpha\gamma\beta} &= \frac{1}{\bar{\eta}_\alpha - \bar{\eta}_\beta}\frac{1}{\eta_\alpha - \eta_\gamma} - \frac{1}{\eta_\alpha - \eta_\beta}\frac{1}{\bar{\eta}_\alpha - \bar{\eta}_\gamma}\\
&= \left(-\eta_\alpha\eta_\beta - \eta_\alpha\eta_\gamma\right)\frac{1}{\eta_\alpha - \eta_\beta}\frac{1}{\eta_\alpha - \eta_\gamma}\\
&= \eta_\alpha\left((\eta_\alpha - \eta_\beta) - (\eta_\alpha - \eta_\gamma)\right)\frac{1}{\eta_\alpha - \eta_\beta}\frac{1}{\eta_\alpha - \eta_\gamma}\\
&= \frac{\eta_\alpha}{\eta_\alpha - \eta_\gamma} - \frac{1}{2} - \left(\frac{\eta_\alpha}{\eta_\alpha - \eta_\beta} - \frac{1}{2}\right)\\
&= -\frac{1}{2}\left(\frac{\eta_\alpha + \eta_\beta}{\eta_\alpha - \eta_\beta} - \frac{\eta_\alpha + \eta_\gamma}{\eta_\alpha - \eta_\gamma}\right)
\end{aligned} \tag{3.7.6}$$

and $S_\beta^+ S_\gamma^- - S_\beta^- S_\gamma^+ = -2\mathrm{i}(\boldsymbol{S}_\beta \times \boldsymbol{S}_\gamma)^z$ [cf. (D.3.3)], we obtain

$$\begin{aligned} H_\alpha^{\bar{\mathrm{P}}\mathrm{T}} &= \frac{\mathrm{i}}{4} \sum_{\substack{\beta \neq \gamma \\ \beta,\gamma \neq \alpha}} \frac{\eta_\alpha + \eta_\beta}{\eta_\alpha - \eta_\beta} (\boldsymbol{S}_\beta \times \boldsymbol{S}_\gamma)^z \\ &= \frac{\mathrm{i}}{4} \sum_{\substack{\beta \\ \beta \neq \alpha}} \frac{\eta_\alpha + \eta_\beta}{\eta_\alpha - \eta_\beta} \left(\boldsymbol{S}_\beta \times (\boldsymbol{S}_{\mathrm{tot}} - \boldsymbol{S}_\alpha - \boldsymbol{S}_\beta)\right)^z \\ &= \frac{\mathrm{i}}{4} \sum_{\substack{\beta \\ \beta \neq \alpha}} \frac{\eta_\alpha + \eta_\beta}{\eta_\alpha - \eta_\beta} \left((\boldsymbol{S}_\alpha \times \boldsymbol{S}_\beta) - \mathrm{i}\boldsymbol{S}_\beta\right)^z + \frac{\mathrm{i}}{4} \sum_{\substack{\beta \\ \beta \neq \alpha}} \frac{\eta_\alpha + \eta_\beta}{\eta_\alpha - \eta_\beta} \left(\boldsymbol{S}_\beta \times \boldsymbol{S}_{\mathrm{tot}}\right)^z, \end{aligned} \tag{3.7.7}$$

where we have used $\boldsymbol{S}_\beta \times \boldsymbol{S}_\beta = i\boldsymbol{S}_\beta$. Since $|\psi_0^{\mathrm{HS}}\rangle$ is a spin singlet, it is trivially annihilated by the second term in the last line of (3.7.7), and hence also annihilated by the first term, which is the z component of a vector. The singlet property of the ground state implies that $|\psi_0^{\mathrm{HS}}\rangle$ is annihilated by all the components of this vector, i.e.,

$$\boldsymbol{D}_\alpha = \frac{\mathrm{i}}{2} \sum_{\substack{\beta=1 \\ \beta \neq \alpha}}^{N} \frac{\eta_\alpha + \eta_\beta}{\eta_\alpha - \eta_\beta} \left[(\boldsymbol{S}_\alpha \times \boldsymbol{S}_\beta) - \mathrm{i}\boldsymbol{S}_\beta\right], \quad \boldsymbol{D}_\alpha|\psi_0^{\mathrm{HS}}\rangle = 0 \quad \forall \alpha. \tag{3.7.8}$$

This is exactly the auxiliary operator (2.2.42) we introduced in (2.2.5), where we have further shown that

$$\frac{2}{9} \sum_{\alpha=1}^{N} \boldsymbol{D}_\alpha^\dagger \boldsymbol{D}_\alpha + \frac{N+1}{12} \boldsymbol{S}_{\mathrm{tot}}^2 = \sum_{\alpha \neq \beta}^{N} \frac{\boldsymbol{S}_\alpha \boldsymbol{S}_\beta}{|\eta_\alpha - \eta_\beta|^2} + \frac{N(N^2+5)}{48}.$$

This proofs once more that $|\psi_0^{\mathrm{HS}}\rangle$ is a ground state of (3.6.17) with energy (3.6.18).

Equation 3.7.8 implies that the Haldane–Shastry ground state is further annihilated by

$$\boldsymbol{\Lambda} = \sum_{\alpha=1}^{N} \boldsymbol{D}_\alpha = \frac{\mathrm{i}}{2} \sum_{\alpha \neq \beta}^{N} \frac{\eta_\alpha + \eta_\beta}{\eta_\alpha - \eta_\beta} (\boldsymbol{S}_\alpha \times \boldsymbol{S}_\beta), \tag{3.7.9}$$

where we have used (B.16). This is the rapidity operator (2.2.8) from Sect. 2.2.2, which together with the total spin operator generates the Yangian symmetry algebra of the Haldane–Shastry model.

For completeness, we further wish to mention the scalar operator we can construct from (3.7.2), which transforms even under P, and which yields the Hamiltonian (3.6.16) when we sum over α. This operator is given by

$$H_\alpha^{\mathrm{PT}} = \frac{1}{2}\left(H_\alpha^{\mathrm{T}} + \Pi H_\alpha^{\mathrm{T}} \Pi\right) = H_\alpha^{\mathrm{PT}=} + H_\alpha^{\mathrm{PT}\neq}, \tag{3.7.10}$$

where

$$H_\alpha^{\text{PT}=} = H_\alpha^{\text{T}=}, \quad H_\alpha^{\text{PT}\neq} = \frac{1}{4} \sum_{\substack{\beta \neq \gamma \\ \beta,\gamma \neq \alpha}} \omega_{\alpha\beta\gamma} \left(S_\beta^+ S_\gamma^- + S_\beta^- S_\gamma^+ \right). \tag{3.7.11}$$

The scalar component of this operator is with (D.2.5) and (D.2.4) given by

$$\left\{ H_\alpha^{\text{PT}} \right\}_0 = \frac{1}{3} \sum_{\substack{\beta \\ \beta \neq \alpha}} \frac{\boldsymbol{S}_\alpha \boldsymbol{S}_\beta}{|\eta_\alpha - \eta_\beta|^2} + \frac{1}{3} \sum_{\substack{\beta \neq \gamma \\ \beta,\gamma \neq \alpha}} \frac{\boldsymbol{S}_\beta \boldsymbol{S}_\gamma}{(\bar{\eta}_\alpha - \bar{\eta}_\beta)(\eta_\alpha - \eta_\gamma)} + \frac{N^2 - 1}{48}, \tag{3.7.12}$$

and annihilates the Gutzwiller state,

$$\left\{ H_\alpha^{\text{PT}} \right\}_0 \left| \psi_0^{\text{HS}} \right\rangle = 0 \quad \forall \alpha.$$

We do not believe that this operator is useful.

3.7.2 Annihilation Operators Which Transform Odd Under T

Finally, we consider annihilation operators we can construct from (3.7.1), and which transform odd under T,

$$H_\alpha^{\bar{\text{T}}} = \frac{1}{2} \left(H_\alpha - \Theta H_\alpha \Theta \right) = H_\alpha^{\bar{\text{T}}=} + H_\alpha^{\bar{\text{T}}\neq} \tag{3.7.13}$$

with

$$\begin{aligned} H_\alpha^{\bar{\text{T}}=} &= \sum_{\substack{\beta \\ \beta \neq \alpha}} \omega_{\alpha\beta\beta} \left(\frac{1}{2} S_\alpha^z \{ S_\beta^+, S_\beta^- \} + \frac{1}{4} [S_\beta^+, S_\beta^-] \right) \\ &= \frac{1}{2} \sum_{\substack{\beta \\ \beta \neq \alpha}} \omega_{\alpha\beta\beta} \left(S_\alpha^z + S_\beta^z \right) \\ &= \frac{N^2 - 1}{24} S_\alpha^z + \frac{1}{2} \sum_{\substack{\beta \\ \beta \neq \alpha}} \omega_{\alpha\beta\beta} S_\beta^z, \end{aligned} \tag{3.7.14}$$

$$H_\alpha^{\bar{\text{T}}\neq} = \sum_{\substack{\beta \neq \gamma \\ \beta,\gamma \neq \alpha}} \omega_{\alpha\beta\gamma} S_\alpha^z S_\beta^+ S_\gamma^-, \tag{3.7.15}$$

where we have used (B.15). $|\psi_0^{\mathrm{HS}}\rangle$ is hence annihilated by all the tensor components of (3.7.13), which are readily obtained with (D.3.11), (D.3.1), and (D.3.3). Let us consider first the scalar operator

$$\left\{H_\alpha^{\bar{T}}\right\}_0 = -\frac{\mathrm{i}}{3}\sum_{\substack{\beta\neq\gamma\\ \beta,\gamma\neq\alpha}} \frac{\boldsymbol{S}_\alpha(\boldsymbol{S}_\beta\times\boldsymbol{S}_\gamma)}{(\bar{\eta}_\alpha-\bar{\eta}_\beta)(\eta_\alpha-\eta_\gamma)}, \tag{3.7.16}$$

which is odd under P. With (3.7.6), we obtain

$$\begin{aligned}
\left\{H_\alpha^{\bar{\mathrm{T}}}\right\}_0 &= \frac{\mathrm{i}}{6}\sum_{\substack{\beta\neq\gamma\\ \beta,\gamma\neq\alpha}} \frac{\eta_\alpha+\eta_\beta}{\eta_\alpha-\eta_\beta}\boldsymbol{S}_\alpha\left(\boldsymbol{S}_\beta\times\boldsymbol{S}_\gamma\right)\\
&= \frac{\mathrm{i}}{6}\sum_{\substack{\beta\\ \beta\neq\alpha}} \frac{\eta_\alpha+\eta_\beta}{\eta_\alpha-\eta_\beta}\boldsymbol{S}_\alpha\left(\boldsymbol{S}_\beta\times(\boldsymbol{S}_{\mathrm{tot}}-\boldsymbol{S}_\alpha-\boldsymbol{S}_\beta)\right)\\
&= \frac{\mathrm{i}}{6}\sum_{\substack{\beta\\ \beta\neq\alpha}} \frac{\eta_\alpha+\eta_\beta}{\eta_\alpha-\eta_\beta}\boldsymbol{S}_\alpha\left(\boldsymbol{S}_\beta\times\boldsymbol{S}_{\mathrm{tot}}\right),
\end{aligned} \tag{3.7.17}$$

where we have used

$$\boldsymbol{S}_\alpha\left(\boldsymbol{S}_\beta\times(-\boldsymbol{S}_\alpha-\boldsymbol{S}_\beta)\right) = \boldsymbol{S}_\beta\left(\boldsymbol{S}_\alpha\times\boldsymbol{S}_\alpha\right)-\boldsymbol{S}_\alpha\left(\boldsymbol{S}_\beta\times\boldsymbol{S}_\beta\right)=0. \tag{3.7.18}$$

The operator (3.7.17) annihilates every spin singlet, and is therefore useless in the present context.

The vector component of (3.7.13), however, constitutes a viable annihilation operator for the Haldane–Shastry ground state,

$$\begin{aligned}
\boldsymbol{A}_\alpha &\equiv 5\left(\left\{H_\alpha^{\bar{\mathrm{T}}=}\right\}_1+\left\{H_\alpha^{\bar{\mathrm{T}}\neq}\right\}_1\right)\\
&= \frac{5}{2}\sum_{\substack{\beta\\ \beta\neq\alpha}}\frac{\boldsymbol{S}_\alpha+\boldsymbol{S}_\beta}{|\eta_\alpha-\eta_\beta|^2}+\sum_{\substack{\beta\neq\gamma\\ \beta,\gamma\neq\alpha}}\frac{4\boldsymbol{S}_\alpha(\boldsymbol{S}_\beta\boldsymbol{S}_\gamma)-\boldsymbol{S}_\beta(\boldsymbol{S}_\alpha\boldsymbol{S}_\gamma)-\boldsymbol{S}_\gamma(\boldsymbol{S}_\alpha\boldsymbol{S}_\beta)}{(\bar{\eta}_\alpha-\bar{\eta}_\beta)(\eta_\alpha-\eta_\gamma)},\\
\boldsymbol{A}_\alpha&\left|\psi_0^{\mathrm{HS}}\right\rangle = 0\quad\forall\alpha.
\end{aligned} \tag{3.7.19}$$

This operator is even under P. Summing over α, we find that the first term annihilates every singlet, since

$$\frac{1}{2}\sum_{\substack{\alpha,\beta\\ \alpha\neq\beta}}\frac{\boldsymbol{S}_\alpha+\boldsymbol{S}_\beta}{|\eta_\alpha-\eta_\beta|^2}=\sum_\alpha\boldsymbol{S}_\alpha\sum_{\substack{\beta\\ \beta\neq\alpha}}\omega_{\alpha\beta\beta}=\frac{N^2-1}{12}\boldsymbol{S}_{\mathrm{tot}}.$$

This implies that $|\psi_0^{\mathrm{HS}}\rangle$ is further annihilated by the vector operator

Table 3.1 Annihilation operators for the Haldane–Shastry ground state

Annihilation operators for $\lvert\psi_0^{\mathrm{HS}}\rangle$					
Operator	Equation	Symmetry transformation properties			
		T	P	Order of tensor	Transl. inv.
$\boldsymbol{S}_{\mathrm{tot}}$	(2.2.6)	−	+	Vector	Yes
$\Omega_\alpha^{\mathrm{HS}}$	(3.4.6)	No	No	2nd	No
$\{H_\alpha^{\mathrm{PT}}\}_0$	(3.7.12)	+	+	Scalar	No
$H^{\mathrm{HS}} - E_0^{\mathrm{HS}}$	(3.6.17)	+	+	Scalar	Yes
$\boldsymbol{D}_\alpha$	(3.7.8)	+	−	Vector	No
$\boldsymbol{\Lambda}$	(3.7.9)	+	−	Vector	Yes
$\boldsymbol{A}_\alpha$	(3.7.19)	−	+	Vector	No
$\boldsymbol{\Upsilon}$	(3.7.20)	−	+	Vector	Yes

With the exception of the defining operator $\Omega_\alpha^{\mathrm{HS}}$, which is the $m=2$ component of a 2nd order tensor, we have only included scalar and vector annihilation operators

$$\boldsymbol{\Upsilon} = 5\sum_\alpha \{H_\alpha^{\bar{\mathrm{T}}\neq}\}_1 = \sum_{\substack{\alpha,\beta,\gamma \\ \alpha\neq\beta\neq\gamma\neq\alpha}} \frac{4\boldsymbol{S}_\alpha(\boldsymbol{S}_\beta\boldsymbol{S}_\gamma) - \boldsymbol{S}_\beta(\boldsymbol{S}_\alpha\boldsymbol{S}_\gamma) - \boldsymbol{S}_\gamma(\boldsymbol{S}_\alpha\boldsymbol{S}_\beta)}{(\bar{\eta}_\alpha - \bar{\eta}_\beta)(\eta_\alpha - \eta_\gamma)}. \tag{3.7.20}$$

This is a three spin operator, and has to our knowledge not been considered before.

3.8 Concluding Remarks

The various annihilation operators for the Haldane–Shastry model are summarized in Table 3.1.

The Haldane–Shastry model, including the operators presented in Sect. 3.7.1, have been known for a long time. In the work of Haldane and Shastry, however, the model was discovered, while we derived it here. Unlike the discovery, the derivation we presented here lends itself to a generalization to higher spins, which is what we will pursue in the following chapter.

It is worth noting that the derivation of the model presented in Sect. 3.6.1, which only assumes the defining condition (3.4.6), is significantly simpler than the previously established verification of the model reviewed in Sect. 2.2.4 with Appendix B. The disadvantage of the present derivation, however, is that it is not clear how to extract information regarding excitations via the formalism employed.

References

1. R. Thomale, D.P. Arovas, B.A. Bernevig, Nonlocal order in gapless systems: entanglement spectrum in spin chains. Phys. Rev. Lett. **105**, 116805 (2010)

2. G. Baym, *Lectures on Quantum Mechanics* (Benjamin/Addison Wesley, New York, 1969)
3. F.D.M. Haldane, Fractional quantization of the Hall effect: a hierarchy of incompressible quantum fluid states. Phys. Rev. Lett. **51**, 605 (1983)
4. K. Gottfried, *Quantum Mechanics, vol. I: Fundamentals* (Benjamin/Addison Wesley, New York, 1966)

Chapter 4
From a Bosonic Pfaffian State to an $S = 1$ Spin Chain

4.1 General Considerations

In this section, we wish to use the bosonic Pfaffian state at Landau level filling fraction $\nu = 1$ and its parent Hamiltonian (see Sect. 2.3), to construct a parent Hamiltonian for the critical $S = 1$ spin liquid state introduced in Sect. 2.4. The Hamiltonian we construct should be invariant under all the trivial symmetries of the spin liquid ground state described in Sect. 2.4.2, i.e., under space translations, P and T, and SU(2) spin rotations. This task would probably be beyond our means if we had not established a suitable technique in Chap. 3, when we derived the Haldane–Shastry Hamiltonian from a bosonic Laughlin state and its parent Hamiltonian. The purpose of this derivation was really to establish the technique which we will fruitfully use in the present analysis.

To begin with, we briefly recall the quantum Hall model and the spin liquid ground state.

4.1.1 A Model and a Ground State

The wave function for the bosonic $m = 1$ Pfaffian Hall state [1–3]

$$\psi_0(z_1, z_2, \ldots, z_N) = \mathrm{Pf}\left(\frac{1}{z_i - z_j}\right) \prod_{i<j}^{N} (z_i - z_j) \prod_{i=1}^{N} e^{-\frac{1}{4}|z_i|^2}, \tag{4.1.1}$$

where the particle number N is even, and the Pfaffian is is given by the fully antisymmetrized sum over all possible pairings of the N particle coordinates,

$$\mathrm{Pf}\left(\frac{1}{z_i - z_j}\right) \equiv \mathcal{A}\left\{\frac{1}{z_1 - z_2} \cdot \ldots \cdot \frac{1}{z_{N-1} - z_N}\right\}. \tag{4.1.2}$$

It is the exact ground state of the three-body Hamiltonian [2, 3]

M. Greiter, *Mapping of Parent Hamiltonians*, Springer Tracts in Modern Physics 244,
DOI: 10.1007/978-3-642-24384-4_4, © Springer-Verlag Berlin Heidelberg 2011

$$V = \sum_{i,j<k}^{N} \delta^{(2)}(z_i - z_j)\delta^{(2)}(z_i - z_k). \tag{4.1.3}$$

In Sect. 2.4, we introduced an $S = 1$ spin liquid state described by a Pfaffian. We considered a one-dimensional lattice with periodic boundary conditions and an even number of sites N on a unit circle embedded in the complex plane, $\eta_\alpha = e^{\mathrm{i}\frac{2\pi}{N}\alpha}$ with $\alpha = 1, \ldots, N$. The wave function is given by a bosonic Pfaffian state in the complex lattice coordinates z_i supplemented by a phase factor,

$$\psi_0^{S=1}(z_1, z_2, \ldots, z_N) = \mathrm{Pf}\left(\frac{1}{z_i - z_j}\right)\prod_{i<j}^{N}(z_i - z_j)\prod_{i=1}^{N} z_i. \tag{4.1.4}$$

The "particles" z_i represent re-normalized spin flips

$$\tilde{S}_\alpha^+ \equiv \frac{S_\alpha^z + 1}{2} S_\alpha^+, \tag{4.1.5}$$

which act on a vacuum with all spins in the $S^z = -1$ state,

$$\left|\psi_0^{S=1}\right\rangle = \sum_{\{z_1,\ldots,z_N\}} \psi_0^{S=1}(z_1, \ldots, z_N)\, \tilde{S}_{z_1}^+ \cdots \tilde{S}_{z_N}^+ |-1\rangle_N, \tag{4.1.6}$$

where the sum runs over all possibilities of distributing the N "particles" over the N lattice sites allowing for double occupation, and

$$|-1\rangle_N \equiv \otimes_{\alpha=1}^{N} |1, -1\rangle_\alpha. \tag{4.1.7}$$

As for the Laughlin state in Sect. 3.1.1, the circular droplet described by the quantum Hall wave function (4.1.1) has a boundary, while the $S = 1$ ground state (4.1.6) with (4.1.4) describes a spin liquid on a compact surface. To circumvent this problem, we formulate the quantum Hall model on the sphere (see Sect. 2.1.6). Then the bosonic $m = 1$ Pfaffian state for N particles on a sphere with $2s = N - 2$ flux quanta is given by

$$\psi_0[u, v] = \mathrm{Pf}\left(\frac{1}{u_i v_j - u_j v_i}\right)\prod_{i<j}^{N}(u_i v_j - u_j v_i). \tag{4.1.8}$$

Within the lowest Landau level, it is the exact and unique zero-energy ground state of the interaction Hamiltonian

$$\begin{aligned} V^{\mathrm{qh}} = &\sum_{m_1=-s}^{s}\sum_{m_2=-s}^{s}\sum_{m_3=-s}^{s}\sum_{m_4=-s}^{s}\sum_{m_5=-s}^{s}\sum_{m_6=-s}^{s} \\ &\cdot a_{m_1}^\dagger a_{m_2}^\dagger a_{m_3}^\dagger a_{m_4} a_{m_5} a_{m_6} \delta_{m_1+m_2+m_3, m_4+m_5+m_6} \\ &\cdot \langle s, m_1; s, m_2 | 2s, m_1 + m_2\rangle\langle 2s, m_1 + m_2; s, m_3 | 3s, m_1 + m_2 + m_3\rangle \\ &\cdot \langle 3s, m_4 + m_5 + m_6 | s, m_4; 2s, m_5 + m_6\rangle\langle 2s, m_5 + m_6 | s, m_5; s, m_6\rangle, \end{aligned} \tag{4.1.9}$$

where a_m annihilates a boson in the properly normalized single particle state

$$\psi^s_{m,0}(u,v) = \langle u, v|a^\dagger_m|0\rangle = \frac{1}{g_m} u^{s+m} v^{s-m}, \tag{4.1.10}$$

with

$$g_m = \sqrt{\frac{4\pi(s+m)!(s-m)!}{(2s+1)!}}, \tag{4.1.11}$$

and $\langle s, m_1; s, m_2|2s-l, m_1+m_2\rangle$ etc. are Clebsch–Gordan coefficients [4].

The differences between the Pfaffian Hall state (4.1.8) and the spin liquid state (4.1.4) are almost in exact correspondence to the differences between the Laughlin state (3.1.6) and the Haldane–Shastry ground state (3.1.2). We will employ the same techniques to adapt the quantum Hall model to the spin chain.

4.1.2 Creation of a Quasihole

The wave function of the spin liquid state (4.1.4) differs from the quantum Hall state in that it contains an additional factor $\prod_i z_i$. We can adapt the quantum Hall state by insertion of a quasihole at the south pole of the sphere. This yields

$$\psi^{\mathrm{qH}}_0[u,v] = \mathrm{Pf}\left(\frac{1}{u_i v_j - u_j v_i}\right) \prod_{i<j}^{N} (u_i v_j - u_j v_i) \prod_{i=1}^{N} u_i, \tag{4.1.12}$$

on a sphere with $2s = N - 1$. It is the exact and unique ground state of

$$H^{\mathrm{qH}} = V^{\mathrm{qH}} + U^{\mathrm{qH}} \tag{4.1.13}$$

with

$$U^{\mathrm{qH}} = U_0 a^\dagger_{-s} a_{-s} \tag{4.1.14}$$

for $U_0 > 0$ if we restrict our Hilbert space again to the lowest Landau level. Note that both V^{qh} and U^{qh} annihilate the ground state (4.1.12) individually. The single particle Hilbert space dimension of the bosons on the sphere is now equal to the dimension dimension of the single particle Hilbert space for the spin flips on the unit circle, $2s + 1 = N$. The expansion coefficients $C_{q_1,\ldots,q_N}$ for the polynomials

$$\psi^{S=1}_0[z] = \sum_{\{q_1,\ldots,q_N\}} C_{q_1,\ldots,q_N} z_1^{q_1} \ldots z_N^{q_N} \tag{4.1.15}$$

and

$$\psi_0^{\mathrm{qH}}[u,v] = \sum_{\{q_1,\ldots,q_N\}} C_{q_1,\ldots,q_N} u_1^{q_1} v_1^{2s-q_1} \ldots u_N^{q_N} v_N^{2s-q_N} \tag{4.1.16}$$

are identical.

4.2 Hilbert Space Renormalization

While the coefficients in the polynomial expansions of the ground states (4.1.15) and (4.1.16) are identical, the expansions of both states in terms of single particle states are not. The state vector for the quantum Hall state is given by

$$\left|\psi_0^{\mathrm{qH}}\right\rangle = \sum_{\{m_1,\ldots,m_N\}} C_{m_1+s,\ldots,m_N+s}\; g_{m_1} \cdots g_{m_M}\; a_{m_1}^{\dagger} \cdots a_{m_N}^{\dagger} |0\rangle \tag{4.2.1}$$

where g_m are the normalizations (4.1.11) of the polynomials $u^{s+m} v^{s-m}$ in (4.1.10). In the spin chain, the polynomials z^q require no such normalization factors, as discussed in Sect. 3.2.

To adjust the quantum Hall state, we renormalize the Hilbert space using Theorem 3.2 of Sect. 3.2 with the same operators G given in (3.2.7). This yields that

$$G^{-1}\left|\psi_0^{\mathrm{qH}}\right\rangle = \sum_{\{m_1,\ldots,m_N\}} C_{m_1+s,\ldots,m_N+s}\; a_{m_1}^{\dagger} \cdots a_{m_N}^{\dagger} |0\rangle \tag{4.2.2}$$

is an exact zero-energy eigenstate of

$$\begin{aligned} V = &\sum_{m_1=-s}^{s} \sum_{m_2=-s}^{s} \sum_{m_3=-s}^{s} \sum_{m_4=-s}^{s} \sum_{m_5=-s}^{s} \sum_{m_6=-s}^{s} \\ &\cdot a_{m_1}^{\dagger} a_{m_2}^{\dagger} a_{m_3}^{\dagger} a_{m_4} a_{m_5} a_{m_6} \delta_{m_1+m_2+m_3,m_4+m_5+m_6} g_{m_1} g_{m_2} g_{m_3} \\ &\cdot \langle s,m_1;s,m_2|2s,m_1+m_2\rangle\langle 2s,m_1+m_2;s,m_3|3s,m_1+m_2+m_3\rangle \\ &\cdot \langle 3s,m_4+m_5+m_6|s,m_4;2s,m_5+m_6\rangle\langle 2s,m_5+m_6|s,m_5;s,m_6\rangle, \\ &\cdot g_{m_4} g_{m_5} g_{m_6} \end{aligned} \tag{4.2.3}$$

Since (4.2.2) is likewise annihilated by (4.1.14), it is also a zero energy state of

$$H = V + U^{\mathrm{qH}}. \tag{4.2.4}$$

With (3.2.5, 3.2.14) and the explicit formula

$$\langle 2s, m_1+m_2; s, m_3|3s, m_1+m_2+m_3\rangle$$
$$= \frac{\sqrt{(3s-m_1-m_2-m_3)!(3s+m_1+m_2+m_3)!}}{\sqrt{(2s-m_1+m_2)!(2s+m_1+m_2)!(s-m_3)!(s+m_3)!}} \cdot 2\sqrt{\frac{s(2s-1)!(4s-1)!}{3\cdot(6s-1)!}} \tag{4.2.5}$$

for the second set of Clebsch–Gordan coefficients [4], we obtain

$$g_{m_1}g_{m_2}g_{m_3}\langle s, m_1; s, m_2|2s, m_1+m_2\rangle\langle 2s, m_1+m_2; s, m_3|3s, m_1+m_2+m_3\rangle$$
$$= \sqrt{(3s-m_1-m_2-m_3)!(3s+m_1+m_2+m_3)!}\frac{2}{\sqrt{3s(6s-1)!}}\left(\frac{2\pi}{2s+1}\right)^{\frac{3}{2}}. \tag{4.2.6}$$

The last two factors in (4.2.6) do not depend on any m_i and can hence be absorbed by rescaling V accordingly. This yields

$$V = \sum_{m_1=-s}^{s}\sum_{m_2=-s}^{s}\sum_{m_3=-s}^{s}\sum_{m_4=-s}^{s}\sum_{m_5=-s}^{s}\sum_{m_6=-s}^{s} a^\dagger_{m_1}a^\dagger_{m_2}a^\dagger_{m_3}a_{m_4}a_{m_5}a_{m_6}$$
$$\cdot V_{m_1,m_2,m_3,m_4,m_5,m_6} \tag{4.2.7}$$

with

$$V_{m_1,m_2,m_3,m_4,m_5,m_6} = V_{m_1+m_2+m_3}\cdot\delta_{m_1+m_2+m_3,m_4+m_5+m_6}, \tag{4.2.8}$$

$$V_m = (3s-m)!(3s+m)!. \tag{4.2.9}$$

Note that the scattering matrix elements $V_{m_1,m_2,m_3,m_4,m_5,m_6}$ in (4.2.7) depend once again only on the conserved total value of L^z, $m_1+m_2+m_3 = m_4+m_5+m_6$, and not on any of the (angular) momentum transfers. This constitutes an enormous simplification.

4.3 Fourier Transformation

4.3.1 Particle Creation and Annihilation Operators

We proceed by transforming the Hamiltonian (4.2.7) into Fourier space, using the transformations

$$a_m = \frac{1}{\sqrt{N}}\sum_{\alpha=1}^{N}(\bar{\eta}_\alpha)^{s+m}a_\alpha, \quad a^\dagger_m = \frac{1}{\sqrt{N}}\sum_{\alpha=1}^{N}(\eta_\alpha)^{s+m}a^\dagger_\alpha, \tag{4.3.1}$$

where $N = 2s + 1$, and $\eta_\alpha = e^{\mathrm{i}\frac{2\pi}{N}\alpha}$. We again interpret α as site indices of a periodic chain with N sites, and η_α as the positions of these sites when the periodic chain is embedded as a unit circle in the complex plane.

The Fourier transformation yields

$$V = \frac{1}{N^3} \sum_{\{\alpha_1,\alpha_2,\alpha_3,\alpha_4,\alpha_5,\alpha_6\}} a^\dagger_{\alpha_6} a^\dagger_{\alpha_5} a^\dagger_{\alpha_4} a^\dagger_{\alpha_3} a^\dagger_{\alpha_2} a^\dagger_{\alpha_1} V_{\alpha_1,\alpha_2,\alpha_3,\alpha_4,\alpha_5,\alpha_6} \tag{4.3.2}$$

with

$$\begin{aligned} &V_{\alpha_1,\alpha_2,\alpha_3,\alpha_4,\alpha_5,\alpha_6} \\ &= \sum_{m_1=-s}^{s} \sum_{m_2=-s}^{s} \sum_{m_3=-s}^{s} \sum_{m_4=-s}^{s} \sum_{m_5=-s}^{s} \sum_{m_6=-s}^{s} V_{m_1+m_2+m_3} \delta_{m_1+m_2+m_3,m_4+m_5+m_6} \\ &\quad \cdot (\eta_{\alpha 6})^{s+m_6} (\eta_{\alpha 5})^{s+m_5} (\eta_{\alpha 4})^{s+m_4} (\bar{\eta}_{\alpha 3})^{s+m_3} (\bar{\eta}_{\alpha 2})^{s+m_2} (\bar{\eta}_{\alpha 1})^{s+m_1} \end{aligned} \tag{4.3.3}$$

and V_m given by (4.2.9) for the interaction Hamiltonian, and

$$\begin{aligned} |\psi_0\rangle &= G^{-1} \left|\psi_0^{\mathrm{qH}}\right\rangle \\ &= \sum_{\{\alpha_1,\ldots,\alpha_N\}} \frac{1}{\sqrt{N}^N} \sum_{\{m_1,\ldots,m_N\}} C_{m_1+s,\ldots,m_N+s} (\eta_{\alpha_1})^{s+m_1} \ldots (\eta_{\alpha_N})^{s+m_N} \\ &\quad \cdot a^\dagger_{\alpha_1} \ldots a^\dagger_{\alpha_N} |0\rangle \\ &= \sum_{\{\alpha_1,\ldots,\alpha_N\}} \psi_0^{S=1}(\eta_{\alpha_1}, \ldots, \eta_{\alpha_N}) a^\dagger_{\alpha_1} \ldots a^\dagger_{\alpha_N} |0\rangle, \end{aligned} \tag{4.3.4}$$

where $\psi_0^{S=1}(\eta_{\alpha_1}, \ldots, \eta_{\alpha_N})$ is given by (4.1.4), for the ground state it annihilates. In (4.3.4), we have used the definition of the coefficients $C_{m_1+s,\ldots,m_N+s}$ from (4.1.15).

4.3.2 Substitution of Spin Flip Operators for Boson Operators

The formulation of the model in terms of position space operators allows us to substitute spin flip operators for the creation and annihilation operators, and thus to turn our boson model into a spin model. For the $S = 1$ model, this step is not as trivial as for the $S = \frac{1}{2}$ model treated in Chap. 3, as the usual spin flip operators do not obey the same commutation relations as bosonic ladder operators in the subspace where each site can be doubly occupied at most. The relation

$$S^-_\alpha (\tilde{S}^+_\alpha)^n |1, -1\rangle_\alpha = n (\tilde{S}^+_\alpha)^{n-1} |1, -1\rangle_\alpha, \quad \text{for } n = 0, 1, 2, \tag{4.3.5}$$

which follows directly form the definition (4.1.5), instructs us how to proceed. Since

$$a(a^{\dagger})^n|0\rangle = n(a^{\dagger})^{n-1}|0\rangle,$$

we may substitute $a^{\dagger}_{\alpha_i} \to S^{+}_{\alpha_i}$, $a_{\alpha_i} \to S^{-}_{\alpha_i}$, in the Hamiltonian and $a^{\dagger}_{\alpha_i} \to \tilde{S}^{+}_{\alpha_i}$, $|0\rangle \to |-1\rangle_N$ in the ground state. In other words, the non-Abelian $S = 1$ spin liquid state (4.1.6) with (4.1.4) introduced in Sect. 2.4, is annihilated by

$$V = \frac{1}{N^3} \sum_{\{\alpha_1,\alpha_2,\alpha_3,\alpha_4,\alpha_5,\alpha_6\}} S^{+}_{\alpha_6} S^{+}_{\alpha_5} S^{+}_{\alpha_4} S^{-}_{\alpha_3} S^{-}_{\alpha_2} S^{-}_{\alpha_1} V_{\alpha_1,\alpha_2,\alpha_3,\alpha_4,\alpha_5,\alpha_6} \tag{4.3.6}$$

with the matrix elements (4.3.3). For the on-site potential term (4.1.14), Fourier transformation and subsequent substitution yields again

$$U^{\mathrm{qH}} = \frac{1}{N} U_0 S^{+}_{\mathrm{tot}} S^{-}_{\mathrm{tot}}. \tag{4.3.7}$$

This term annihilates any singlet state, and will not be helpful in constructing a parent Hamiltonian. We will keep in mind, however, that the original term was required to single out the ground state wave function (4.1.12) on the quantum Hall sphere.

Note that this substitution does not just amount to a renaming of operators, as it did for the spin $\frac{1}{2}$ chain discussed in Chap. 3. In the present case, it effectively renormalizes the single particle Hilbert spaces once more, and hence leads to a different model. To see this, compare the normalizations of "unoccupied", "singly occupied", and "doubly occupied" sites in the $S = 1$ spin chain,

$$\begin{aligned} \langle 1,-1|\tilde{S}^{-}_{\alpha} \tilde{S}^{+}_{\alpha}|1,-1\rangle &= \frac{1}{2}, \\ \langle 1,-1|(\tilde{S}^{-}_{\alpha})^2 (\tilde{S}^{+}_{\alpha})^2|1,-1\rangle &= 1, \end{aligned}$$

to those of bosons,

$$\langle 0|a^n a^{\dagger^n}|0\rangle = n!. \tag{4.3.8}$$

The difference does not just amount to a different overall normalizations of the states. If we were, for example, to renormalize the already renormalized spin operators $\tilde{S}_{\alpha} \longrightarrow \sqrt{2}\tilde{S}_{\alpha}$, we would obtain

$$\begin{aligned} \langle 1,-1|\tilde{S}^{-}_{\alpha} \tilde{S}^{+}_{\alpha}|1,-1\rangle &= 1, \\ \langle 1,-1|(\tilde{S}^{-}_{\alpha})^2 (\tilde{S}^{+}_{\alpha})^2|1,-1\rangle &= 4. \end{aligned}$$

This would match (4.3.8) for $n = 1$, but not for $n = 2$. The amplitudes of the individual spin configurations in the spin state vector are hence different from those of the corresponding amplitudes in the boson state vector.

4.3.3 *Many Body Annihilation Operators*

Since the scatting elements (4.3.3) depend only on the total angular momentum quantum number m, we can rewrite (4.3.6) as

$$V = \sum_{m=-3s}^{3s} V_m B_m^{\dagger} B_m, \tag{4.3.9}$$

where V_m is given by (4.2.9), and

$$B_m = B_m^{\neq} + B_m^{=} \tag{4.3.10}$$

with

$$B_m^{\neq} = \frac{1}{\sqrt{N^3}} \sum_{\substack{\alpha_1,\alpha_2,\alpha_3=1\\ \alpha_1\neq\alpha_2\neq\alpha_3\neq\alpha_1}}^{N} B_{m;\alpha_1,\alpha_2,\alpha_3}^{\neq} S_{\alpha_3}^{-} S_{\alpha_2}^{-} S_{\alpha_1}^{-}, \tag{4.3.11}$$

$$B_m^{=} = \frac{3}{\sqrt{N^3}} \sum_{\substack{\alpha_1,\alpha_2=1\\ \alpha_1\neq\alpha_2}}^{N} B_{m;\alpha_1,\alpha_2}^{=} \left(S_{\alpha_2}^{-}\right)^2 S_{\alpha_1}^{-}. \tag{4.3.12}$$

The coefficients in (4.3.11) and (4.3.12) are given by

$$B_{m;\alpha_1,\alpha_2,\alpha_3}^{\neq} = \sum_{m_1=-s}^{s} \sum_{m_2=-s}^{s} \sum_{m_3=-s}^{s} (\bar{\eta}_{\alpha_3})^{s+m_3} (\bar{\eta}_{\alpha_2})^{s+m_2} (\bar{\eta}_{\alpha_1})^{s+m_1} \cdot \delta_{m,m_1+m_2+m_3}, \tag{4.3.13}$$

$$B_{m;\alpha_1,\alpha_2}^{=} = \sum_{m_1=-s}^{s} \sum_{m_2=-s}^{s} \sum_{m_3=-s}^{s} (\bar{\eta}_{\alpha_2})^{s+m_3} (\bar{\eta}_{\alpha_2})^{s+m_2} (\bar{\eta}_{\alpha_1})^{s+m_1} \cdot \delta_{m,m_1+m_2+m_3}. \tag{4.3.14}$$

The factor 3 in the definition (4.3.12) of $B_m^{=}$ stems from the three possibilities of two coordinates being equal.

4.3.4 Evaluation of $B_{m;\alpha_1,\alpha_2,\alpha_3}^{\neq}$

In this section, we evaluate

$$B_{m;\alpha_1,\alpha_2,\alpha_3}^{\neq} = \sum_{m_1=-s}^{s} \sum_{m_2=-s}^{s} \sum_{m_3=-s}^{s} (\bar{\eta}_{\alpha_3})^{s+m_3} (\bar{\eta}_{\alpha_2})^{s+m_2} (\bar{\eta}_{\alpha_1})^{s+m_1} \cdot \delta_{m,m_1+m_2+m_3} \tag{4.3.15}$$

subject to the condition that none the coordinates α_1, α_2, and α_3 coincide.

To begin with, we carry out the sum over m_3, and obtain

$$B^{\neq}_{m;\alpha_1,\alpha_2,\alpha_3} = \sum_{m_1}{}' (\bar{\eta}_{\alpha_1})^{s+m_1} \underbrace{\sum_{m_2}{}' (\bar{\eta}_{\alpha_3})^{s+m-m_1-m_2} (\bar{\eta}_{\alpha_2})^{s+m_2}}_{\equiv I_{m_1}}, \tag{4.3.16}$$

where the primed sums are restricted such that all the exponents of the $\bar{\eta}_\alpha$'s are between 0 and $2s$. With $-s \leq m - m_1 - m_2 \leq s$, we have

$$\begin{aligned}
I_{m_1} &= (\bar{\eta}_{\alpha_3})^{2s+m-m_1} \sum_{m_2=\max\{-s,-s+m-m_1\}}^{\min\{s,s+m-m_1\}} (\bar{\eta}_{23})^{s+m_2} \\
&= \begin{cases} A_{m-m_1;\alpha_2,\alpha_3} & \text{for } -2s \leq m - m_1 \leq 2s, \\ 0 & \text{otherwise}, \end{cases} \\
&= \begin{cases} \dfrac{\bar{\eta}_{\alpha_2}^{m-m_1} - \bar{\eta}_{\alpha_3}^{m-m_1}}{\bar{\eta}_{\alpha_2} - \bar{\eta}_{\alpha_3}} & \text{for } m \leq m_1 \leq 2s + m, \\ -\dfrac{\bar{\eta}_{\alpha_2}^{m-m_1} - \bar{\eta}_{\alpha_3}^{m-m_1}}{\bar{\eta}_{\alpha_2} - \bar{\eta}_{\alpha_3}} & \text{for } -2s + m \leq m_1 \leq m - 1, \\ 0 & \text{otherwise}, \end{cases}
\end{aligned} \tag{4.3.17}$$

where we have defined $\bar{\eta}_{23} \equiv \bar{\eta}_{\alpha_2-\alpha_3} = \bar{\eta}_{\alpha_2}\eta_{\alpha_3}$ and used the result (3.3.19) for the sum (3.3.18) from Sect. 3.3.3.

For the evaluation of the sum over m_1, we consider three different regimes for m.

(a) $-s < m \leq s$. In this regime, $m - m_1$ changes sign as we sum over m_1. We obtain

$$\begin{aligned}
B^{\neq}_{m;\alpha_1,\alpha_2,\alpha_3} &= \frac{\bar{\eta}_{\alpha_2}^{s+m}}{\bar{\eta}_{\alpha_2} - \bar{\eta}_{\alpha_3}} \left(-\sum_{m_1=-s}^{m-1} \bar{\eta}_{12}^{s+m_1} + \sum_{m_1=m}^{s} \bar{\eta}_{12}^{s+m_1} \right) \\
&\quad + \text{same term with } \bar{\eta}_{\alpha_2} \leftrightarrow \bar{\eta}_{\alpha_3} \\
&= \frac{\bar{\eta}_{\alpha_2}^{s+m}}{\bar{\eta}_{\alpha_2} - \bar{\eta}_{\alpha_3}} \left(-\frac{\bar{\eta}_{12}^{s+m} - 1}{\bar{\eta}_{12} - 1} + \frac{1 - \bar{\eta}_{12}^{s+m}}{\bar{\eta}_{12} - 1} \right) \\
&\quad + \text{same term with } \bar{\eta}_{\alpha_2} \leftrightarrow \bar{\eta}_{\alpha_3}
\end{aligned}$$

$$
\begin{aligned}
&= -\frac{2\bar{\eta}_{\alpha_2}}{\bar{\eta}_{\alpha_2}-\bar{\eta}_{\alpha_3}} \cdot \frac{\bar{\eta}_{\alpha_1}^{s+m}-\bar{\eta}_{\alpha_2}^{s+m}}{\bar{\eta}_{\alpha_1}-\bar{\eta}_{\alpha_2}} + \text{ same term with } \bar{\eta}_{\alpha_2} \leftrightarrow \bar{\eta}_{\alpha_3} \\
&= \frac{2\bar{\eta}_{\alpha_2}^{s+m+1}}{(\bar{\eta}_{\alpha_1}-\bar{\eta}_{\alpha_2})(\bar{\eta}_{\alpha_2}-\bar{\eta}_{\alpha_3})} + \frac{2\bar{\eta}_{\alpha_3}^{s+m+1}}{(\bar{\eta}_{\alpha_2}-\bar{\eta}_{\alpha_3})(\bar{\eta}_{\alpha_3}-\bar{\eta}_{\alpha_1})} \\
&\quad - \frac{2\bar{\eta}_{\alpha_1}^{s+m}}{\bar{\eta}_{\alpha_2}-\bar{\eta}_{\alpha_3}} \left(\frac{\bar{\eta}_{\alpha_2}}{\bar{\eta}_{\alpha_1}-\bar{\eta}_{\alpha_2}} - \frac{\bar{\eta}_{\alpha_3}}{\bar{\eta}_{\alpha_1}-\bar{\eta}_{\alpha_3}} \right) \\
&= \frac{2\bar{\eta}_{\alpha_2}^{s+m+1}}{(\bar{\eta}_{\alpha_1}-\bar{\eta}_{\alpha_2})(\bar{\eta}_{\alpha_2}-\bar{\eta}_{\alpha_3})} + \frac{2\bar{\eta}_{\alpha_3}^{s+m+1}}{(\bar{\eta}_{\alpha_2}-\bar{\eta}_{\alpha_3})(\bar{\eta}_{\alpha_3}-\bar{\eta}_{\alpha_1})} \\
&\quad + \frac{2\bar{\eta}_{\alpha_1}^{s+m+1}}{(\bar{\eta}_{\alpha_3}-\bar{\eta}_{\alpha_1})(\bar{\eta}_{\alpha_1}-\bar{\eta}_{\alpha_2})} \\
&\equiv 2Q^{\neq}_{m;\alpha_1,\alpha_2,\alpha_3},
\end{aligned} \tag{4.3.18}
$$

where $Q^{\neq}_{m;\alpha_1,\alpha_2,\alpha_3}$ is strictly periodic under $m \longrightarrow m+N$ with $N=2s+1$.

(b) $-3s \leq m \leq -s$. Since $-s \leq m_1 \leq s$, this implies that we are always in the first regime in (4.3.17), $m \leq m_1 \leq 2s+m$. This yields

$$
\begin{aligned}
B^{\neq}_{m;\alpha_1,\alpha_2,\alpha_3} &= \frac{\bar{\eta}_{\alpha_2}^{s+m}}{\bar{\eta}_{\alpha_2}-\bar{\eta}_{\alpha_3}} \sum_{m_1=-s}^{2s+m} \bar{\eta}_{12}^{s+m_1} + \text{ same term with } \bar{\eta}_{\alpha_2} \leftrightarrow \bar{\eta}_{\alpha_3} \\
&= \frac{\bar{\eta}_{\alpha_2}^{s+m}}{\bar{\eta}_{\alpha_2}-\bar{\eta}_{\alpha_3}} \frac{\bar{\eta}_{12}^{s+m}-1}{\bar{\eta}_{12}-1} + \text{ same term with } \bar{\eta}_{\alpha_2} \leftrightarrow \bar{\eta}_{\alpha_3} \\
&= -Q^{\neq}_{m;\alpha_1,\alpha_2,\alpha_3}.
\end{aligned} \tag{4.3.19}
$$

(c) $s < m \leq 3s$. Since $-s \leq m_1 \leq s$, this implies that we are always in the second regime in (4.3.17), $-2s+m \leq m_1 \leq m-1$. This yields

$$
\begin{aligned}
B^{\neq}_{m;\alpha_1,\alpha_2,\alpha_3} &= -\frac{\bar{\eta}_{\alpha_2}^{s+m}}{\bar{\eta}_{\alpha_2}-\bar{\eta}_{\alpha_3}} \sum_{m_1=-2s+m}^{s} \bar{\eta}_{12}^{s+m_1} + \text{ same term with } \bar{\eta}_{\alpha_2} \leftrightarrow \bar{\eta}_{\alpha_3} \\
&= -\frac{\bar{\eta}_{\alpha_2}^{s+m}}{\bar{\eta}_{\alpha_2}-\bar{\eta}_{\alpha_3}} \sum_{m_1=-2s+m-1}^{s} \bar{\eta}_{12}^{s+m_1} \\
&\quad + \text{ same term with } \bar{\eta}_{\alpha_2} \leftrightarrow \bar{\eta}_{\alpha_3} \\
&= -\frac{\bar{\eta}_{\alpha_2}^{s+m}}{\bar{\eta}_{\alpha_2}-\bar{\eta}_{\alpha_3}} \frac{1-\bar{\eta}_{12}^{s+m}}{\bar{\eta}_{12}-1} + \text{ same term with } \bar{\eta}_{\alpha_2} \leftrightarrow \bar{\eta}_{\alpha_3} \\
&= -Q^{\neq}_{m;\alpha_1,\alpha_2,\alpha_3}.
\end{aligned} \tag{4.3.20}
$$

Note that since $Q^{\neq}_{\pm s;\alpha_1,\alpha_2,\alpha_3} = 0$, it does not matter with which regime we associate the cases $m = \pm s$. As a curiosity, note further that $Q^{\neq}_{s+2;\alpha_1,\alpha_2,\alpha_3} = 1$ [cf. (B.7) of Appendix B].

4.3.5 Evaluation of $B^{=}_{m;\alpha_1,\alpha_2}$

We now evaluate

$$B^{=}_{m;\alpha_1,\alpha_2} = \sum_{m_1=-s}^{s} \sum_{m_2=-s}^{s} \sum_{m_3=-s}^{s} (\bar{\eta}_{\alpha_2})^{s+m_3} (\bar{\eta}_{\alpha_2})^{s+m_2} (\bar{\eta}_{\alpha_1})^{s+m_1} \cdot \delta_{m,m_1+m_2+m_3} \tag{4.3.21}$$

subject to the condition $\alpha_1 \neq \alpha_2$.

To begin with, we carry out the sum over m_3, and obtain

$$\begin{aligned} B^{=}_{m;\alpha_1,\alpha_2} &= \sum_{m_1}{}' (\bar{\eta}_{\alpha_1})^{s+m_1} \sum_{m_2}{}' (\bar{\eta}_{\alpha_2})^{2s+m-m_1}, \\ &= (\bar{\eta}_{\alpha_2})^{s+m-1} \sum_{m_1}{}' (\bar{\eta}_{12})^{s+m_1} \sum_{m_2}{}' 1, \end{aligned} \tag{4.3.22}$$

where the primed sums are restricted such that all the exponents of the original $\bar{\eta}_\alpha$'s in (4.3.21) are between 0 and $2s$. With $-s \leq m - m_1 - m_2 \leq s$, we have

$$\begin{aligned} \sum_{m_2}{}' 1 &= \sum_{m_2=\max\{-s,-s+m-m_1\}}^{\min\{s,s+m-m_1\}} 1 \\ &= \begin{cases} N + m - m_1 & \text{for } m \leq m_1 \leq 2s + m, \\ N - m + m_1 & \text{for } -2s + m \leq m_1 \leq m - 1, \\ 0 & \text{otherwise,} \end{cases} \end{aligned} \tag{4.3.23}$$

For the evaluation of the sum over m_1, we again consider three different regimes for m.

(a) $-s < m \leq s$. In this regime, $m - m_1$ changes sign as we sum over m_1. We obtain

$$\begin{aligned}
\frac{B^{=}_{m;\alpha_1,\alpha_2}}{\bar{\eta}_{\alpha_2}^{s+m-1}} &= \sum_{m_1=-s}^{m-1} (N-m+m_1)\bar{\eta}_{12}^{s+m_1} + \sum_{m_1=m}^{s} (N+m-m_1)\bar{\eta}_{12}^{s+m_1} \\
&= N \underbrace{\sum_{m_1=-s}^{s} \bar{\eta}_{12}^{s+m_1}}_{=0} - m\left(\frac{\bar{\eta}_{12}^{s+m}-1}{\bar{\eta}_{12}-1} - \frac{1-\bar{\eta}_{12}^{s+m}}{\bar{\eta}_{12}-1}\right) \\
&\quad + \sum_{m_1=-s}^{m-1} m_1\bar{\eta}_{12}^{s+m_1} - \sum_{m_1=m}^{s} m_1\bar{\eta}_{12}^{s+m_1}.
\end{aligned} \tag{4.3.24}$$

With the formula

$$\sum_{q=a}^{b} q x^q = \frac{(b+1)x^{b+1} - a x^a}{x-1} - \frac{x^{b+2}-x^{a+1}}{(x-1)^2}, \quad b \geq a, \tag{4.3.25}$$

we obtain for the last two sums in (4.3.24),

$$\begin{aligned}
\sum_{m_1=-s}^{m-1} m_1\bar{\eta}_{12}^{s+m_1} &= \frac{m\bar{\eta}_{12}^{s+m}+s}{\bar{\eta}_{12}-1} - \frac{\bar{\eta}_{12}^{s+m+1}-\bar{\eta}_{12}}{(\bar{\eta}_{12}-1)^2} \\
-\sum_{m_1=m}^{s} m_1\bar{\eta}_{12}^{s+m_1} &= -\frac{(s+1)-m\bar{\eta}_{12}^{s+m}}{\bar{\eta}_{12}-1} + \frac{\bar{\eta}_{12}-\bar{\eta}_{12}^{s+m+1}}{(\bar{\eta}_{12}-1)^2}.
\end{aligned}$$

Summing up all the terms we find

$$\begin{aligned}
B^{=}_{m;\alpha_1,\alpha_2} &= \bar{\eta}_{\alpha_2}^{s+m-1}\left(\frac{2m-1}{\bar{\eta}_{12}-1} - 2\frac{\bar{\eta}_{12}^{s+m+1}-\bar{\eta}_{12}}{(\bar{\eta}_{12}-1)^2}\right) \\
&= \bar{\eta}_{\alpha_2}^{s+m-1}\left(\frac{2m+1}{\bar{\eta}_{12}-1} - 2\frac{\bar{\eta}_{12}^{s+m+1}-1}{(\bar{\eta}_{12}-1)^2}\right) \\
&= (2m+1)\frac{\bar{\eta}_{\alpha_2}^{s+m}}{\bar{\eta}_{\alpha_1}-\bar{\eta}_{\alpha_2}} - 2\frac{\bar{\eta}_{\alpha_1}^{s+m+1}-\bar{\eta}_{\alpha_2}^{s+m+1}}{(\bar{\eta}_{\alpha_1}-\bar{\eta}_{\alpha_2})^2} \\
&= (2m+1)P_{m;\alpha_1,\alpha_2} + 2Q^{=}_{m;\alpha_1,\alpha_2},
\end{aligned} \tag{4.3.26}$$

where we have defined

$$P_{m;\alpha_1,\alpha_2} \equiv \frac{\bar{\eta}_{\alpha_2}^{s+m}}{\bar{\eta}_{\alpha_1}-\bar{\eta}_{\alpha_2}}, \quad Q^{=}_{m;\alpha_1,\alpha_2} \equiv -\frac{\bar{\eta}_{\alpha_1}^{s+m+1}-\bar{\eta}_{\alpha_2}^{s+m+1}}{(\bar{\eta}_{\alpha_1}-\bar{\eta}_{\alpha_2})^2}. \tag{4.3.27}$$

(b) $-3s \leq m \leq -s$. Since $-s \leq m_1 \leq s$, this implies that we are always in the first regime in (4.3.23), $m \leq m_1 \leq 2s+m$. This yields

$$\begin{aligned}
\frac{B^{=}_{m;\alpha_1,\alpha_2}}{\bar{\eta}_{\alpha_2}^{s+m-1}} &= \sum_{m_1=-s}^{2s+m} (N+m-m_1)\bar{\eta}_{12}^{s+m_1} \\
&= (N+m)\frac{\bar{\eta}_{12}^{s+m}-1}{\bar{\eta}_{12}-1} - \frac{(N+m)\bar{\eta}_{12}^{s+m}+s}{\bar{\eta}_{12}-1} + \frac{\bar{\eta}_{12}^{s+m+1}-\bar{\eta}_{12}}{(\bar{\eta}_{12}-1)^2} \\
&= -\frac{N+m+s+1}{\bar{\eta}_{12}-1} + \frac{\bar{\eta}_{12}^{s+m+1}-1}{(\bar{\eta}_{12}-1)^2},
\end{aligned} \tag{4.3.28}$$

and

$$\begin{aligned}
B^{=}_{m;\alpha_1,\alpha_2} &= -(N+m+s+1)\frac{\bar{\eta}_{\alpha_2}^{s+m}}{\bar{\eta}_{\alpha_1}-\bar{\eta}_{\alpha_2}} + \frac{\bar{\eta}_{\alpha_1}^{s+m+1}-\bar{\eta}_{\alpha_2}^{s+m+1}}{(\bar{\eta}_{\alpha_1}-\bar{\eta}_{\alpha_2})^2} \\
&= -(N+m+s+1)P_{m;\alpha_1,\alpha_2} - Q^{=}_{m;\alpha_1,\alpha_2}.
\end{aligned} \tag{4.3.29}$$

(c) $s < m \leq 3s$. Since $-s \leq m_1 \leq s$, this implies that we are always in the second regime in (4.3.23), $-2s+m \leq m_1 \leq m-1$. This yields

$$\begin{aligned}
\frac{B^{=}_{m;\alpha_1,\alpha_2}}{\bar{\eta}_{\alpha_2}^{s+m-1}} &= \sum_{m_1=-2s+m}^{s} (N-m+m_1)\bar{\eta}_{12}^{s+m_1} \\
&= \sum_{m_1=-2s+m-1}^{s} (N-m+m_1)\bar{\eta}_{12}^{s+m_1} \\
&= (N-m)\frac{1-\bar{\eta}_{12}^{s+m}}{\bar{\eta}_{12}-1} \\
&\quad + \frac{(s+1)+(N-m)\bar{\eta}_{12}^{s+m}}{\bar{\eta}_{12}-1} - \frac{\bar{\eta}_{12}-\bar{\eta}_{12}^{s+m+1}}{(\bar{\eta}_{12}-1)^2} \\
&= \frac{N-m+s}{\bar{\eta}_{12}-1} + \frac{\bar{\eta}_{12}^{s+m+1}-1}{(\bar{\eta}_{12}-1)^2},
\end{aligned} \tag{4.3.30}$$

and

$$B^{=}_{m;\alpha_1,\alpha_2} = (N-m+s)\frac{\bar{\eta}_{\alpha_2}^{s+m}}{\bar{\eta}_{\alpha_1}-\bar{\eta}_{\alpha_2}} + \frac{\bar{\eta}_{\alpha_1}^{s+m+1}-\bar{\eta}_{\alpha_2}^{s+m+1}}{(\bar{\eta}_{\alpha_1}-\bar{\eta}_{\alpha_2})^2} \tag{4.3.31}$$

$$= (N-m+s)P_{m;\alpha_1,\alpha_2} - Q^{=}_{m;\alpha_1,\alpha_2}. \tag{4.3.32}$$

Note that since $Q^{=}_{s;\alpha_1,\alpha_2} = 0$ and

$$Q^{=}_{-s;\alpha_1,\alpha_2} = -\frac{\bar{\eta}_{\alpha_1}-\bar{\eta}_{\alpha_2}}{(\bar{\eta}_{\alpha_1}-\bar{\eta}_{\alpha_2})^2} = -\frac{1}{(\bar{\eta}_{\alpha_1}-\bar{\eta}_{\alpha_2})} = -P_{-s;\alpha_1,\alpha_2},$$

it does not matter with which regimes we associate the cases $m = \pm s$. The expressions (4.3.26) and (4.3.29) are equal for $m = -s$, and (4.3.26) and (4.3.31) are equal for m=s.

4.4 The Defining Condition for the $S = 1$ Pfaffian Chain

4.4.1 Derivation

In Sect. 4.3, we have shown that the non-Abelian $S = 1$ spin liquid state (4.1.6) with (4.1.4) introduced in Sect. 2.4, is annihilated by

$$V = \sum_{m=-3s}^{3s} V_m B_m^{\dagger} B_m, \tag{4.4.1}$$

where

$$V_m = (3s - m)!(3s + m)! \tag{4.4.2}$$

and

$$B_m = B_m^{\neq} + B_m^{=} \tag{4.4.3}$$

with

$$B_m^{\neq} = \frac{1}{\sqrt{N^3}} \sum_{\substack{\alpha_1,\alpha_2,\alpha_3=1 \\ \alpha_1\neq\alpha_2\neq\alpha_3\neq\alpha_1}}^{N} B_{m;\alpha_1,\alpha_2,\alpha_3}^{\neq} S_{\alpha_3}^{-} S_{\alpha_2}^{-} S_{\alpha_1}^{-}, \tag{4.4.4}$$

$$B_m^{=} = \frac{3}{\sqrt{N^3}} \sum_{\substack{\alpha_1,\alpha_2=1 \\ \alpha_1\neq\alpha_2}}^{N} B_{m;\alpha_1,\alpha_2}^{=} \left(S_{\alpha_2}^{-}\right)^2 S_{\alpha_1}^{-}. \tag{4.4.5}$$

We calculated the coefficients in (4.4.4) and (4.4.5) in Sects. 4.3.4 and 4.3.5, respectively, and found

$$B_{m;\alpha_1,\alpha_2,\alpha_3}^{\neq} = \begin{cases} -Q_{m;\alpha_1,\alpha_2,\alpha_3}^{\neq} & \text{for } \quad s < m \leq 3s, \\ 2Q_{m;\alpha_1,\alpha_2,\alpha_3}^{\neq} & \text{for } -s < m \leq s, \\ -Q_{m;\alpha_1,\alpha_2,\alpha_3}^{\neq} & \text{for } -3s < m \leq -s, \end{cases} \tag{4.4.6}$$

and

$$B_{m;\alpha_1,\alpha_2}^{=} = \begin{cases} (N - m + s)\, P_{m;\alpha_1,\alpha_2} - Q_{m;\alpha_1,\alpha_2}^{=} & \text{for } \quad s < m \leq 3s, \\ (2m + 1)\, P_{m;\alpha_1,\alpha_2} + 2Q_{m;\alpha_1,\alpha_2}^{=} & \text{for } -s < m \leq s, \\ -(N + m + s + 1)\, P_{m;\alpha_1,\alpha_2} - Q_{m;\alpha_1,\alpha_2}^{=} & \text{for } -3s < m \leq -s. \end{cases} \tag{4.4.7}$$

$Q^{\neq}_{m;\alpha_1,\alpha_2,\alpha_3}$ is defined in (4.3.18), and $P_{m;\alpha_1,\alpha_2}$ and $Q^{=}_{m;\alpha_1,\alpha_2}$ are defined in (4.3.27). All three are periodic functions of m, i.e.,

$$\begin{aligned} Q^{\neq}_{m+N;\alpha_1,\alpha_2,\alpha_3} &= Q^{\neq}_{m;\alpha_1,\alpha_2,\alpha_3}, \\ P_{m+N;\alpha_1,\alpha_2} &= P_{m;\alpha_1,\alpha_2}, \\ Q^{=}_{m+N;\alpha_1,\alpha_2} &= Q^{=}_{m;\alpha_1,\alpha_2}. \end{aligned} \tag{4.4.8}$$

The property that $\left|\psi_0^{S=1}\right\rangle$ is annihilated by V implies with (4.4.1) that

$$\begin{aligned} \left\langle\psi_0^{S=1}\middle|V\middle|\psi_0^{S=1}\right\rangle &= \sum_{m=-3s}^{3s} V_m \left\langle\psi_0^{S=1}\middle|B_m^{\dagger}B_m\middle|\psi_0^{S=1}\right\rangle \\ &= \sum_{m=-3s}^{3s} V_m \left\| B_m\left|\psi_0^{S=1}\right\rangle\right\|^2 = 0. \end{aligned} \tag{4.4.9}$$

Since all the values V_m for $-3s \le m \le 3s$ are positive, and the norms of the vectors by definition non-negative, (4.4.9) implies that the vectors $B_m\left|\psi_0^{S=1}\right\rangle$ must vanish for all allowed values of m. In other words,

$$B_m\left|\psi_0^{S=1}\right\rangle = 0 \quad \forall\, m \in [-3s, 3s]. \tag{4.4.10}$$

This implies that $\left|\psi_0^{S=1}\right\rangle$ is further annihilated by any linear combination of the B_m's, and in particular also those in which the terms involving $Q^{\neq}_{m+N;\alpha_1,\alpha_2,\alpha_3}$ and $Q^{=}_{m+N;\alpha_1,\alpha_2}$ cancel. These include for $-s < m \le s$

$$\begin{aligned} B_m + 2B_{m-N} &= \left[(2m+1) - 2(s+m+1)\right] \sum_{\alpha_1\neq\alpha_2}^{N} P_{m;\alpha_1,\alpha_2} \left(S^-_{\alpha_2}\right)^2 S^-_{\alpha_1} \\ &= -N \sum_{\alpha_1\neq\alpha_2}^{N} P_{m;\alpha_1,\alpha_2} \left(S^-_{\alpha_2}\right)^2 S^-_{\alpha_1}, \end{aligned} \tag{4.4.11}$$

and for $m = s + 1$

$$B_{s+1} - B_{-s} = 2N \sum_{\alpha_1\neq\alpha_2}^{N} P_{s+1;\alpha_1,\alpha_2} \left(S^-_{\alpha_2}\right)^2 S^-_{\alpha_1}. \tag{4.4.12}$$

Given the periodicity of $P_{m;\alpha_1,\alpha_2}$ in m, (4.4.11) and (4.4.12) imply that

$$P_m\left|\psi_0^{S=1}\right\rangle = 0 \quad \forall\, m, \tag{4.4.13}$$

where we have defined

$$\begin{aligned} P_m &\equiv \sum_{\alpha_1 \neq \alpha_2}^{N} P_{m;\alpha_1,\alpha_2} \left(S^-_{\alpha_2}\right)^2 S^-_{\alpha_1} \\ &= \sum_{\alpha_1 \neq \alpha_2}^{N} \frac{\bar{\eta}_{\alpha_2}^{s+m}}{\bar{\eta}_{\alpha_1} - \bar{\eta}_{\alpha_2}} \left(S^-_{\alpha_2}\right)^2 S^-_{\alpha_1}. \end{aligned} \tag{4.4.14}$$

Since the spin liquid state $\left|\psi_0^{S=1}\right\rangle$ is invariant under parity, i.e., under $\eta\alpha \to \bar{\eta}_\alpha$ (see Sect. 2.4.2), it is also annihilated by the complex conjugates $\bar{P}_m$ of P_m for all m.

The non-Abelian $S = 1$ spin liquid state (4.1.6) with (4.1.4) is further annihilated by the operators

$$\begin{aligned} \Omega_\alpha^{S=1} &\equiv -\frac{1}{N} \sum_{m=0}^{N} \bar{\eta}_\alpha^{s+m} \bar{P}_m \\ &= \sum_{\substack{\beta=1 \\ \beta \neq \alpha}}^{N} \frac{1}{\eta_\alpha - \eta_\beta} (S_\alpha^-)^2 S_\beta^-, \quad \Omega_\alpha^{S=1}\left|\psi_0^{S=1}\right\rangle = 0 \quad \forall\, \alpha, \end{aligned} \tag{4.4.15}$$

which are obtained from the complex conjugate of (4.4.14) by Fourier transformation, as well as their complex conjugates,

$$\bar{\Omega}_\alpha^{S=1} = \sum_{\substack{\beta=1 \\ \beta \neq \alpha}}^{N} \frac{1}{\bar{\eta}_\alpha - \bar{\eta}_\beta} (S_\alpha^-)^2 S_\beta^-, \quad \bar{\Omega}_\alpha^{S=1}\left|\psi_0^{S=1}\right\rangle = 0 \quad \forall\, \alpha. \tag{4.4.16}$$

Note that we would not need to exclude configurations with $\beta = \alpha$, as the spin operators take care of this automatically.

In Sect. 4.5, we will use the operators $\Omega_\alpha^{S=1}$ to construct a parent Hamiltonian, which is translationally invariant, invariant under P and T, and invariant under SU(2) spin rotations, for the non-Abelian $S = 1$ spin liquid state $\left|\psi_0^{S=1}\right\rangle$. The analysis will imply that $\left|\psi_0^{S=1}\right\rangle$ is completely specified by the condition (4.4.15) plus the the mentioned symmetries. Therefore, we will refer to (4.4.15) as the *defining condition* of non-Abelian $S = 1$ spin chain we introduce in Sect. 2.4.

4.4.2 A Second Condition

It is worth noting that the condition (4.4.13) with (4.4.14) implies that the remaining terms in B_m annihilate $\left|\psi_0^{S=1}\right\rangle$ as well. In particular, we have

$$Q_m\left|\psi_0^{S=1}\right\rangle = \bar{Q}_m\left|\psi_0^{S=1}\right\rangle = 0 \quad \forall\, m, \tag{4.4.17}$$

where we have defined

$$
\begin{aligned}
Q_m \equiv & \frac{1}{3} \sum_{\substack{\alpha_1,\alpha_2,\alpha_3=1 \\ \alpha_1\neq\alpha_2\neq\alpha_3\neq\alpha_1}}^{N} Q^{\neq}_{m;\alpha_1,\alpha_2,\alpha_3} S^-_{\alpha_3} S^-_{\alpha_2} S^-_{\alpha_1} + \sum_{\alpha_1\neq\alpha_2}^{N} Q^{=}_{m;\alpha_1,\alpha_2} \left(S^-_{\alpha_2}\right)^2 S^-_{\alpha_1} \\
= & \sum_{\substack{\alpha_1,\alpha_2,\alpha_3=1 \\ \alpha_1\neq\alpha_2\neq\alpha_3\neq\alpha_1}}^{N} \frac{\bar{\eta}_{\alpha_1}^{s+m+1}}{(\bar{\eta}_{\alpha_3}-\bar{\eta}_{\alpha_1})(\bar{\eta}_{\alpha_1}-\bar{\eta}_{\alpha_2})} S^-_{\alpha_3} S^-_{\alpha_2} S^-_{\alpha_1} \\
& - \sum_{\alpha_1\neq\alpha_2}^{N} \frac{\bar{\eta}_{\alpha_1}^{s+m+1} - \bar{\eta}_{\alpha_2}^{s+m+1}}{(\bar{\eta}_{\alpha_1}-\bar{\eta}_{\alpha_2})^2} \left(S^-_{\alpha_2}\right)^2 S^-_{\alpha_1} \\
= & \sum_{\substack{\alpha_1,\alpha_2,\alpha_3=1 \\ \alpha_1\neq\alpha_2 \\ \alpha_1\neq\alpha_3}}^{N} \frac{\bar{\eta}_{\alpha_1}^{s+m+1}}{(\bar{\eta}_{\alpha_3}-\bar{\eta}_{\alpha_1})(\bar{\eta}_{\alpha_1}-\bar{\eta}_{\alpha_2})} S^-_{\alpha_3} S^-_{\alpha_2} S^-_{\alpha_1} \\
& + \sum_{\alpha_1\neq\alpha_2}^{N} \frac{\bar{\eta}_{\alpha_2}^{s+m+1}}{(\bar{\eta}_{\alpha_1}-\bar{\eta}_{\alpha_2})^2} \left(S^-_{\alpha_2}\right)^2 S^-_{\alpha_1}.
\end{aligned}
\tag{4.4.18}
$$

The non-Abelian $S = 1$ spin liquid state $\left|\psi_0^{S=1}\right\rangle$ is further annihilated by the operators

$$
\begin{aligned}
\Xi_\alpha \equiv & -\frac{1}{N} \sum_{m=0}^{N} \bar{\eta}_\alpha^{s+m+1} \bar{Q}_m \\
= & \sum_{\substack{\beta,\gamma=1 \\ \beta,\gamma\neq\alpha}}^{N} \frac{S^-_\alpha S^-_\beta S^-_\gamma}{(\eta_\alpha-\eta_\beta)(\eta_\alpha-\eta_\gamma)} - \sum_{\substack{\beta=1 \\ \beta\neq\alpha}}^{N} \frac{(S^-_\alpha)^2 S^-_\beta}{(\eta_\alpha-\eta_\beta)^2}, \quad \Xi_\alpha \left|\psi_0^{S=1}\right\rangle = 0 \quad \forall\, \alpha,
\end{aligned}
\tag{4.4.19}
$$

which are obtained from the complex conjugate of (4.4.18) by Fourier transformation, as well as their complex conjugates $\bar{\Xi}_\alpha$. These operators, however, do not appear promising for the construction of a simple parent Hamiltonian for the state.

4.4.3 Direct Verification

In this section, we wish to verify the defining condition (4.4.15) directly for the $S = 1$ ground state (4.1.4). The method will be similar to the proof of the singlet property in Sect. 2.4.2. To begin with, we again notice that when we substitute (4.1.4) with (4.1.2) into (4.1.6), we may replace theantisymmetrization $\mathcal{A}$ in (4.1.2) by an overall normalization factor 9 which we ignore, as it is taken care by the commutativity of the bosonic operators $\tilde{S}_\alpha$. Let $\tilde{\psi}_0$ be $\psi_0^{S=1}$ without the antisymmetrization in (4.1.2),

$$\tilde{\psi}_0(z_1,\dots,z_N) = \left\{\frac{1}{z_1-z_2}\cdot\ldots\cdot\frac{1}{z_{N-1}-z_N}\right\}\cdot\prod_{i<j}^{N}(z_i-z_j)\prod_{i=1}^{N}z_i. \quad (4.4.20)$$

Since $\tilde{\psi}_0(z_1,z_2,\dots,z_N)$ is still symmetric under interchange of pairs, we may assume that the spin flip operators $(S_\alpha^-)^2$ and S_β^- of (4.4.15) will act on the pairs (z_1,z_2) and (z_3,z_4), respectively:

$$\begin{aligned}(S_\alpha^-)^2 S_\beta^- \left|\psi_0^{S=1}\right\rangle = \sum_{\{z_5,\dots,z_N\}} &(S_\alpha^-)^2(S_\alpha^+)^2 \\ &\left\{\sum_{z_4(\neq\eta_\beta)} \tilde{\psi}_0(\eta_\alpha,\eta_\alpha,\eta_\beta,z_4,z_5,\dots,z_N) S_\beta^- \tilde{S}_\beta^+ \tilde{S}_{z_4}^+\right.\\ &+\sum_{z_3(\neq\eta_\beta)} \tilde{\psi}_0(\eta_\alpha,\eta_\alpha,z_3,\eta_\beta,z_5,\dots,z_N) S_\beta^- \tilde{S}_{z_3}^+ \tilde{S}_\beta^+ \\ &\left.+\tilde{\psi}_0(\eta_\alpha,\eta_\alpha,\eta_\beta,\eta_\beta,z_5,\dots,z_N) S_\beta^- (\tilde{S}_\beta^+)^2\right\} \\ &\cdot \tilde{S}_{z_5}^+\ldots\tilde{S}_{z_N}^+|-1\rangle_N \\ = 4\sum_{\{z_5,\dots,z_N\}} &\left\{\sum_{z_4}\tilde{\psi}_0(\eta_\alpha,\eta_\alpha,\eta_\beta,z_4,z_5,\dots,z_N)\tilde{S}_{z_4}^+\right\}\tilde{S}_{z_5}^+\ldots\tilde{S}_{z_N}^+|-1\rangle_N, \end{aligned} \quad (4.4.21)$$

where we have used

$$S_\alpha^-(\tilde{S}_\alpha^+)^n|1,-1\rangle_\alpha = n(\tilde{S}_\alpha^+)^{n-1}|1,-1\rangle_\alpha, \quad (4.4.22)$$

which follows directly form the definition (4.1.5). We hence obtain

$$\Omega_\alpha^{S=1}\left|\psi_0^{S=1}\right\rangle = \sum_{\{z_4,\dots,z_N\}} \underbrace{\sum_{\substack{\beta=1\\ \beta\neq\alpha}}^{N} \frac{\tilde{\psi}_0(\eta_\alpha,\eta_\alpha,\eta_\beta,z_4,\dots,z_N)}{\eta_\alpha-\eta_\beta}}_{=0} \tilde{S}_{z_4}^+\ldots\tilde{S}_{z_N}^+|-1\rangle_N, \quad (4.4.23)$$

since

$$\begin{aligned}\frac{\tilde{\psi}_0(\eta_\alpha,\eta_\alpha,\eta_\beta,z_4,\dots,z_N)}{\eta_\alpha-\eta_\beta} &= (\eta_\alpha-\eta_\beta)\eta_\alpha^2\eta_\beta \\ &\cdot\prod_{i=4}^{N}(\eta_\alpha-z_i)^2 z_i\prod_{i=5}^{N}(\eta_\beta-z_i)\prod_{4\le i<j}^{N}(z_i-z_j) \\ &\cdot\frac{1}{z_5-z_6}\cdot\ldots\cdot\frac{1}{z_{N-1}-z_N}\end{aligned}$$

vanishes for $\beta=\alpha$ and contains only powers $\eta_\beta^1,\eta_\beta^2,\dots,\eta_\beta^{N-2}$. Note that the calculation for $\bar{\Omega}_\alpha^{S=1}$ is almost identical, since

$$\frac{\tilde{\psi}_0(\eta_\alpha, \eta_\alpha, \eta_\beta, z_4, \ldots, z_N)}{\bar{\eta}_\alpha - \eta_\beta b} = -\eta_\alpha \eta_\beta \frac{\tilde{\psi}_0(\eta_\alpha, \eta_\alpha, \eta_\beta, z_4, \ldots, z_N)}{\eta_\alpha - \eta_\beta}$$

vanishes also for $\beta = \alpha$ and contains only powers $\eta_\beta^2, \eta_\beta^3, \ldots, \eta_\beta^{N-1}$.

4.5 Construction of a Parent Hamiltonian

We will now construct a parent Hamiltonian for the non-Abelian $S = 1$ spin liquid state (4.1.6) with (4.1.4) using the annihilation operator (4.4.15), i.e.,

$$\Omega_\alpha^{S=1} \left|\psi_0^{S=1}\right\rangle = 0 \quad \forall\, \alpha, \quad \text{where} \quad \Omega_\alpha^{S=1} = \sum_{\substack{\beta=1 \\ \beta\neq\alpha}}^{N} \frac{1}{\eta_\alpha - \eta_\beta} \left(S_\alpha^-\right)^2 S_\beta^- . \tag{4.5.1}$$

The Hamiltonian has to be Hermitian, and we wish it to be invariant under translations, time reversal (T), parity (P), and SU(2) spin rotations.

4.5.1 *Translational, Time Reversal, and Parity Symmetry*

The operator $\Omega_\alpha^{S=1\dagger} \Omega_\alpha^{S=1}$ is Hermitian and positive semi-definite, meaning that all the eigenvalues are non-negative. A translationally invariant operator is given by

$$\begin{aligned} H_0 = \frac{1}{2} \sum_{\alpha=1}^{N} \Omega_\alpha^{S=1\dagger} \Omega_\alpha^{S=1} &= \frac{1}{2} \sum_{\substack{\alpha,\beta,\gamma \\ \alpha\neq\beta,\gamma}} \frac{1}{\bar{\eta}_\alpha - \bar{\eta}_\beta} \frac{1}{\eta_\alpha - \eta_\gamma} \left(S_\alpha^+\right)^2 \left(S_\alpha^-\right)^2 S_\beta^+ S_\gamma^- \\ &= \sum_{\substack{\alpha,\beta,\gamma \\ \alpha\neq\beta,\gamma}} \omega_{\alpha\beta\gamma} S_\alpha^z \left(S_\alpha^z + 1\right) S_\beta^+ S_\gamma^- , \end{aligned} \tag{4.5.2}$$

where $\omega_{\alpha\beta\gamma}$ is defined in (3.6.3),and we have used that

$$\left(S_\alpha^+\right)^2 \left(S_\alpha^-\right)^2 = 2 S_\alpha^z \left(S_\alpha^z + 1\right)$$

for $S = 1$, which is readily verified with (C.6). With the transformation properties under time reversal,

$$\mathrm{T}: \quad \eta_\alpha \to \Theta \eta_\alpha \Theta = \bar{\eta}_\alpha, \quad \boldsymbol{S} \to \Theta \boldsymbol{S} \Theta = -\boldsymbol{S},$$

and hence

$$\omega_{\alpha\beta\gamma} \to \omega_{\alpha\gamma\beta}, \quad S^+ \to -S^-, \quad S^- \to -S^+, \quad S^z \to -S^z,$$

the operator (4.5.2) transforms into

$$\Theta H_0 \Theta = \sum_{\substack{\alpha,\beta,\gamma \\ \alpha\neq\beta,\gamma}} \omega_{\alpha\beta\gamma} S^{\mathrm{z}}_\alpha \left(S^{\mathrm{z}}_\alpha - 1\right) S^-_\gamma S^+_\beta . \tag{4.5.3}$$

We proceed with the T invariant operator

$$H_0^{\mathrm{T}} = \frac{1}{2}\left(H_0 + \Theta H_0 \Theta\right) = H_0^{\mathrm{T}=} + H_0^{\mathrm{T}\neq}, \tag{4.5.4}$$

where

$$\begin{aligned} H_0^{\mathrm{T}=} &= \frac{1}{2}\sum_{\substack{\alpha,\beta \\ \alpha\neq\beta}} \omega_{\alpha\beta\beta}\left(S^{\mathrm{z}2}_\alpha\{S^+_\beta, S^-_\beta\} + S^{\mathrm{z}}_\alpha[S^+_\beta, S^-_\beta]\right) \\ &= \frac{1}{2}\sum_{\alpha\neq\beta} \omega_{\alpha\beta\beta}\left(S^{\mathrm{z}^2}_\alpha\{S^+_\beta, S^-_\beta\} + 2S^{\mathrm{z}}_\alpha S^{\mathrm{z}}_\beta\right), \end{aligned} \tag{4.5.5}$$

$$H_0^{\mathrm{T}\neq} = \sum_{\substack{\alpha,\beta,\gamma \\ \alpha\neq\beta\neq\gamma\neq\alpha}} \omega_{\alpha\beta\gamma}\, S^{\mathrm{z}2}_\alpha S^+_\beta S^-_\gamma . \tag{4.5.6}$$

With the transformation properties under parity,

$$\mathrm{P}: \quad \eta_\alpha \to \Pi\eta_\alpha\Pi = \bar{\eta}_\alpha, \quad \boldsymbol{S} \to \Theta\boldsymbol{S}\Theta = \boldsymbol{S}, \tag{4.5.7}$$

and hence $\omega_{\alpha\beta\gamma} \to \omega_{\alpha\gamma\beta}$, we obtain the P and T invariant operator

$$H_0^{\mathrm{PT}} = \frac{1}{2}\left(H_0^{\mathrm{T}} + \Pi H_0^{\mathrm{T}} \Pi\right) = H_0^{\mathrm{PT}=} + H_0^{\mathrm{PT}\neq}, \tag{4.5.8}$$

where

$$H_0^{\mathrm{PT}=} = H_0^{\mathrm{T}=}, \quad H_0^{\mathrm{PT}\neq} = \frac{1}{2}\sum_{\substack{\alpha,\beta,\gamma \\ \alpha\neq\beta\neq\gamma\neq\alpha}} \omega_{\alpha\beta\gamma} S^{\mathrm{z}2}_\alpha\left(S^+_\beta S^-_\gamma + S^-_\beta S^+_\gamma\right). \tag{4.5.9}$$

4.5.2 Spin Rotation Symmetry

Since the non-Abelian spin liquid state $\left|\psi_0^{S=1}\right\rangle$ is a spin singlet, the property that it is annihilated by (4.5.8) with (4.5.9) and (4.5.5) implies that it is annihilated by each tensor component of (4.5.8) individually.

With the tensor decompositions (D.2.4), (D.2.5), and $S^2 = 2$ for $S = 1$, we can rewrite the two contributions as

$$H_0^{\mathrm{PT}=} = \sum_{\alpha \neq \beta} \omega_{\alpha\beta\beta} \left[\left(\frac{2}{3} + \frac{1}{\sqrt{6}} T^0_{\alpha\alpha} \right) \left(\frac{4}{3} - \frac{1}{\sqrt{6}} T^0_{\beta\beta} \right) + \frac{1}{3} \boldsymbol{S}_\alpha \boldsymbol{S}_\beta + \frac{1}{\sqrt{6}} T^0_{\alpha\beta} \right], \tag{4.5.10}$$

$$H_0^{\mathrm{PT}\neq} = \sum_{\substack{\alpha,\beta,\gamma \\ \alpha \neq \beta \neq \gamma \neq \alpha}} \omega_{\alpha\beta\gamma} \left(\frac{2}{3} + \frac{1}{\sqrt{6}} T^0_{\alpha\alpha} \right) \left(\frac{2}{3} \boldsymbol{S}_\beta \boldsymbol{S}_\gamma - \frac{1}{\sqrt{6}} T^0_{\beta\gamma} \right). \tag{4.5.11}$$

Projecting out the scalar components under SU(2) spin rotations yields

$$\left\{ H_0^{\mathrm{PT}=} \right\}_0 = \sum_{\alpha \neq \beta} \omega_{\alpha\beta\beta} \left(\frac{8}{9} - \frac{1}{6} \left\{ T^0_{\alpha\alpha} T^0_{\beta\beta} \right\}_0 + \frac{1}{3} \boldsymbol{S}_\alpha \boldsymbol{S}_\beta \right), \tag{4.5.12}$$

$$\left\{ H_0^{\mathrm{PT}\neq} \right\}_0 = \sum_{\substack{\alpha,\beta,\gamma \\ \alpha \neq \beta \neq \gamma \neq \alpha}} \omega_{\alpha\beta\gamma} \left(\frac{4}{9} \boldsymbol{S}_\beta \boldsymbol{S}_\gamma - \frac{1}{6} \left\{ T^0_{\alpha\alpha} T^0_{\beta\gamma} \right\}_0 \right). \tag{4.5.13}$$

The next step is to calculate the scalar component of the tensor products in (4.5.12).

4.5.3 Evaluation of $\{T^0_{\alpha\alpha} T^0_{\beta\gamma}\}_0$

We evaluate the scalar component of the tensor product of $T^0_{\alpha\alpha}$ and $T^0_{\beta\gamma}$ with $\alpha \neq \beta, \gamma$ using (3.5.20),

$$\left\{ T^0_{\alpha\alpha} T^0_{\beta\gamma} \right\}_j = \langle j, 0 | 2, 0; 2, 0 \rangle \sum_{m=-2}^{2} T^m_{\alpha\alpha} T^{-m}_{\beta\gamma} \langle 2, m; 2, -m | j, 0 \rangle. \tag{4.5.14}$$

With (D.2.3) and the Clebsch–Gordan coefficients

$$\langle 2, m; 2, -m | 0, 0 \rangle = \frac{(-1)^m}{\sqrt{5}}, \tag{4.5.15}$$

we obtain

$$
\begin{aligned}
5\left\{T^0_{\alpha\alpha}T^0_{\alpha\gamma}\right\}_0 &= \sum_{m=-2}^{2}(-1)^m T^m_{\alpha\alpha}T^{-m}_{\alpha\gamma} \\
&= S^-_\alpha S^-_\alpha S^+_\beta S^+_\gamma \\
&\quad + \left(S^z_\alpha S^-_\alpha + S^-_\alpha S^z_\alpha\right)\left(S^z_\beta S^+_\gamma + S^+_\beta S^z_\gamma\right) \\
&\quad + \frac{1}{6}\left(4S^z_\alpha S^z_\alpha - S^+_\alpha S^-_\alpha - S^-_\alpha S^+_\alpha\right)\left(4S^z_\beta S^z_\gamma - S^+_\beta S^-_\gamma - S^-_\beta S^+_\gamma\right) \\
&\quad + \left(S^z_\alpha S^+_\alpha + S^+_\alpha S^z_\alpha\right)\left(S^z_\beta S^-_\gamma + S^-_\beta S^z_\gamma\right) \\
&\quad + S^+_\alpha S^+_\alpha S^-_\beta S^-_\gamma .
\end{aligned}
\tag{4.5.16}
$$

We wish to write this in a more convenient form, which directly displays that it transforms as a scalar under spin rotations. Since

$$
\mathbf{1}\otimes\mathbf{1}\otimes\mathbf{1}\otimes\mathbf{1} = 3\cdot\mathbf{0}\oplus 6\cdot\mathbf{1}\oplus 6\cdot\mathbf{2}\oplus 3\cdot\mathbf{3}\oplus\mathbf{4},
$$

we can only form three scalars from four spin operators. For $\alpha\neq\beta\neq\gamma\neq\alpha$, three such scalars are

$$
\boldsymbol{S}^2_\alpha(\boldsymbol{S}_\beta\boldsymbol{S}_\gamma),\quad (\boldsymbol{S}_\alpha\boldsymbol{S}_\beta)(\boldsymbol{S}_\alpha\boldsymbol{S}_\gamma),\quad \text{and}\quad (\boldsymbol{S}_\alpha\boldsymbol{S}_\gamma)(\boldsymbol{S}_\alpha\boldsymbol{S}_\beta).
$$

For $\alpha\neq\beta=\gamma$, the latter two are identical, but we have the additional scalar $\boldsymbol{S}_\alpha\boldsymbol{S}_\beta$. For $\alpha\neq\beta,\gamma$ in general, we write

$$
\begin{aligned}
5\left\{T^0_{\alpha\alpha}T^0_{\alpha\gamma}\right\}_0 = a\,\boldsymbol{S}^2_\alpha(\boldsymbol{S}_\beta\boldsymbol{S}_\gamma) + b\left[(\boldsymbol{S}_\alpha\boldsymbol{S}_\beta)(\boldsymbol{S}_\alpha\boldsymbol{S}_\gamma) + (\boldsymbol{S}_\alpha\boldsymbol{S}_\gamma)(\boldsymbol{S}_\alpha\boldsymbol{S}_\beta)\right] \\
+ c\,\delta_{\beta\gamma}\,\boldsymbol{S}_\alpha\boldsymbol{S}_\beta,
\end{aligned}
\tag{4.5.17}
$$

where we have used the invariance of the tensor product under interchange of β and γ. The coefficients a and b may depend on whether $\beta=\gamma$ or not.

Since the $S^-_\alpha S^-_\alpha$ term in (4.5.16) has to come form the second term in (4.5.17), we can immediately infer $b=2$. To obtain a and c, we first write out the second term in (4.5.17) for $\alpha\neq\beta,\gamma$,

$$
\begin{aligned}
&2\left[(\boldsymbol{S}_\alpha\boldsymbol{S}_\beta)(\boldsymbol{S}_\alpha\boldsymbol{S}_\gamma) + (\boldsymbol{S}_\alpha\boldsymbol{S}_\gamma)(\boldsymbol{S}_\alpha\boldsymbol{S}_\beta)\right] \\
&\quad = \frac{1}{2}\left(2S^z_\alpha S^z_\beta + S^+_\alpha S^-_\beta + S^-_\alpha S^+_\beta\right)\left(2S^z_\alpha S^z_\gamma + S^+_\alpha S^-_\gamma + S^-_\alpha S^+_\gamma\right) \\
&\qquad + \text{same with } \beta\leftrightarrow\gamma \\
&\quad = S^+_\alpha S^+_\alpha S^-_\beta S^-_\gamma + S^-_\alpha S^-_\alpha S^+_\beta S^+_\gamma \\
&\qquad + S^z_\alpha S^z_\beta\left(S^+_\alpha S^-_\gamma + S^-_\alpha S^+_\gamma\right) + S^z_\alpha S^z_\gamma\left(S^+_\alpha S^-_\beta + S^-_\alpha S^+_\beta\right) \\
&\qquad + \left(S^+_\alpha S^-_\beta + S^-_\alpha S^+_\beta\right)S^z_\alpha S^z_\gamma + \left(S^+_\alpha S^-_\gamma + S^-_\alpha S^+_\gamma\right)S^z_\alpha S^z_\beta \\
&\qquad + \frac{1}{2}S^+_\alpha S^-_\alpha\left(S^-_\beta S^+_\gamma + S^-_\gamma S^+_\beta\right) + \frac{1}{2}S^-_\alpha S^+_\alpha\left(S^+_\beta S^-_\gamma + S^+_\gamma S^-_\beta\right) \\
&\qquad + 4S^z_\alpha S^z_\alpha S^z_\beta S^z_\gamma,
\end{aligned}
\tag{4.5.18}
$$

and order the terms such that the S_β operators are to the left of the S_γ operators,

$$\begin{aligned}
&2\left[(\boldsymbol{S}_\alpha \boldsymbol{S}_\beta)(\boldsymbol{S}_\alpha \boldsymbol{S}_\gamma) + (\boldsymbol{S}_\alpha \boldsymbol{S}_\gamma)(\boldsymbol{S}_\alpha \boldsymbol{S}_\beta)\right]\\
&\quad = S_\alpha^+ S_\alpha^+ S_\beta^- S_\gamma^- + S_\alpha^- S_\alpha^- S_\beta^+ S_\gamma^+\\
&\quad\quad + S_\alpha^z S_\alpha^+ S_\beta^z S_\gamma^- + S_\alpha^z S_\alpha^- S_\beta^z S_\gamma^+ + S_\alpha^z S_\alpha^+ S_\beta^- S_\gamma^z + S_\alpha^z S_\alpha^- S_\beta^+ S_\gamma^z\\
&\quad\quad - S_\alpha^z S_\alpha^+ [S_\beta^-, S_\gamma^z] - S_\alpha^z S_\alpha^- [S_\beta^+, S_\gamma^z]\\
&\quad\quad + S_\alpha^+ S_\alpha^z S_\beta^- S_\gamma^z + S_\alpha^- S_\alpha^z S_\beta^+ S_\gamma^z + S_\alpha^+ S_\alpha^z S_\beta^z S_\gamma^- + S_\alpha^- S_\alpha^z S_\beta^z S_\gamma^+\\
&\quad\quad - S_\alpha^+ S_\alpha^z [S_\beta^z, S_\gamma^-] - S_\alpha^- S_\alpha^z [S_\beta^z, S_\gamma^+]\\
&\quad\quad + \frac{1}{2} S_\alpha^+ S_\alpha^- (S_\beta^- S_\gamma^+ + S_\beta^+ S_\gamma^-) + \frac{1}{2} S_\alpha^- S_\alpha^+ (S_\beta^+ S_\gamma^- + S_\beta^- S_\gamma^+)\\
&\quad\quad - \frac{1}{2}[S_\alpha^+, S_\alpha^-][S_\beta^+, S_\gamma^-]\\
&\quad\quad + 4 S_\alpha^z S_\alpha^z S_\beta^z S_\gamma^z .
\end{aligned}\tag{4.5.19}$$

With

$$\begin{aligned}
&[S_\alpha^z, S_\alpha^+][S_\beta^-, S_\beta^z] + [S_\alpha^z, S_\alpha^-][S_\beta^+, S_\beta^z] + \frac{1}{2}[S_\alpha^+, S_\alpha^-][S_\beta^+, S_\beta^-]\\
&\quad = S_\alpha^+ S_\beta^- + S_\alpha^- S_\beta^+ + 2 S_\alpha^z S_\beta^z = 2 \boldsymbol{S}_\alpha \boldsymbol{S}_\beta ,
\end{aligned}$$

we finally obtain

$$\begin{aligned}
&2\left[(\boldsymbol{S}_\alpha \boldsymbol{S}_\beta)(\boldsymbol{S}_\alpha \boldsymbol{S}_\gamma) + (\boldsymbol{S}_\alpha \boldsymbol{S}_\gamma)(\boldsymbol{S}_\alpha \boldsymbol{S}_\beta)\right] + 2\delta_{\alpha\gamma}\, \boldsymbol{S}_\alpha \boldsymbol{S}_\beta\\
&\quad = S_\alpha^+ S_\alpha^+ S_\beta^- S_\gamma^- + S_\alpha^- S_\alpha^- S_\beta^+ S_\gamma^+\\
&\quad\quad + S_\alpha^z S_\alpha^+ S_\beta^z S_\gamma^- + S_\alpha^z S_\alpha^- S_\beta^z S_\gamma^+ + S_\alpha^z S_\alpha^+ S_\beta^- S_\gamma^z + S_\alpha^z S_\alpha^- S_\beta^+ S_\gamma^z\\
&\quad\quad + S_\alpha^+ S_\alpha^z S_\beta^- S_\gamma^z + S_\alpha^- S_\alpha^z S_\beta^+ S_\gamma^z + S_\alpha^+ S_\alpha^z S_\beta^z S_\gamma^- + S_\alpha^- S_\alpha^z S_\beta^z S_\gamma^+\\
&\quad\quad + \frac{1}{2}(S_\alpha^+ S_\alpha^- + S_\alpha^- S_\alpha^+)(S_\beta^+ S_\gamma^- + S_\beta^- S_\gamma^+)\\
&\quad\quad + 4\, S_\alpha^z S_\alpha^z S_\beta^z S_\gamma^z .
\end{aligned}\tag{4.5.20}$$

Subtracting this from (4.5.16), we obtain

$$\begin{aligned}
5\left\{T_{\alpha\alpha}^0 T_{\beta\gamma}^0\right\}_0 &- 2\left[(\boldsymbol{S}_\alpha \boldsymbol{S}_\beta)(\boldsymbol{S}_\alpha \boldsymbol{S}_\gamma) + (\boldsymbol{S}_\alpha \boldsymbol{S}_\gamma)(\boldsymbol{S}_\alpha \boldsymbol{S}_\beta)\right] - 2\delta_{\beta\gamma}\, \boldsymbol{S}_\alpha \boldsymbol{S}_\beta\\
&= \frac{1}{6}(4 S_\alpha^z S_\alpha^z - S_\alpha^+ S_\alpha^- - S_\alpha^- S_\alpha^+)(4 S_\beta^z S_\gamma^z - S_\beta^+ S_\gamma^- - S_\beta^- S_\gamma^+)\\
&\quad - \frac{1}{2}(S_\alpha^+ S_\alpha^- + S_\alpha^- S_\alpha^+)(S_\beta^+ S_\gamma^- + S_\beta^- S_\gamma^+) - 4\, S_\alpha^z S_\alpha^z S_\beta^z S_\gamma^z .\\
&= -\frac{4}{3} \boldsymbol{S}_\alpha^2 (\boldsymbol{S}_\beta \boldsymbol{S}_\gamma),
\end{aligned}\tag{4.5.21}$$

or

$$5\left\{T^0_{\alpha\alpha}T^0_{\beta\gamma}\right\}_0 = -\frac{4}{3}\boldsymbol{S}_\alpha^2(\boldsymbol{S}_\beta\boldsymbol{S}_\gamma) + 2\left[(\boldsymbol{S}_\alpha\boldsymbol{S}_\beta)(\boldsymbol{S}_\alpha\boldsymbol{S}_\gamma) + (\boldsymbol{S}_\alpha\boldsymbol{S}_\gamma)(\boldsymbol{S}_\alpha\boldsymbol{S}_\beta)\right] + 2\delta_{\beta\gamma}\,\boldsymbol{S}_\alpha\boldsymbol{S}_\beta \tag{4.5.22}$$

As an aside, since the Clebsch–Gordan coefficient

$$\langle 2,0;2,0|1,0\rangle = 0, \tag{4.5.23}$$

(4.5.14) implies that the tensor product of $T^0_{\alpha\alpha}$ and $T^0_{\beta\gamma}$ has no vector component, i.e.,

$$\left\{T^0_{\alpha\alpha}T^0_{\beta\gamma}\right\}_1 = 0. \tag{4.5.24}$$

It follows that we cannot obtain a vector annihilation operator which is even under P and T from the operator H_0 defined in (4.5.2).

4.5.4 Writing Out the Hamiltonian

Substitution of (4.5.22) into (4.5.12) and (4.5.13) yields

$$\left\{H_0^{\mathrm{PT}=}\right\}_0 = \frac{1}{15}\sum_{\alpha\neq\beta}\omega_{\alpha\beta\beta}\left[16 - 2\left(\boldsymbol{S}_\alpha\boldsymbol{S}_\beta\right)^2 + 4\,\boldsymbol{S}_\alpha\boldsymbol{S}_\beta\right], \tag{4.5.25}$$

$$\left\{H_0^{\mathrm{PT}\neq}\right\}_0 = \frac{1}{15}\sum_{\substack{\alpha,\beta,\gamma\\ \alpha\neq\beta\neq\gamma\neq\alpha}}\omega_{\alpha\beta\gamma}\left[8\,\boldsymbol{S}_\beta\boldsymbol{S}_\gamma - (\boldsymbol{S}_\alpha\boldsymbol{S}_\beta)(\boldsymbol{S}_\alpha\boldsymbol{S}_\gamma) - (\boldsymbol{S}_\alpha\boldsymbol{S}_\gamma)(\boldsymbol{S}_\alpha\boldsymbol{S}_\beta)\right]. \tag{4.5.26}$$

With (B.20), we rewrite the first term in (4.5.26) as

$$\begin{aligned} 8\sum_{\substack{\alpha,\beta,\gamma\\ \alpha\neq\beta\neq\gamma\neq\alpha}}\omega_{\alpha\beta\gamma}\boldsymbol{S}_\beta\boldsymbol{S}_\gamma &= 16\sum_{\alpha\neq\beta}\omega_{\alpha\beta\beta}\boldsymbol{S}_\alpha\boldsymbol{S}_\beta - 4\sum_{\alpha\neq\beta}\boldsymbol{S}_\alpha\boldsymbol{S}_\beta \\ &= 16\sum_{\alpha\neq\beta}\omega_{\alpha\beta\beta}\boldsymbol{S}_\alpha\boldsymbol{S}_\beta - 4\boldsymbol{S}_{\mathrm{tot}}^2 + 8N. \end{aligned}$$

Collecting all the terms we obtain

$$\begin{aligned} 15\{H_0^{\mathrm{PT}}\}_0 = &\sum_{\alpha\neq\beta}\frac{1}{|\eta_\alpha-\eta_\beta|^2}\left[20\,\boldsymbol{S}_\alpha\boldsymbol{S}_\beta - 2\left(\boldsymbol{S}_\alpha\boldsymbol{S}_\beta\right)^2\right] \\ &- \sum_{\substack{\alpha,\,\beta,\,\gamma\\ \alpha\neq\beta\neq\gamma\neq\alpha}}\frac{1}{\bar\eta_\alpha-\bar\eta_\beta}\frac{1}{\eta_\alpha-\eta_\gamma}\left[(\boldsymbol{S}_\alpha\boldsymbol{S}_\beta)(\boldsymbol{S}_\alpha\boldsymbol{S}_\gamma) + (\boldsymbol{S}_\alpha\boldsymbol{S}_\gamma)(\boldsymbol{S}_\alpha\boldsymbol{S}_\beta)\right] \\ &- 4\boldsymbol{S}_{\mathrm{tot}}^2 + \frac{4N(N^2-1)}{3} + 8N, \end{aligned} \tag{4.5.27}$$

and finally

$$\frac{3}{4}\{H_0^{\mathrm{PT}}\}_0 = \sum_{\alpha\neq\beta} \frac{1}{|\eta_\alpha-\eta_\beta|^2}\left[\boldsymbol{S}_\alpha\boldsymbol{S}_\beta - \frac{1}{10}(\boldsymbol{S}_\alpha\boldsymbol{S}_\beta)^2\right]$$
$$-\frac{1}{20}\sum_{\substack{\alpha,\beta,\gamma\\ \alpha\neq\beta\neq\gamma\neq\alpha}} \frac{1}{(\bar\eta_\alpha-\bar\eta_\beta)(\eta_\alpha-\eta_\gamma)}\left[(\boldsymbol{S}_\alpha\boldsymbol{S}_\beta)(\boldsymbol{S}_\alpha\boldsymbol{S}_\gamma)+(\boldsymbol{S}_\alpha\boldsymbol{S}_\gamma)(\boldsymbol{S}_\alpha\boldsymbol{S}_\beta)\right]$$
$$-\frac{1}{5}\boldsymbol{S}_{\mathrm{tot}}^2 + \frac{N(N^2+5)}{15}. \tag{4.5.28}$$

Note that the second term in the first line of (4.5.28) is equal to what we would get if we were to take $\beta=\gamma$ on the term in the second line.

In conclusion, we have derived that the non-Abelian $S=1$ Pfaffian spin liquid state $\left|\psi_0^{S=1}\right\rangle$ introduced in Sect. 2.4 is an exact eigenstate of

$$H^{S=1} = \frac{2\pi^2}{N^2}\left[\sum_{\alpha\neq\beta}\frac{\boldsymbol{S}_\alpha\boldsymbol{S}_\beta}{|\eta_\alpha-\eta_\beta|^2} - \frac{1}{20}\sum_{\substack{\alpha,\beta,\gamma\\ \alpha\neq\beta,\gamma}}\frac{(\boldsymbol{S}_\alpha\boldsymbol{S}_\beta)(\boldsymbol{S}_\alpha\boldsymbol{S}_\gamma)+(\boldsymbol{S}_\alpha\boldsymbol{S}_\gamma)(\boldsymbol{S}_\alpha\boldsymbol{S}_\beta)}{(\bar\eta_\alpha-\bar\eta_\beta)(\eta_\alpha-\eta_\gamma)}\right] \tag{4.5.29}$$

with energy eigenvalue

$$E_0^{S=1} = -\frac{2\pi^2}{N^2}\frac{N(N^2+5)}{15} = -\frac{2\pi^2}{15}\left(N+\frac{5}{N}\right). \tag{4.5.30}$$

The information regarding the positive semi-definiteness of H_0^{PT}, which was still intact on the level of (4.5.10) and (4.5.11), has unfortunately been lost as we carried out the projection onto the scalar components (4.5.12) and (4.5.13). We will recover this information in Sect. 5.5.1. Exact diagonalization studies [5] carried out numerically for up to $N=18$ sites further show that $\left|\psi_0^{S=1}\right\rangle$ is the unique ground states of (4.5.29), and that the model is gapless.

4.6 Vector Annihilation Operators

4.6.1 Annihilation Operators Which Transform Even Under T

We can use the defining condition (4.5.1) further to construct a vector annihilation operator. First note that since

$$\boldsymbol{\Omega}_\alpha^{S=1}\left|\psi_0^{S=1}\right\rangle = 0 \quad \forall\alpha,$$

$\left|\psi_0^{S=1}\right\rangle$ is also annihilated by the Hermitian operator

$$H_\alpha = \frac{1}{2}\Omega_\alpha^{S=1\,\dagger}\Omega_\alpha^{S=1} = \sum_{\substack{\beta,\gamma\\ \alpha\neq\beta,\gamma}} \omega_{\alpha\beta\gamma}\, S_\alpha^{\mathrm{z}}\left(S_\alpha^{\mathrm{z}}+1\right) S_\beta^{+} S_\gamma^{-}, \tag{4.6.1}$$

which is just the operator (4.5.2) without the sum over α. Constructing an operator which is even under T,

$$H_\alpha^{\mathrm{T}} = \frac{1}{2}\left(H_\alpha + \Theta H_\alpha \Theta\right) = H_\alpha^{\mathrm{T}=} + H_\alpha^{\mathrm{T}\neq}, \tag{4.6.2}$$

with

$$H_\alpha^{\mathrm{T}=} = \frac{1}{2}\sum_{\substack{\beta\\ \beta\neq\alpha}} \omega_{\alpha\beta\beta}\left(S_\alpha^{\mathrm{z}\,2}\{S_\beta^{+}, S_\beta^{-}\} + 2S_\alpha^{\mathrm{z}} S_\beta^{\mathrm{z}}\right), \tag{4.6.3}$$

$$H_\alpha^{\mathrm{T}\neq} = \sum_{\substack{\beta\neq\gamma\\ \beta,\gamma\neq\alpha}} \omega_{\alpha\beta\gamma}\, S_\alpha^{\mathrm{z}\,2} S_\beta^{+} S_\gamma^{-}. \tag{4.6.4}$$

and odd under P, we obtain

$$H_\alpha^{\bar{\mathrm{P}}\mathrm{T}} = \frac{1}{2}\left(H_\alpha^{\mathrm{T}} - \Pi H_\alpha^{\mathrm{T}} \Pi\right) = H_\alpha^{\bar{\mathrm{P}}\mathrm{T}=} + H_\alpha^{\bar{\mathrm{P}}\mathrm{T}\neq}, \tag{4.6.5}$$

where

$$H_\alpha^{\bar{\mathrm{P}}\mathrm{T}=} = 0, \quad H_\alpha^{\bar{\mathrm{P}}\mathrm{T}\neq} = \frac{1}{2}\sum_{\substack{\beta\neq\gamma\\ \beta,\gamma\neq\alpha}} \omega_{\alpha\beta\gamma}\, S_\alpha^{\mathrm{z}\,2}\left(S_\beta^{+} S_\gamma^{-} - S_\beta^{-} S_\gamma^{+}\right). \tag{4.6.6}$$

With (3.7.6) and $S_\beta^{+} S_\gamma^{-} - S_\beta^{-} S_\gamma^{+} = -2\mathrm{i}(\boldsymbol{S}_\beta \times \boldsymbol{S}_\gamma)^{\mathrm{z}}$ (cf. (D.3.3)), we obtain

$$H_\alpha^{\bar{\mathrm{P}}\mathrm{T}} = \frac{1}{2}\sum_{\substack{\beta\neq\gamma\\ \beta,\gamma\neq\alpha}} \frac{\eta_\alpha + \eta_\beta}{\eta_\alpha - \eta_\beta}\, S_\alpha^{\mathrm{z}\,2}(\boldsymbol{S}_\beta \times \boldsymbol{S}_\gamma)^{\mathrm{z}}. \tag{4.6.7}$$

With (D.3.8) and (D.3.3) we find that the scalar component of the product of the z-components of three vectors vanishes identically, while the vector component is given by

$$5\left\{S_\alpha^{\mathrm{z}\,2}(\boldsymbol{S}_\beta \times \boldsymbol{S}_\gamma)^{\mathrm{z}}\right\}_1 = S_\alpha^{\mathrm{z}}\left(\boldsymbol{S}_\alpha(\boldsymbol{S}_\beta \times \boldsymbol{S}_\gamma)\right) + \left(\boldsymbol{S}_\alpha(\boldsymbol{S}_\beta \times \boldsymbol{S}_\gamma)\right)S_\alpha^{\mathrm{z}} + 2(\boldsymbol{S}_\beta \times \boldsymbol{S}_\gamma)^{\mathrm{z}},$$

where we have used $\alpha\neq\beta,\gamma$ and $\boldsymbol{S}_\alpha^2 = 2$. The Pfaffian spin liquid state $\left|\psi_0^{S=1}\right\rangle$ is hence annihilated by the vector operator

$$5\{H_\alpha^{\bar{\text{P}}\text{T}}\}_1 = \mathrm{i} \sum_{\substack{\beta \neq \gamma \\ \beta,\gamma \neq \alpha}} \frac{\eta_\alpha + \eta_\beta}{\eta_\alpha - \eta_\beta} \Big[(\boldsymbol{S}_\beta \times \boldsymbol{S}_\gamma) + \frac{1}{2} \boldsymbol{S}_\alpha \big(\boldsymbol{S}_\alpha (\boldsymbol{S}_\beta \times \boldsymbol{S}_\gamma) \big) + \frac{1}{2} \big(\boldsymbol{S}_\alpha (\boldsymbol{S}_\beta \times \boldsymbol{S}_\gamma) \big) \boldsymbol{S}_\alpha \Big]. \tag{4.6.8}$$

With

$$\sum_{\substack{\gamma \\ \gamma \neq \alpha,\beta}} \boldsymbol{S}_\gamma = \boldsymbol{S}_{\text{tot}} - \boldsymbol{S}_\alpha - \boldsymbol{S}_\beta,$$

$\boldsymbol{S}_\beta \times \boldsymbol{S}_\beta = i\boldsymbol{S}_\beta$, and (3.7.18), we find from (4.6.8) that $|\psi_0^{S=1}\rangle$ is also annihilated by

$$\mathrm{i} \sum_{\substack{\beta \\ \beta \neq \alpha}} \frac{\eta_\alpha + \eta_\beta}{\eta_\alpha - \eta_\beta} \Big[(\boldsymbol{S}_\alpha \times \boldsymbol{S}_\beta) - i\boldsymbol{S}_\beta + \frac{1}{2} \big(\boldsymbol{S}_\alpha (\boldsymbol{S}_\beta \times \boldsymbol{S}_{\text{tot}}) \big) \boldsymbol{S}_\alpha \Big]. \tag{4.6.9}$$

We can rewrite the product of the four spin operators in the last term as

$$\big(\boldsymbol{S}_\alpha (\boldsymbol{S}_\beta \times \boldsymbol{S}_{\text{tot}}) \big) S_\alpha^d = \sum_{\gamma=1}^{N} \varepsilon^{abc} S_\alpha^a S_\beta^b \big[S_\gamma^c, S_\alpha^d \big] + \text{something} \cdot S_{\text{tot}}^c, \tag{4.6.10}$$

where the second term annihilates every singlet. The first term yields

$$\begin{aligned} \varepsilon^{abc} S_\alpha^a S_\beta^b \big[S_\alpha^c, S_\alpha^d \big] &= \mathrm{i} \varepsilon^{abc} \varepsilon^{cde} S_\alpha^a S_\beta^b S_\alpha^e \\ &= \mathrm{i} \big(\delta^{ad} \delta^{be} - \delta^{ae} \delta^{bd} \big) S_\alpha^a S_\beta^b S_\alpha^e \\ &= \mathrm{i} S_\alpha^d (\boldsymbol{S}_\alpha \boldsymbol{S}_\beta) - \mathrm{i} S_\beta^d \boldsymbol{S}_\alpha^2, \end{aligned} \tag{4.6.11}$$

where we have used $\alpha \neq \beta$. The Pfaffian spin liquid state (4.1.6) with (4.1.4) is therefore also annihilated by

$$\begin{aligned} \boldsymbol{D}_\alpha^{S=1} &= \frac{1}{2} \sum_{\substack{\beta \\ \beta \neq \alpha}} \frac{\eta_\alpha + \eta_\beta}{\eta_\alpha - \eta_\beta} \Big[\mathrm{i} (\boldsymbol{S}_\alpha \times \boldsymbol{S}_\beta) + 2\boldsymbol{S}_\beta - \frac{1}{2} \boldsymbol{S}_\alpha (\boldsymbol{S}_\alpha \boldsymbol{S}_\beta) \Big], \\ \boldsymbol{D}_\alpha^{S=1} |\psi_0^{S=1}\rangle &= 0 \quad \forall \alpha. \end{aligned} \tag{4.6.12}$$

This is the analog of the auxiliary operator (2.2.42) or (3.7.8) of the Haldane–Shastry model.

Equation (4.6.12) implies that $|\psi_0^{S=1}\rangle$ is further annihilated by

$$\mathbf{\Lambda}^{S=1} = \sum_{\alpha=1}^{N} \mathbf{D}_{\alpha}^{S=1} = \frac{1}{2} \sum_{\alpha \neq \beta} \frac{\eta_\alpha + \eta_\beta}{\eta_\alpha - \eta_\beta} \left[\mathrm{i}(\mathbf{S}_\alpha \times \mathbf{S}_\beta) - \frac{1}{2} \mathbf{S}_\alpha (\mathbf{S}_\alpha \mathbf{S}_\beta) \right], \quad (4.6.13)$$

where we have used (B.16). This is the analog of the rapidity operator (2.2.8) or (3.7.9) of the Haldane-Shastry model. In contrast to the Haldane–Shastry model, however, the operator (4.6.13) does not commute with the Hamiltonian (4.5.29).

4.6.2 Annihilation Operators Which Transform Odd Under T

Finally, we consider annihilation operators we can construct from (4.6.1), and which transform odd under T,

$$H_\alpha^{\bar{\mathrm{T}}} = \frac{1}{2} (H_\alpha - \Theta H_\alpha \Theta) = H_\alpha^{\bar{\mathrm{T}}=} + H_\alpha^{\bar{\mathrm{T}}\neq} \quad (4.6.14)$$

with

$$\begin{aligned} H_\alpha^{\bar{\mathrm{T}}=} &= \frac{1}{2} \sum_{\substack{\beta \\ \beta \neq \alpha}} \omega_{\alpha\beta\beta} \left(S_\alpha^{\mathrm{z}\,2} [S_\beta^+, S_\beta^-] + S_\alpha^{\mathrm{z}} \{ S_\beta^+, S_\beta^- \} \right) \\ &= \sum_{\substack{\beta \\ \beta \neq \alpha}} \omega_{\alpha\beta\beta} \left(S_\alpha^{\mathrm{z}\,2} S_\beta^{\mathrm{z}} + S_\alpha^{\mathrm{z}} (\mathbf{S}_\beta^2 - S_\beta^{\mathrm{z}\,2}) \right), \quad (4.6.15) \\ H_\alpha^{\bar{\mathrm{T}}\neq} &= \sum_{\substack{\beta \neq \gamma \\ \beta, \gamma \neq \alpha}} \omega_{\alpha\beta\gamma} S_\alpha^{\mathrm{z}} S_\beta^+ S_\gamma^-. \quad (4.6.16) \end{aligned}$$

Let us first look at the component which transforms odd under P,

$$H_\alpha^{\bar{\mathrm{P}}\bar{\mathrm{T}}} = \frac{1}{2} \left(H_\alpha^{\bar{\mathrm{T}}} - \Pi H_\alpha^{\bar{\mathrm{T}}} \Pi \right) = H_\alpha^{\bar{\mathrm{P}}\bar{\mathrm{T}}=} + H_\alpha^{\bar{\mathrm{P}}\bar{\mathrm{T}}\neq}, \quad (4.6.17)$$

where

$$H_\alpha^{\bar{\mathrm{P}}\bar{\mathrm{T}}=} = 0, \quad H_\alpha^{\bar{\mathrm{P}}\bar{\mathrm{T}}\neq} = \frac{1}{2} \sum_{\substack{\beta \neq \gamma \\ \beta, \gamma \neq \alpha}} \omega_{\alpha\beta\gamma} S_\alpha^{\mathrm{z}} (S_\beta^+ S_\gamma^- - S_\beta^- S_\gamma^+). \quad (4.6.18)$$

This operator has no vector component. With (D.3.10), we obtain the scalar component

$$\{H_\alpha^{\bar{\mathrm{T}}}\}_0 = -\frac{\mathrm{i}}{3} \sum_{\substack{\beta \neq \gamma \\ \beta, \gamma \neq \alpha}} \frac{\mathbf{S}_\alpha (\mathbf{S}_\beta \times \mathbf{S}_\gamma)}{(\bar{\eta}_\alpha - \bar{\eta}_\beta)(\eta_\alpha - \eta_\gamma)}, \quad (4.6.19)$$

It is identical to (3.7.16) in the Haldane–Shastry model, and annihilates every spin singlet. We will not consider it further.

We will now turn the component which transforms even under P,

$$H_\alpha^{\mathrm{P}\bar{\mathrm{T}}} = \frac{1}{2}\left(H_\alpha^{\bar{\mathrm{T}}} + \Pi H_\alpha^{\bar{\mathrm{T}}}\Pi\right) = H_\alpha^{\mathrm{P}\bar{\mathrm{T}}=} + H_\alpha^{\mathrm{P}\bar{\mathrm{T}}\neq}, \tag{4.6.20}$$

where

$$H_\alpha^{\mathrm{P}\bar{\mathrm{T}}=} = H_\alpha^{\bar{\mathrm{T}}=},\ H_\alpha^{\mathrm{P}\bar{\mathrm{T}}\neq} = \frac{1}{2}\sum_{\substack{\beta\neq\gamma\\ \beta,\gamma\neq\alpha}} \omega_{\alpha\beta\gamma} S_\alpha^{\mathrm{z}}\left(S_\beta^+ S_\gamma^- + S_\beta^- S_\gamma^+\right), \tag{4.6.21}$$

which has no scalar, but a vector component. With (D.3.8) and (D.3.3), we write

$$\begin{aligned}\left\{H_\alpha^{\bar{\mathrm{T}}=}\right\}_1 = \frac{1}{5}\sum_{\substack{\beta\\ \beta\neq\alpha}} \omega_{\alpha\beta\beta}\Big[&S_\alpha^{\mathrm{z}}(\boldsymbol{S}_\alpha\boldsymbol{S}_\beta) + \boldsymbol{S}_\alpha(S_\alpha^{\mathrm{z}})\boldsymbol{S}_\beta \\ &+ \boldsymbol{S}_\alpha^2 S_\beta^{\mathrm{z}} + 4S_\alpha^{\mathrm{z}}\boldsymbol{S}_\beta^2 - \boldsymbol{S}_\alpha(S_\beta^{\mathrm{z}})\boldsymbol{S}_\beta - (\boldsymbol{S}_\alpha\boldsymbol{S}_\beta)S_\beta^{\mathrm{z}}\Big].\end{aligned} \tag{4.6.22}$$

Writing out the second term, we obtain

$$\begin{aligned}\boldsymbol{S}_\alpha(S_\alpha^{\mathrm{z}})\boldsymbol{S}_\beta &= \frac{1}{2}\left(S_\alpha^- S_\alpha^{\mathrm{z}} S_\beta^+ + S_\alpha^+ S_\alpha^{\mathrm{z}} S_\beta^-\right) + S_\alpha^{\mathrm{z}} S_\alpha^{\mathrm{z}} S_\beta^{\mathrm{z}}, \\ &= S_\alpha^{\mathrm{z}}(\boldsymbol{S}_\alpha\boldsymbol{S}_\beta) + \frac{1}{2}\left(\left[S_\alpha^-, S_\alpha^{\mathrm{z}}\right]S_\beta^+ + \left[S_\alpha^+, S_\alpha^{\mathrm{z}}\right]S_\beta^-\right) \\ &= S_\alpha^{\mathrm{z}}(\boldsymbol{S}_\alpha\boldsymbol{S}_\beta) + \mathrm{i}(\boldsymbol{S}_\alpha \times \boldsymbol{S}_\beta)^{\mathrm{z}}.\end{aligned} \tag{4.6.23}$$

Similarly, the fifth term gives

$$\boldsymbol{S}_\alpha(S_\beta^{\mathrm{z}})\boldsymbol{S}_\beta = (\boldsymbol{S}_\alpha\boldsymbol{S}_\beta)S_\beta^{\mathrm{z}} + \mathrm{i}(\boldsymbol{S}_\alpha \times \boldsymbol{S}_\beta)^{\mathrm{z}}. \tag{4.6.24}$$

Collecting the terms, we obtain

$$\left\{H_\alpha^{\bar{\mathrm{T}}=}\right\}_1 = \frac{2}{5}\sum_{\substack{\beta\\ \beta\neq\alpha}} \omega_{\alpha\beta\beta}\Big[4S_\alpha^{\mathrm{z}} + S_\beta^{\mathrm{z}} + S_\alpha^{\mathrm{z}}(\boldsymbol{S}_\alpha\boldsymbol{S}_\beta) - (\boldsymbol{S}_\alpha\boldsymbol{S}_\beta)S_\beta^{\mathrm{z}}\Big], \tag{4.6.25}$$

where we have used $\boldsymbol{S}_\alpha^2 = \boldsymbol{S}_\beta^2 = 2$. With (D.3.9), we further obtain

$$\left\{H_\alpha^{\bar{\mathrm{T}}\neq}\right\}_1 = \frac{1}{5}\sum_{\substack{\beta\neq\gamma\\ \beta,\gamma\neq\alpha}} \omega_{\alpha\beta\gamma}\Big[4S_\alpha^{\mathrm{z}}(\boldsymbol{S}_\beta\boldsymbol{S}_\gamma) - S_\beta^{\mathrm{z}}(\boldsymbol{S}_\alpha\boldsymbol{S}_\gamma) - S_\gamma^{\mathrm{z}}(\boldsymbol{S}_\alpha\boldsymbol{S}_\beta)\Big]. \tag{4.6.26}$$

Combining (4.6.25) and (4.6.26), we finally obtain the vector annihilation operator

$$
\begin{aligned}
\boldsymbol{A}_\alpha^{S=1} &\equiv 5\Big(\big\{H_\alpha^{\bar{\mathrm{T}}=}\big\}_1 + \big\{H_\alpha^{\bar{\mathrm{T}}\neq}\big\}_1\Big) \\
&= 2\sum_{\substack{\beta \\ \beta\neq\alpha}} \frac{4\boldsymbol{S}_\alpha + \boldsymbol{S}_\beta + \boldsymbol{S}_\alpha(\boldsymbol{S}_\alpha\boldsymbol{S}_\beta) - (\boldsymbol{S}_\alpha\boldsymbol{S}_\beta)\boldsymbol{S}_\beta}{|\eta_\alpha - \eta_\beta|^2} \\
&\quad + \sum_{\substack{\beta\neq\gamma \\ \beta,\gamma\neq\alpha}} \frac{4\boldsymbol{S}_\alpha(\boldsymbol{S}_\beta\boldsymbol{S}_\gamma) - \boldsymbol{S}_\beta(\boldsymbol{S}_\alpha\boldsymbol{S}_\gamma) - \boldsymbol{S}_\gamma(\boldsymbol{S}_\alpha\boldsymbol{S}_\beta)}{(\bar\eta_\alpha - \bar\eta_\beta)(\eta_\alpha - \eta_\gamma)}, \\
\boldsymbol{A}_\alpha^{S=1}\big|\psi_0^{S=1}\big\rangle &= 0 \quad \forall\,\alpha.
\end{aligned} \tag{4.6.27}
$$

This operator is rather complicated, but does simplify as we sum over α. From (4.6.15), we obtain

$$
\sum_\alpha \big\{H_\alpha^{\bar{\mathrm{T}}=}\big\}_1 = 2\sum_\alpha \boldsymbol{S}_\alpha \sum_{\substack{\beta \\ \beta\neq\alpha}} \omega_{\alpha\beta\beta} = \frac{N^2-1}{6}\boldsymbol{S}_{\text{tot}}.
$$

This implies that $\big|\psi_0^{S=1}\big\rangle$ is also annihilated by

$$
\boldsymbol{\Upsilon}^{S=1} = 5\sum_\alpha \big\{H_\alpha^{\bar{\mathrm{T}}\neq}\big\}_1 = \sum_{\substack{\alpha,\beta,\gamma \\ \alpha\neq\beta\neq\gamma\neq\alpha}} \frac{4\boldsymbol{S}_\alpha(\boldsymbol{S}_\beta\boldsymbol{S}_\gamma) - \boldsymbol{S}_\beta(\boldsymbol{S}_\alpha\boldsymbol{S}_\gamma) - \boldsymbol{S}_\gamma(\boldsymbol{S}_\alpha\boldsymbol{S}_\beta)}{(\bar\eta_\alpha - \bar\eta_\beta)(\eta_\alpha - \eta_\gamma)}. \tag{4.6.28}
$$

4.7 Concluding Remarks

The various annihilation operators for the $S = 1$ model derived in this section are summarized in Table 4.1.

The main result, of course, is the Hamiltonian $H^{S=1}$ given by (4.5.29). It is a three-spin operator. The three-body interaction terms fall off as $1/(r_{12}\, r_{13})$, which makes the model long-ranged. Since the wave function (2.4.1) introduced in Sect. 2.4 is critical, i.e., has algebraically decaying correlations, it is not surprising that we need a Hamiltonian with long-ranged interaction to single it out as unique and exact ground states. Hamiltonians with only short-ranged interactions, like the Heisenberg model, tend to single out states with exponentially decaying correlations, and a Haldane gap in the excitation spectrum [6–9].

The most intriguing feature of the $S = 1$ Pfaffian spin liquid state we have elevated into an exactly soluble model here is that the spinon excitations obey a novel form of quantum statistics, which is presumably the closest analog to non-Abelian statistics one can define in one dimensions. As explained in Sect. 2.4.5, there is an internal, topological Hilbert space of dimension 2^n associated with a state with $2n$ spinons. In

Table 4.1 Annihilation operators for the $S = 1$spin liquid ground state

Annihilation operators for $\vert\psi_0^{S=1}\rangle$					
Operator	Equation	Symmetry transformation properties			
		T	P	Order of tensor	Translationally invarient
$\boldsymbol{S}_{\text{tot}}$	(2.2.6)	$-$	$+$	Vector	yes
$\Omega_\alpha^{S=1}$	(4.4.15)	No	No	3rd	No
Ξ_α	(4.4.19)	No	No	3rd	No
$H^{S=1} - E_0^{S=1}$	(4.5.29)	$+$	$+$	Scalar	Yes
$\boldsymbol{D}_\alpha^{S=1}$	(4.6.12)	$+$	$-$	Vector	No
$\boldsymbol{\Lambda}^{S=1}$	(4.6.13)	$+$	$-$	Vector	Yes
$\boldsymbol{A}_\alpha^{S=1}$	(4.6.27)	$-$	$+$	Vector	No
$\boldsymbol{\Upsilon}^{S=1}$	(4.6.28)	$-$	$+$	Vector	Yes

With the exception of the defining operator $\Omega_{\alpha^{S=1}}$ and Ξ_α, which are the $m = 3$ components of 3rd order tensors, we have only included scalar and vector annihilation operators

the thermodynamic limit, all the states in this internal Hilbert space become degenerate. We assume that the information regarding the internal state is encoded in fractional shifts in the momentum spacings between the individual the spinons (see Sect. 2.4.5). These shifts are topological quantum numbers, and are hence insensitive against local, external perturbations. This makes this model, and presumably a range of models of critical $S = 1$ spin chains, suited for applications as protected qubits in quantum computing.

Preliminary numerical work [5] indicates that the rapidity operator $\boldsymbol{\Lambda}$ given in (4.6.13) does not commute with $H^{S=1}$. The model hence does not appear to share the integrability structure of the spin $\frac{1}{2}$ Haldane–Shastry model. We conjecture that the reason for this is related to the rich internal structure of the Hilbert space, which makes the universality class of the states we introduce here both much less accessible and much more interesting than the Abelian $S = \frac{1}{2}$ Heisenberg model.

In the following chapter, we will employ the theoretical method developed here to generalize the model to arbitrary spin, i.e., to identify a parent Hamiltonian for the state (2.4.37) introduced in Sect. 2.4.6.

References

1. G. Moore, N. Read, Nonabelions in the fractional quantum Hall effect. Nucl. Phys. B **360**, 362 (1991)
2. M. Greiter, X.G. Wen, F. Wilczek, Paired Hall state at half filling. Phys. Rev. Lett. **66**, 3205 (1991)
3. M. Greiter, X.G. Wen, F. Wilczek, Paired Hall states. Nucl. Phys. B **374**, 567 (1992)
4. G. Baym, *Lectures on Quantum Mechanics* (Benjamin/Addison Wesley, New York, 1969)
5. R. Thomale, S. Rachel, P. Schmitteckert, M. Greiter, manuscript in preparation
6. F.D.M. Haldane, Contiuum dynamics of the 1-D Heisenberg antiferromagnet: identification with the O(3) nonlinear sigma model. Phys. Lett. **93**, 464 (1983)

7. F.D.M. Haldane, Nonlinear field theory of large-spin Heisenberg antiferromagnets: semiclassically quantized solitons of the one-dimensional easy-axis Néel state, Phys. Rev. Lett. **50**, 1153 (1983)
8. I. Affleck, Field theory methods and quantum critical phenomena. In: E. Brézin, J. Zinn-Justin (eds) *Fields Strings and Critical Phenomena*, Vol. XLIX of Les Houches Lectures (Elsevier, Amsterdam, 1990)
9. E. Fradkin, *Field Theories of Condensed Matter Systems*, Number 82 in Frontiers in Physics (Addison Wesley, Redwood City, 1991)

Chapter 5
Generalization to Arbitrary Spin S

5.1 A Critical Spin Liquid State With Spin S

5.1.1 Generation Through Projection of Gutzwiller States

In this section, we wish to generalize the model introduced and derived in the previous section for spin $S = 1$ to arbitrary spin $S = s$. The generalization of the $S = 1$ ground state (2.4.3) with (2.4.1) was introduced in Sect. 2.4.6. In essence, we combine $2s$ identical copies of the Gutzwiller or Haldane–Shastry ground state with spin $\frac{1}{2}$, and project the spin on each site onto spin s,

$$\underbrace{\mathbf{\frac{1}{2}} \otimes \mathbf{\frac{1}{2}} \otimes \ldots \otimes \mathbf{\frac{1}{2}}}_{2s} = \boldsymbol{s} \oplus (2s-1) \cdot \boldsymbol{s-1} \oplus \ldots$$

The projection onto the completely symmetric representation can be carried out conveniently using Schwinger bosons (see Sect. 2.4.3). In particular, if we write the Haldane–Shastry ground state as

$$\begin{aligned} \left|\psi_0^{\text{HS}}\right\rangle &= \sum_{\{z_1, z_2, \ldots, z_M\}} \psi_0^{\text{HS}}(z_1, \ldots, z_M)\, S_{z_1}^{+} \cdot \ldots \cdot S_{z_M}^{+} \left|\downarrow\downarrow \ldots \downarrow\right\rangle \\ &= \sum_{\{z_1, \ldots, z_M; w_1, \ldots, w_M\}} \psi_0^{\text{HS}}(z_1, \ldots, z_M)\, a_{z_1}^{+} \ldots a_{z_M}^{\dagger} b_{w_1}^{+} \ldots b_{w_M}^{\dagger} \left|0\right\rangle \\ &\equiv \Psi_0^{\text{HS}}\big[a^{\dagger}, b^{\dagger}\big] \left|0\right\rangle, \end{aligned} \tag{5.1.1}$$

where $\psi_0^{\text{HS}}(z_1, \ldots, z_M)$ is given by (2.2.3), $M = \frac{N}{2}$ and the w_k's are those lattice sites which are not occupied by any of the z_i's, we can write the spin S state obtained by the mentioned projection as (cf 2.4.37)

$$\left|\psi_0^{S}\right\rangle = \left(\Psi_0^{\text{HS}}\big[a^{\dagger}, b^{\dagger}\big]\right)^{2s} \left|0\right\rangle. \tag{5.1.2}$$

M. Greiter, *Mapping of Parent Hamiltonians*, Springer Tracts in Modern Physics 244,
DOI: 10.1007/978-3-642-24384-4_5, © Springer-Verlag Berlin Heidelberg 2011

In Sect. 2.4.6, we mentioned that the state can alternatively be written as

$$\left|\psi_0^S\right\rangle = \sum_{\{z_1,\dots,z_{SN}\}} \psi_0^S(z_1,\dots,z_{SN})\tilde{S}_{z_1}^+ \cdot \dots \cdot \tilde{S}_{z_{SN}}^+ \left|-s\right\rangle_N, \tag{5.1.3}$$

where N is the number of lattice sites,

$$\left|-s\right\rangle_N \equiv \otimes_{\alpha=1}^N \left|s,-s\right\rangle_\alpha \tag{5.1.4}$$

is the "vacuum" state in which all the spins are maximally polarized in the negative $\hat{z}$-direction, and $\tilde{S}^+$ are re-normalized spin flip operators $\tilde{S}^+$ which satisfy

$$\frac{1}{\sqrt{(2s)!}}(a^\dagger)^n(b^\dagger)^{(2s-n)}\left|0\right\rangle = (\tilde{S}^+)^n\left|s,-s\right\rangle. \tag{5.1.5}$$

In a basis in which S^{z} is diagonal, we may write

$$\tilde{S}^+ \equiv \frac{1}{b^\dagger b+1}a^\dagger b = \frac{1}{S-S^{\mathrm{z}}+1}S^+. \tag{5.1.6}$$

Note that (5.1.5) implies

$$\begin{aligned} S^-(\tilde{S}^+)^n\left|s,-s\right\rangle &= b^\dagger a\frac{1}{\sqrt{(2s)!}}(a^\dagger)^n(b^\dagger)^{(2s-n)}\left|0\right\rangle \\ &= n(\tilde{S}^+)^{n-1}\left|s,-s\right\rangle. \end{aligned} \tag{5.1.7}$$

The wave function for the spin S state (5.1.2) are then given by

$$\psi_0^S(z_1,\dots,z_{SN}) = \prod_{m=1}^{2s}\left(\prod_{\substack{i,j=(m-1)M+1\\ i<j}}^{mM}(z_i-z_j)^2\right)\prod_{i=1}^{sN} z_i. \tag{5.1.8}$$

Note the similarity to Read–Rezayi states [1] in the quantized Hall effect.

For the purposes in Sects. 5.1.2 and 5.2.2, it is convenient to write the state in the form

$$\left|\psi_0^S\right\rangle = \left[\sum_{\{z_1,\dots,z_M\}} \psi_0^{\mathrm{HS}}(z_1,\dots,z_M)\,\tilde{S}_{z_1}^+ \cdot \dots \cdot \tilde{S}_{z_M}^+\right]^{2s}\left|0\right\rangle. \tag{5.1.9}$$

5.1.2 Direct Verification of the Singlet Property

The singlet property of $\left|\psi_0^S\right\rangle$ is manifest from the method we employed to construct it by combining $2s$ copies of states which are singlets, and in particular through (5.1.2). It is nonetheless instructive to proof it directly from (5.1.9), as the proof of the defining condition for the state in Sect. 5.2.2 will proceed along similar lines.

Since the S^{z}_{tot} component of (5.1.9) is trivially equal to zero, it is sufficient to show that $\left|\psi_0^S\right\rangle$ is annihilated by S^-_{tot}. As we act with S^-_α on (5.1.9), we have to distinguish between configurations with $n = 0, 1, 2, \ldots, 2s$ re-normalized spin flips $\tilde{S}^+_\alpha$ at site α. Since the state is symmetric under interchange of the $2s$ copies of ψ_0^{HS}, we may assume that the n spin flips are present in the first n copies, and account for the restriction through ordering by a combinatorial factor. This yields

$$
\begin{aligned}
\sum_\alpha S^-_\alpha \left|\psi_0^S\right\rangle
&= \sum_{\alpha=1}^{N} S^-_\alpha \sum_{n=0}^{2s} \binom{2s}{n} \left[\sum_{\{z_2,\ldots,z_M\}} \psi_0^{\text{HS}}(\eta_\alpha, z_2, \ldots, z_M) \tilde{S}^+_\alpha \tilde{S}^+_{z_2} \cdot \ldots \cdot \tilde{S}^+_{z_M} \right]^n \\
&\qquad \cdot \left[\sum_{\{z_1,\ldots,z_M\} \neq \eta_\alpha} \psi_0^{\text{HS}}(z_1, \ldots, z_M) \tilde{S}^+_{z_1} \cdot \ldots \cdot \tilde{S}^+_{z_M} \right]^{2s-n} |0\rangle \\
&= 2s \sum_{\alpha=1}^{N} \left[\sum_{\{z_2,\ldots,z_M\}} \psi_0^{\text{HS}}(\eta_\alpha, z_2, \ldots, z_M) \tilde{S}^+_{z_2} \cdot \ldots \cdot \tilde{S}^+_{z_M} \right] \\
&\qquad \cdot \sum_{n=1}^{2s} \binom{2s-1}{n-1} \left[\sum_{\{z_2,\ldots,z_M\}} \psi_0^{\text{HS}}(\eta_\alpha, z_2, \ldots, z_M) \tilde{S}^+_\alpha \tilde{S}^+_{z_2} \cdot \ldots \cdot \tilde{S}^+_{z_M} \right]^{n-1} \\
&\qquad \cdot \left[\sum_{\{z_1,\ldots,z_M\} \neq \eta_\alpha} \psi_0^{\text{HS}}(z_1, \ldots, z_M) \tilde{S}^+_{z_1} \cdot \ldots \cdot \tilde{S}^+_{z_M} \right]^{2s-n} |0\rangle \\
&= 2s \left[\sum_{\{z_2,\ldots,z_M\}} \underbrace{\sum_{\alpha=1}^{N} \psi_0^{\text{HS}}(\eta_\alpha, z_2, \ldots, z_M)}_{=0} \tilde{S}^+_{z_2} \cdot \ldots \cdot \tilde{S}^+_{z_M} \right] \\
&\qquad \cdot \left[\sum_{\{z_1,\ldots,z_M\}} \psi_0^{\text{HS}}(z_1, \ldots, z_M) \tilde{S}^+_{z_1} \cdot \ldots \cdot \tilde{S}^+_{z_M} \right]^{2s-1} |0\rangle ,
\end{aligned}
\tag{5.1.10}
$$

where we have used (5.1.7) and that $\psi_0^{\text{HS}}(\eta_\alpha, z_2, \ldots, z_M)$ contains only powers $\eta_\beta^1, \eta_\beta^2, \ldots, \eta_\beta^{N-1}$.

5.2 The Defining Condition for the Spin S Chain

5.2.1 Statement

The defining condition for the spin $\boldsymbol{S}$ state is by direct generalization of (3.4.6) and (4.4.15) given by

$$\Omega_\alpha^S = \sum_{\substack{\beta=1\\ \beta\neq\alpha}}^N \frac{1}{\eta_\alpha - \eta_\beta}(S_\alpha^-)^{2s} S_\beta^-, \qquad \Omega_\alpha^S \left|\psi_0^S\right\rangle = 0 \quad \forall\alpha. \tag{5.2.1}$$

Since the state is real, it is also annihilated by the complex conjugate of Ω_α^S,

$$\bar{\Omega}_\alpha^S = \sum_{\substack{\beta=1\\ \beta\neq\alpha}}^N \frac{1}{\bar{\eta}_\alpha - \bar{\eta}_\beta}(S_\alpha^-)^{2s} S_\beta^-, \qquad \bar{\Omega}_\alpha^S \left|\psi_0^S\right\rangle = 0 \quad \forall\alpha. \tag{5.2.2}$$

5.2.2 Direct Verification

Unlike for the cases of spin $\frac{1}{2}$ and spin one, we have not derived the defining condition (5.2.1) from the parent Hamiltonian of a quantized Hall state. The direct and explicit verification presented here does therefore not just serve to check the validity of the previous analysis, but is an essential part of the entire argument we present.

Let us consider the action of $(S_\alpha^-)^{2s} S_\beta^-$ on $\left|\psi_0^S\right\rangle$ written in the form (5.1.9). Since $\psi_0^{\mathrm{HS}}(z_1,\ldots,z_M)$ vanishes whenever two arguments z_i coincide, one of the z_i's in each of the $2s$ copies in (5.1.9) must equal η_α; since $\psi_0^{\mathrm{HS}}(z_1,\ldots,z_M)$ is symmetric under interchange of the z_i's and we count each distinct configuration in the sums over $\{z_1,\ldots,z_M\}$ only once, we may take $z_1 = \eta_\alpha$. Regarding the action of S_β^- on (5.1.9), we have to distinguish between configurations with $n = 0, 1, 2, \ldots, 2s$ re-normalized spin flips $\tilde{S}_\beta^+$ at site β. Since the state is symmetric under interchange of the $2s$ copies, we may assume that the n spin flips are present in the first n copies, and account for the restriction through ordering by a combinatorial factor. This yields

$$\begin{aligned}
&(S_\alpha^-)^{2s} S_\beta^- \left|\psi_0^S\right\rangle \\
&= (S_\alpha^-)^{2s} S_\beta^- \sum_{n=0}^{2s} \binom{2s}{n} \left[\sum_{\{z_3,\ldots,z_M\}} \psi_0^{\mathrm{HS}}(\eta_\alpha,\eta_\beta,z_3,\ldots)\tilde{S}_\alpha^+ \tilde{S}_\beta^+ \tilde{S}_{z_3}^+ \cdot \ldots \cdot \tilde{S}_{z_M}^+\right]^n \\
&\qquad \cdot \left[\sum_{\{z_2,\ldots,z_M\}\neq\eta_\beta} \psi_0^{\mathrm{HS}}(\eta_\alpha,z_2,\ldots)\tilde{S}_\alpha^+ \tilde{S}_{z_2}^+ \cdot \ldots \cdot \tilde{S}_{z_M}^+\right]^{2s-n} \left|0\right\rangle
\end{aligned}$$

$$
\begin{aligned}
&= (2s)!\,2s \left[\sum_{\{z_2,\ldots,z_M\}} \psi_0^{\mathrm{HS}}(\eta_\alpha, \eta_\beta, z_3, \ldots, z_M) \tilde{S}_{z_3}^{+} \cdot \ldots \cdot \tilde{S}_{z_M}^{+} \right] \\
&\quad \cdot \sum_{n=1}^{2s} \binom{2s-1}{n-1} \left[\sum_{\{z_3,\ldots,z_M\}} \psi_0^{\mathrm{HS}}(\eta_\alpha, \eta_\beta, z_3, \ldots, z_M) \tilde{S}_{\beta}^{+} \tilde{S}_{z_3}^{+} \cdot \ldots \cdot \tilde{S}_{z_M}^{+} \right]^{n-1} \\
&\quad \cdot \left[\sum_{\{z_2,\ldots,z_M\} \neq \eta_\beta} \psi_0^{\mathrm{HS}}(\eta_\alpha, z_2, \ldots, z_M) \tilde{S}_{z_2}^{+} \cdot \ldots \cdot \tilde{S}_{z_M}^{+} \right]^{2s-n} |0\rangle \\
&= (2s)!\,2s \left[\sum_{\{z_3,\ldots,z_M\}} \psi_0^{\mathrm{HS}}(\eta_\alpha, \eta_\beta, z_3, \ldots, z_M) \tilde{S}_{z_3}^{+} \cdot \ldots \cdot \tilde{S}_{z_M}^{+} \right] \\
&\quad \cdot \left[\sum_{\{z_2,\ldots,z_M\}} \psi_0^{\mathrm{HS}}(\eta_\alpha, z_2, \ldots, z_M) \tilde{S}_{z_2}^{+} \cdot \ldots \cdot \tilde{S}_{z_M}^{+} \right]^{2s-1} |0\rangle,
\end{aligned}
\tag{5.2.3}
$$

where we have used (5.1.7). This implies

$$
\begin{aligned}
\Omega_\alpha^S \big| \psi_0^S \big\rangle &= \frac{1}{(2s)!2s} \sum_{\substack{\beta=1 \\ \beta\neq\alpha}}^{N} (S_\alpha^-)^{2s} S_\beta^- \big| \psi_0^S \big\rangle \\
&= (2s)!\,2s \left[\sum_{\{z_3,\ldots,z_M\}} \underbrace{\sum_{\beta=1}^{N} \frac{\psi_0^{\mathrm{HS}}(\eta_\alpha, \eta_\beta, z_3, \ldots, z_M)}{\eta_\alpha - \eta_\beta}}_{=0} \tilde{S}_{z_3}^{+} \cdot \ldots \cdot \tilde{S}_{z_M}^{+} \right] \\
&\quad \cdot \left[\sum_{\{z_2,\ldots,z_M\}} \psi_0^{\mathrm{HS}}(\eta_\alpha, z_2, \ldots, z_M) \tilde{S}_{z_2}^{+} \cdot \ldots \cdot \tilde{S}_{z_M}^{+} \right]^{2s-1} |0\rangle,
\end{aligned}
\tag{5.2.4}
$$

where we have used that

$$
\frac{\psi_0^{\mathrm{HS}}(\eta_\alpha, \eta_\beta, z_3, \ldots z_M)}{\eta_\alpha - \eta_\beta} = (\eta_\alpha - \eta_\beta)\eta_\alpha\eta_\beta \prod_{i=3}^{M} (\eta_\alpha - z_i)^2 (\eta_\beta - z_i)^2 z_i \prod_{3\leq i<j}^{M} (z_i - z_j)^2
$$

vanishes for $\beta = \alpha$ and contains only powers $\eta_\beta^1, \eta_\beta^2, \ldots, \eta_\beta^{N-2}$. Note that the calculation for $\bar{\Omega}_\alpha^S$ is almost identical, since

$$
\frac{\psi_0^{\mathrm{HS}}(\eta_\alpha, \eta_\beta, z_3, \ldots z_M)}{\bar{\eta}_\alpha - \bar{\eta}_\beta} = -\eta_\alpha\eta_\beta \frac{\psi_0^{\mathrm{HS}}(\eta_\alpha, \eta_\beta, z_3, \ldots z_M)}{\eta_\alpha - \eta_\beta}
$$

vanishes also for $\beta = \alpha$ and contains only powers $\eta_\beta^2, \eta_\beta^3, \ldots, \eta_\beta^{N-1}$.

5.3 Construction of a Parent Hamiltonian

5.3.1 Translational Symmetry

A Hermitian and translationally invariant operator which annihilates $\left|\psi_0^S\right\rangle$ is given by

$$H_0 = \frac{1}{2a_0}\sum_{\alpha=1}^{N}\Omega_\alpha^{S\dagger}\Omega_\alpha^S = \frac{1}{2a_0}\sum_{\substack{\alpha,\beta,\gamma\\ \alpha\neq\beta,\gamma}}\frac{1}{\bar{\eta}_\alpha-\bar{\eta}_\beta}\frac{1}{\eta_\alpha-\eta_\gamma}(S_\alpha^+)^{2s}(S_\alpha^-)^{2s}S_\beta^+S_\gamma^-, \tag{5.3.1}$$

where a_0 is a parameter we will conveniently choose below. We wish the Hamiltonian to be further invariant under P, T, and spin rotations. From (D.2.6), the tensor content of $S_\beta^+S_\gamma^-$ is

$$S_\beta^+S_\gamma^- = \frac{2}{3}\boldsymbol{S}_\beta\boldsymbol{S}_\gamma - \mathrm{i}(\boldsymbol{S}_\beta\times\boldsymbol{S}_\gamma)^{\mathrm{z}} - \frac{1}{\sqrt{6}}T_{\beta\gamma}^0, \tag{5.3.2}$$

where

$$\begin{aligned} T_{\beta\gamma}^0 &= \frac{1}{\sqrt{6}}\left(4S_\beta^{\mathrm{z}}S_\gamma^{\mathrm{z}} - S_\beta^+S_\gamma^- - S_\beta^-S_\gamma^+\right)\\ &= \frac{1}{\sqrt{6}}\left(6S_\beta^{\mathrm{z}}S_\gamma^{\mathrm{z}} - 2\boldsymbol{S}_\beta\boldsymbol{S}_\gamma\right). \end{aligned} \tag{5.3.3}$$

This implies that we only have to know the scalar, vector and 2nd order tensor components of $(S_\alpha^+)^{2s}(S_\alpha^-)^{2s}$ in order to obtain the scalar component of H_0.

5.3.2 *Tensor Decomposition of* $(S^+)^{2s}(S^-)^{2s}$

Since $(S^+)^{2s}(S^-)^{2s}$ contains only a single spin operator $\boldsymbol{S}$ with Casimir $\boldsymbol{S}^2 = s(s+1)$, its scalar component U must be a constant, its vector proportional to

$$V^0 = S^{\mathrm{z}} \tag{5.3.4}$$

(cf. (D.1.1)), its 2nd order tensor component proportional to

$$T^0 = \frac{1}{\sqrt{6}}\left(4S^{\mathrm{z}}S^{\mathrm{z}} - S^+S^- - S^-S^+\right) = \frac{2}{\sqrt{6}}\left[3S^{\mathrm{z}2} - s(s+1)\right] \tag{5.3.5}$$

(cf. (D.2.3)), and its 3rd order tensor component proportional to

$$\begin{aligned} W^0 &= -\frac{1}{\sqrt{5}}(S^- S^+ S^z + S^+ S^z S^- + S^z S^- S^+ + S^+ S^- S^z + S^- S^z S^+ \\ &\qquad + S^z S^+ S^-) + \frac{4}{\sqrt{5}} S^z S^z S^z, \\ &= \frac{2}{\sqrt{5}}\Big[5S^{z2} - 3s(s+1) + 1\Big] S^z \end{aligned} \tag{5.3.6}$$

(cf. (D.3.7)). Our task in this section is to calculate the constants of proportionality in the expansion

$$\begin{aligned} (S^+)^{2s}(S^-)^{2s} = a_0\{1 + aV^0 + bT^0 + cW^0\} \\ + \text{tensors of order} > 3. \end{aligned} \tag{5.3.7}$$

To begin with, note that $(S^+)^{2s}$ and $(S^-)^{2s}$ are up to a sign equal to the tensor components with $m = \pm 2s$ of one and the same tensor of order $2s$,

$$\begin{aligned} T^{(2s)^{2s}} &= (-1)^{2s}(S^+)^{2s}, \\ T^{(2s)^{-2s}} &= (S^-)^{2s}. \end{aligned} \tag{5.3.8}$$

Recalling (3.5.17), we write

$$T^{(2s)^{2s}} T^{(2s)^{-2s}} = \sum_{j=1}^{4s} T^{(j)^0} \langle j, 0 | 2s, 2s; 2s, -2s \rangle\,, \tag{5.3.9}$$

where $T^{(j)^0}$ is with (3.5.13) given by

$$T^{(j)^0} = \sum_{m=-2s}^{2s} T^{(2s)^m} T^{(2s)^{-m}} \langle 2s, m; 2s, -m | j, 0 \rangle\,. \tag{5.3.10}$$

With (3.5.9), we can calculate the components $T^{(2s)^m}$ from $T^{(2s)^{\pm 2s}}$,

$$T^{(2s)^{m\pm 1}} = \frac{1}{\sqrt{2s(2s+1) - m(m\pm 1)}} \left[S^\pm, T^{(2s)^m}\right]. \tag{5.3.11}$$

Specifically, $T^{(2s)^{2s-n}}$ is given in terms of $T^{(2s)^{2s}}$ by

$$T^{(2s)^{2s-n}} = \left(\prod_{i=1}^{n} \frac{1}{\sqrt{2s(2s+1) - (2s-i+1)(2s-i)}}\right)$$

$$\underbrace{\cdot\left[S^-,\left[S^-,\ldots\left[S^-,T^{(2s)}{}^{2s}\right]\ldots\right]\right]}_{n\text{ operators }S^-}. \tag{5.3.12}$$

To evaluate the first term, we use

$$\prod_{i=1}^{n}\big(s(s+1)-(s-i+1)(s-i)\big)=\prod_{i=1}^{n}(2s-i+1)i=\frac{(2s)!\cdot n!}{(2s-n)!}, \tag{5.3.13}$$

which holds for $1\le n\le 2s$, $2s$ and n integer. This yields

$$T^{(2s)}{}^{2s-n}=\sqrt{\frac{(4s-n)!}{(4s)!\cdot n!}}\sum_{k=0}^{n}\binom{n}{k}(-1)^{n-k}(S^-)^kT^{(2s)}{}^{2s}(S^-)^{n-k}. \tag{5.3.14}$$

Similarly, we find

$$T^{(2s)}{}^{-2s+n}=\sqrt{\frac{(4s-n)!}{(4s)!\cdot n!}}\sum_{k=0}^{n}\binom{n}{k}(-1)^{k}(S^+)^{n-k}T^{(2s)}{}^{-2s}(S^+)^{k}. \tag{5.3.15}$$

Note that (5.3.14) and (5.3.15) hold for $0\le n\le 4s$. With (5.3.8) and the shorthand $|m\rangle\equiv|s,m\rangle$, we can write

$$\begin{aligned}(S^-)^kT^{(2s)}{}^{2s}(S^-)^{n-k}&=(-1)^{2s}\,|s-k\rangle\,\langle -s+n-k|\\&\quad\cdot\langle s-k|(S^-)^k|s\rangle\langle s|(S^+)^{2s}|-s\rangle\langle -s|(S^-)^{k-n}|-s+n-k\rangle,\end{aligned}$$

and similarly

$$\begin{aligned}(S^+)^{n-k}T^{(2s)}{}^{-2s}(S^+)^k&=(-1)^{2s}\,|-s+n-k\rangle\,\langle s-k|\\&\quad\cdot\langle -s+n-k|\,(S^+)^{n-k}\,|-s\rangle\,\langle -s|\,(S^-)^{2s}\,|s\rangle\,\langle s|\,(S^+)^k\,|s-k\rangle\,.\end{aligned}$$

This implies that in the product $T^{(2s)}{}^{2s-n}T^{(2s)}{}^{-2s+n}$, only terms with matching values of k in the sums (5.3.14) and (5.3.15) contribute. With (5.3.13) we obtain

$$\langle s|\,(S^+)^k(S^-)^k\,|s\rangle=\begin{cases}\dfrac{(2s)!\cdot k!}{(2s-k)!} & \text{for } 0\le k\le 2s,\\ 0 & \text{otherwise,}\end{cases}$$

$$\langle -s|\,(S^-)^{k-n}(S^+)^{n-k}\,|-s\rangle=\begin{cases}\dfrac{(2s)!\cdot (n-k)!}{(2s-n+k)!} & \text{for } 0\le n-k\le 2s,\\ 0 & \text{otherwise,}\end{cases}$$

$$\langle s|\,(S^+)^{2s}(S^-)^{2s}\,|s\rangle=(2s)!^2. \tag{5.3.16}$$

This yields

$$
\begin{aligned}
&T^{(2s)}{}^{2s-n}\,T^{(2s)}{}^{-2s+n} \\
&\quad = (2s)!^2(-1)^{2s+n} \\
&\qquad \cdot \frac{(4s-n)!}{(4s)!\cdot n!} \sum_{k=\max(n-2s,0)}^{\min(n,2s)} \binom{n}{k}\binom{n}{k} \frac{(2s)!\cdot k!}{(2s-k)!}\,\frac{(2s)!\cdot(n-k)!}{(2s-n+k)!}\,|s-k\rangle\langle s-k| \\
&\quad = (2s)!^2(-1)^{2s+n}\binom{4s}{n}^{-1} \sum_{k=\max(n-2s,0)}^{\min(n,2s)} \binom{2s}{k}\binom{2s}{n-k}\,|s-k\rangle\langle s-k|\,.
\end{aligned}
\tag{5.3.17}
$$

Substitution into (5.3.10) yields

$$
T^{(j)}{}^0 = \sum_{n=0}^{4s} T^{(2s)}{}^{2s-n}\,T^{(2s)}{}^{-2s+n}\,\underbrace{\langle 2s, 2s-n; 2s, -2s+n|j,0\rangle\,.}_{\equiv C_j^{2s-n}} \tag{5.3.18}
$$

With

$$
\sum_{n=0}^{4s}\ \sum_{k=\max(n-2s,0)}^{\min(n,2s)} = \sum_{k=0}^{2s}\sum_{n=k}^{2s+k},
$$

we obtain

$$
\begin{aligned}
T^{(j)}{}^0 &= (2s)!^2(-1)^{2s}\sum_{k=0}^{2s}\left\{\sum_{n=k}^{2s+k} C_j^{2s-n}(-1)^n \frac{\binom{2s}{k}\binom{2s}{n-k}}{\binom{4s}{n}}\right\}|s-k\rangle\langle s-k| \\
&= (2s)!^2(-1)^{2s}\sum_{k=0}^{2s}\left\{\sum_{p=0}^{2s} C_j^{2s-k-p}(-1)^{k+p} \frac{\binom{2s}{k}\binom{2s}{p}}{\binom{4s}{k+p}}\right\}|s-k\rangle\langle s-k|\,.
\end{aligned}
\tag{5.3.19}
$$

The individual tensors in the decomposition

$$
(S^+)^{2s}(S^-)^{2s} = \sum_{j=1}^{4s}\left\{(S^+)^{2s}(S^-)^{2s}\right\}_j \tag{5.3.20}
$$

are hence with (5.3.8), (5.3.9), and the definition of C_j^{2s-n} in (5.3.18) given by

$$
\begin{aligned}
\left\{(S^+)^{2s}(S^-)^{2s}\right\}_j &= (-1)^{2s} C_j^{2s}\,T^{(j)}{}^0 \\
&= \frac{(2s)!^2}{2s+1}\sum_{k=0}^{2s} P_j^k\,|s-k\rangle\langle s-k|\,,
\end{aligned}
\tag{5.3.21}
$$

where we have defined

$$P_j^k = (2s+1)C_j^{2s}\sum_{p=0}^{2s} C_j^{2s-k-p}(-1)^{k+p}\frac{\binom{2s}{k}\binom{2s}{p}}{\binom{4s}{k+p}}. \tag{5.3.22}$$

We are not aware of any method to evaluate this sum analytically. We have used *Mathematica* to evaluate it for $k = 0$ and $j = 0, 1, 2, 3$ as a function of s, and then obtained the coefficients in the expansion (5.3.7) from these terms.

With the Clebsch–Gordan coefficients

$$\begin{aligned}
C_0^m &= \frac{(-1)^{2s-m}}{\sqrt{4s+1}},\\
C_1^m &= \frac{\sqrt{3}(-1)^{2s-m}\cdot m}{\sqrt{2s(2s+1)(4s+1)}},\\
C_2^m &= \frac{\sqrt{5}(-1)^{2s-m}\cdot\left(3m^2-2s(2s+1)\right)}{\sqrt{2s(2s+1)(4s-1)(4s+1)(4s+3)}},\\
C_3^m &= \frac{\sqrt{7}(-1)^{2s-m}\cdot m\left(5m^2+1-6s(2s+1)\right)}{2\sqrt{s(s+1)(2s-1)(2s+1)(4s-1)(4s+1)(4s+3)}},
\end{aligned}$$

we find

$$\begin{aligned}
P_0^0 &= 1,\\
P_1^0 &= \frac{3s}{s+1},\\
P_2^0 &= \frac{5s(2s-1)}{(s+1)(2s+3)},\\
P_3^0 &= \frac{7s(2s-1)(s-1)}{(s+1)(2s+3)(s+2)}.
\end{aligned} \tag{5.3.23}$$

Comparing (5.3.21) with (5.3.23) to the coefficients of $|s\rangle\langle s|$ we obtain from (5.3.4–5.3.6),

$$\begin{aligned}
V^0 &= s\,|s\rangle\langle s| + \ldots,\\
T^0 &= \frac{2}{\sqrt{6}}s(2s-1)\,|s\rangle\langle s| + \ldots,\\
W^0 &= \frac{2}{\sqrt{5}}s(2s-1)(s-1)\,|s\rangle\langle s| + \ldots,
\end{aligned} \tag{5.3.24}$$

we find

$$
\begin{aligned}
a_0 &= \frac{(2s)!^2}{2s+1}, \\
a &= \frac{3}{s+1}, \\
b &= \frac{\sqrt{6}}{2} \frac{5}{(s+1)(2s+3)}, \\
c &= \frac{\sqrt{5}}{2} \frac{7}{(s+1)(2s+3)(s+2)}
\end{aligned}
\tag{5.3.25}
$$

for the coefficients in the expansion (5.3.7).

5.3.3 Time Reversal and Parity Symmetry

The for the scalar and vector component relevant part of the operator H_0 introduced (5.3.1) is with (3.6.3) and (5.3.7) given by

$$
H_0' = \frac{1}{2} \sum_{\substack{\alpha,\beta,\gamma \\ \alpha \neq \beta,\gamma}} \omega_{\alpha\beta\gamma} \left\{ 1 + a V_\alpha^0 + b T_{\alpha\alpha}^0 + c W_{\alpha\alpha\alpha}^0 \right\} S_\beta^+ S_\gamma^- . \tag{5.3.26}
$$

From now on, we omit the prime. With the transformation properties under time reversal,

$$
\mathrm{T:} \quad \eta_\alpha \to \Theta \eta_\alpha \Theta = \bar{\eta}_\alpha, \quad \boldsymbol{S} \to \Theta \boldsymbol{S} \Theta = -\boldsymbol{S},
$$

and hence

$$
\begin{gathered}
\omega_{\alpha\,\beta\,\gamma} \to \omega_{\alpha\gamma\beta}, \quad S^+ \to -S^-, \quad S^- \to -S^+, \quad S^{\mathrm{z}} \to -S^{\mathrm{z}}, \\
V^0 \to -V^0, \quad T^0 \to T^0, \quad W^0 \to -W^0,
\end{gathered}
$$

the operator (5.3.26) transforms into

$$
\Theta H_0 \Theta = \frac{1}{2} \sum_{\substack{\alpha,\beta,\gamma \\ \alpha \neq \beta,\gamma}} \omega_{\alpha\beta\gamma} \left\{ 1 - a V_\alpha^0 + b T_{\alpha\alpha}^0 - c W_{\alpha\alpha\alpha}^0 \right\} S_\gamma^- S_\beta^+ , \tag{5.3.27}
$$

We proceed with the T invariant operator

$$
H_0^{\mathrm{T}} = \frac{1}{2} \left(H_0 + \Theta H_0 \Theta \right) = H_0^{\mathrm{T}=} + H_0^{\mathrm{T}\neq}, \tag{5.3.28}
$$

where

$$
\begin{aligned}
H_0^{\mathrm{T}=} &= \frac{1}{2}\sum_{\substack{\alpha,\beta\\ \alpha\neq\beta}} \omega_{\alpha\beta\beta}\left[\left(1+bT^0_{\alpha\alpha}\right)\frac{1}{2}\{S^+_\beta, S^-_\beta\} + \left(aV^0_\alpha + cW^0_{\alpha\alpha\alpha}\right)\frac{1}{2}[S^+_\beta, S^-_\beta]\right] \\
&= \frac{1}{2}\sum_{\alpha\neq\beta} \omega_{\alpha\beta\beta}\left[\left(1+bT^0_{\alpha\alpha}\right)\left(\frac{2s(s+1)}{3} - \frac{1}{\sqrt{6}}T^0_{\beta\beta}\right) + \left(aS^{\mathrm{z}}_\alpha + cW^0_{\alpha\alpha\alpha}\right)S^{\mathrm{z}}_\beta\right]
\end{aligned}
\tag{5.3.29}
$$

$$
\begin{aligned}
H_0^{\mathrm{T}\neq} &= \frac{1}{2}\sum_{\substack{\alpha,\beta,\gamma\\ \alpha\neq\beta\neq\gamma\neq\alpha}} \omega_{\alpha\beta\gamma}\left(1+bT^0_{\alpha\alpha}\right)S^+_\beta S^-_\gamma . \\
&= \frac{1}{2}\sum_{\substack{\alpha,\beta,\gamma\\ \alpha\neq\beta\neq\gamma\neq\alpha}} \omega_{\alpha\beta\gamma}\left(1+bT^0_{\alpha\alpha}\right)\left(\frac{2}{3}\boldsymbol{S}_\beta\boldsymbol{S}_\gamma - \mathrm{i}(\boldsymbol{S}_\beta\times\boldsymbol{S}_\gamma)^{\mathrm{z}} - \frac{1}{\sqrt{6}}T^0_{\beta\gamma}\right).
\end{aligned}
\tag{5.3.30}
$$

With the transformation properties under parity,

$$
\mathrm{P:}\quad \eta_\alpha \to \Pi\eta_\alpha\Pi = \overline{\eta}_\alpha, \quad \boldsymbol{S}\to\Theta\boldsymbol{S}\Theta = \boldsymbol{S}, \tag{5.3.31}
$$

and hence $\omega_{\alpha\beta\gamma} \to \omega_{\alpha\gamma\beta}$, we obtain the P and T invariant operator

$$
H_0^{\mathrm{PT}} = \frac{1}{2}\left(H_0^{\mathrm{T}} + \Pi H_0^{\mathrm{T}}\Pi\right) = H_0^{\mathrm{PT}=} + H_0^{\mathrm{PT}\neq}, \tag{5.3.32}
$$

where

$$
H_0^{\mathrm{PT}=} = H_0^{\mathrm{T}=}, \tag{5.3.33}
$$

$$
H_0^{\mathrm{PT}\neq} = \frac{1}{2}\sum_{\substack{\alpha,\beta,\gamma\\ \alpha\neq\beta\neq\gamma\neq\alpha}} \omega_{\alpha\beta\gamma}\left(1+bT^0_{\alpha\alpha}\right)\left(\frac{2}{3}\boldsymbol{S}_\beta\boldsymbol{S}_\gamma - \frac{1}{\sqrt{6}}T^0_{\beta\gamma}\right). \tag{5.3.34}
$$

5.3.4 Spin Rotation Symmetry

Since the critical spin liquid state $\left|\psi_0^S\right\rangle$ introduced in Sects. 2.4.6 and 5.1.1 is a spin singlet, the property that it is annihilated by (5.3.32) with (5.3.33) and (5.3.34) implies that it is annihilated by each tensor component of (5.3.32) individually.

Since we wish to construct a Hamiltonian which is invariant under SU(2) spin rotations, we proceed by projecting out the scalar component. This yields

$$\left\{H_0^{\mathrm{PT}=}\right\}_0 = \frac{1}{2}\sum_{\alpha\neq\beta}\omega_{\alpha\beta\beta}\left[\frac{2s(s+1)}{3} - \frac{b}{\sqrt{6}}\left\{T^0_{\alpha\alpha}T^0_{\beta\beta}\right\}_0 + \frac{a}{3}\boldsymbol{S}_\alpha\boldsymbol{S}_\beta\right], \tag{5.3.35}$$

$$\left\{H_0^{\mathrm{PT}\neq}\right\}_0 = \frac{1}{2}\sum_{\substack{\alpha,\beta,\gamma\\ \alpha\neq\beta\neq\gamma\neq\alpha}}\omega_{\alpha\beta\gamma}\left[\frac{2}{3}\boldsymbol{S}_\beta\boldsymbol{S}_\gamma - \frac{b}{\sqrt{6}}\left\{T^0_{\alpha\alpha}T^0_{\beta\gamma}\right\}_0\right]. \tag{5.3.36}$$

With (4.5.22), or specifically

$$5\left\{T^0_{\alpha\alpha}T^0_{\beta\beta}\right\}_0 = -\frac{4}{3}s^2(s+1)^2 + 4(\boldsymbol{S}_\alpha\boldsymbol{S}_\beta)^2 + 2\boldsymbol{S}_\alpha\boldsymbol{S}_\beta, \quad \alpha\neq\beta,$$

and (5.3.25), we obtain

$$\begin{aligned}
-\frac{b}{\sqrt{6}}\left\{T^0_{\alpha\alpha}T^0_{\beta\beta}\right\}_0 &= \frac{2s^2(s+1)}{3(2s+3)} - \frac{\boldsymbol{S}_\alpha\boldsymbol{S}_\beta + 2(\boldsymbol{S}_\alpha\boldsymbol{S}_\beta)^2}{(s+1)(2s+3)},\\
\frac{a}{3}\boldsymbol{S}_\alpha\boldsymbol{S}_\beta &= \frac{\boldsymbol{S}_\alpha\boldsymbol{S}_\beta}{(s+1)},
\end{aligned}$$

and hence

$$\left\{H_0^{\mathrm{PT}=}\right\}_0 = \sum_{\alpha\neq\beta}\omega_{\alpha\beta\beta}\frac{1}{2s+3}\left[s(s+1)^2 + \boldsymbol{S}_\alpha\boldsymbol{S}_\beta - \frac{(\boldsymbol{S}_\alpha\boldsymbol{S}_\beta)^2}{(s+1)}\right]. \tag{5.3.37}$$

Similarly, we use

$$\begin{aligned}
5\left\{T^0_{\alpha\alpha}T^0_{\beta\gamma}\right\}_0 = -\frac{4s(s+1)}{3}\boldsymbol{S}_\beta\boldsymbol{S}_\gamma + 2\left[(\boldsymbol{S}_\alpha\boldsymbol{S}_\beta)(\boldsymbol{S}_\alpha\boldsymbol{S}_\gamma) + (\boldsymbol{S}_\alpha\boldsymbol{S}_\gamma)(\boldsymbol{S}_\alpha\boldsymbol{S}_\beta)\right],&\\
\alpha\neq\beta\neq\gamma\neq\alpha,&
\end{aligned}$$

to obtain

$$-\frac{b}{\sqrt{6}}\left\{T^0_{\alpha\alpha}T^0_{\beta\gamma}\right\}_0 = \frac{2s\boldsymbol{S}_\beta\boldsymbol{S}_\gamma}{3(2s+3)} + \frac{(\boldsymbol{S}_\alpha\boldsymbol{S}_\beta)(\boldsymbol{S}_\alpha\boldsymbol{S}_\gamma) + (\boldsymbol{S}_\alpha\boldsymbol{S}_\gamma)(\boldsymbol{S}_\alpha\boldsymbol{S}_\beta)}{(s+1)(2s+3)}$$

and hence

$$\begin{aligned}
&\left\{H_0^{\mathrm{PT}\neq}\right\}_0\\
&\quad = \sum_{\substack{\alpha,\beta,\gamma\\ \alpha\neq\beta\neq\gamma\neq\alpha}}\omega_{\alpha\beta\gamma}\frac{1}{2s+3}\left[(s+1)\boldsymbol{S}_\beta\boldsymbol{S}_\gamma - \frac{(\boldsymbol{S}_\alpha\boldsymbol{S}_\beta)(\boldsymbol{S}_\alpha\boldsymbol{S}_\gamma) + (\boldsymbol{S}_\alpha\boldsymbol{S}_\gamma)(\boldsymbol{S}_\alpha\boldsymbol{S}_\beta)}{2(s+1)}\right].
\end{aligned} \tag{5.3.38}$$

With (B.20), we rewrite the first sum in (5.3.38) as

$$\sum_{\substack{\alpha,\beta,\gamma\\ \alpha\neq\beta\neq\gamma\neq\alpha}} \omega_{\alpha\beta\gamma}\boldsymbol{S}_\beta\boldsymbol{S}_\gamma = 2\sum_{\alpha\neq\beta}\omega_{\alpha\beta\beta}\boldsymbol{S}_\alpha\boldsymbol{S}_\beta - \frac{1}{2}\boldsymbol{S}_{\text{tot}}^2 + \frac{s(s+1)}{2}N.$$

Collecting all the terms we obtain

$$\begin{aligned}\left\{H_0^{\text{PT}}\right\}_0 &= \sum_{\alpha\neq\beta}\omega_{\alpha\beta\beta}\left[\boldsymbol{S}_\alpha\boldsymbol{S}_\beta - \frac{(\boldsymbol{S}_\alpha\boldsymbol{S}_\beta)^2}{(s+1)(2s+3)}\right]\\ &\quad+\sum_{\substack{\alpha,\beta,\gamma\\ \alpha\neq\beta\neq\gamma\neq\alpha}}\omega_{\alpha\beta\gamma}\frac{(\boldsymbol{S}_\alpha\boldsymbol{S}_\beta)(\boldsymbol{S}_\alpha\boldsymbol{S}_\gamma)+(\boldsymbol{S}_\alpha\boldsymbol{S}_\gamma)(\boldsymbol{S}_\alpha\boldsymbol{S}_\beta)}{2(s+1)(2s+3)}\\ &\quad-\frac{s+1}{2(2s+3)}\boldsymbol{S}_{\text{tot}}^2+\frac{s(s+1)^2}{2s+3}\frac{N(N^2+5)}{12}.\end{aligned}\tag{5.3.39}$$

Note that the second term in the first line of (5.3.39) is equal to what we would get if we were to take $\beta=\gamma$ on theterm in the second line.

The spin S spin liquid state $\left|\psi_0^S\right\rangle$ introduced in Sects. 2.4.6 and 5.1.1 is hence an exact eigenstate of

$$\begin{aligned}H^S &= \frac{2\pi^2}{N^2}\left[\sum_{\alpha\neq\beta}\frac{\boldsymbol{S}_\alpha\boldsymbol{S}_\beta}{|\eta_\alpha-\eta_\beta|^2}-\frac{1}{2(s+1)(2s+3)}\right.\\ &\qquad\left.\times\sum_{\substack{\alpha,\beta,\gamma\\ \alpha\neq\beta,\gamma}}\frac{(\boldsymbol{S}_\alpha\boldsymbol{S}_\beta)(\boldsymbol{S}_\alpha\boldsymbol{S}_\gamma)+(\boldsymbol{S}_\alpha\boldsymbol{S}_\gamma)(\boldsymbol{S}_\alpha\boldsymbol{S}_\beta)}{(\bar{\eta}_\alpha-\bar{\eta}_\beta)(\eta_\alpha-\eta_\gamma)}\right]\\ &=\frac{2\pi^2}{N^2}\left[\sum_{\alpha\neq\beta}\frac{\boldsymbol{S}_\alpha\boldsymbol{S}_\beta}{|\eta_\alpha-\eta_\beta|^2}-\sum_{\substack{\alpha,\beta,\gamma\\ \alpha\neq\beta,\gamma}}\Re\left\{\frac{1}{(\bar{\eta}_\alpha-\bar{\eta}_\beta)(\eta_\alpha-\eta_\gamma)}\right\}\frac{(\boldsymbol{S}_\alpha\boldsymbol{S}_\beta)(\boldsymbol{S}_\alpha\boldsymbol{S}_\gamma)}{(s+1)(2s+3)}\right],\end{aligned}\tag{5.3.40}$$

where $\Re$ denotes the real part. The energy eigenvalue is given by

$$E_0^S = -\frac{2\pi^2}{N^2}\frac{s(s+1)^2}{2s+3}\frac{N(N^2+5)}{12} = -\frac{\pi^2}{6}\frac{s(s+1)^2}{2s+3}\left(N+\frac{5}{N}\right).\tag{5.3.41}$$

This is the main result of this work. We will show in Sect. 5.5.1 that $\left|\psi_0^S\right\rangle$ is also a ground state of (5.3.39), i.e., that all the eigenvalues of $H^S - E_0^S$ are non-negative. Exact diagonalization studies [2] carried out numerically for up to $N=16$ sites for the $S=1$ model and for up to $N=10$ sites for the $S=\frac{3}{2}$ model further show that $\left|\psi_0^{S=1}\right\rangle$ and $\left|\psi_0^{S=\frac{3}{2}}\right\rangle$ are the unique ground states of (5.3.40), and that the models are gapless. We assume this property to hold for general spin S.

5.4 Vector Annihilation Operators

5.4.1 Annihilation Operators Which Transform Even Under T

We can use the defining condition (5.2.1) further to construct a vector annihilation operator. First note that since

$$\Omega_\alpha^S \left|\psi_0^S\right\rangle = 0 \quad \forall \alpha,$$

$\left|\psi_0^S\right\rangle$ is also annihilated by the Hermitian operator

$$H_\alpha = \frac{1}{2a_0} \Omega_\alpha^{S\dagger} \Omega_\alpha^S = \frac{1}{2a_0} \sum_{\substack{\beta,\gamma \\ \alpha\neq\beta,\gamma}} \omega_{\alpha\beta\gamma} (S_\alpha^+)^{2s} (S_\alpha^-)^{2s} \boldsymbol{S}_\beta^+ \boldsymbol{S}_\gamma^-, \tag{5.4.1}$$

and therefore also by the scalar and the vector components of

$$H_\alpha' = \frac{1}{2} \sum_{\substack{\beta,\gamma \\ \alpha\neq\beta,\gamma}} \omega_{\alpha\beta\gamma} \left\{1 + a\, V_\alpha^0 + b\, T_{\alpha\alpha}^0 + c\, W_{\alpha\alpha\alpha}^0\right\} \boldsymbol{S}_\beta^+ \boldsymbol{S}_\gamma^-, \tag{5.4.2}$$

which is just the operator (5.3.26) without the sum over α. From now on, we omit the prime. Constructing an operator which is even under T,

$$H_\alpha^{\mathrm{T}} = \frac{1}{2} \left(H_\alpha + \Theta H_\alpha \Theta\right) = H_\alpha^{\mathrm{T}=} + H_\alpha^{\mathrm{T}\neq}, \tag{5.4.3}$$

where

$$H_\alpha^{\mathrm{T}=} = \frac{1}{2} \sum_{\substack{\beta \\ \beta\neq\alpha}} \omega_{\alpha\beta\beta} \left[\left(1 + b\, T_{\alpha\alpha}^0\right) \left(\frac{2s(s+1)}{3} - \frac{1}{\sqrt{6}}\, T_{\beta\gamma}^0\right) + \left(a\, S_\alpha^{\mathrm{z}} + c\, W_{\alpha\alpha\alpha}^0\right) S_\beta^{\mathrm{z}}\right], \tag{5.4.4}$$

$$H_\alpha^{\mathrm{T}\neq} = \frac{1}{2} \sum_{\substack{\beta\neq\gamma \\ \beta,\gamma\neq\alpha}} \omega_{\alpha\beta\gamma} \left(1 + b\, T_{\alpha\alpha}^0\right) \left(\frac{2}{3} \boldsymbol{S}_\beta \boldsymbol{S}_\gamma - \mathrm{i}(\boldsymbol{S}_\beta \times \boldsymbol{S}_\gamma)^{\mathrm{z}} - \frac{1}{\sqrt{6}}\, T_{\beta\gamma}^0\right), \tag{5.4.5}$$

and odd under P, we obtain

$$H_\alpha^{\bar{\mathrm{P}}\mathrm{T}} = \frac{1}{2} \left(H_\alpha^{\mathrm{T}} - \Pi H_\alpha^{\mathrm{T}} \Pi\right) = H_\alpha^{\bar{\mathrm{P}}\mathrm{T}=} + H_\alpha^{\bar{\mathrm{P}}\mathrm{T}\neq}, \tag{5.4.6}$$

where

$$H_\alpha^{\bar{\mathrm{P}}\mathrm{T}=} = 0, \quad H_\alpha^{\bar{\mathrm{P}}\mathrm{T}\neq} = -\frac{\mathrm{i}}{2} \sum_{\substack{\beta\neq\gamma \\ \beta,\gamma\neq\alpha}} \omega_{\alpha\beta\gamma} \left(1 + b\, T_{\alpha\alpha}^{0}\right)(\boldsymbol{S}_\beta \times \boldsymbol{S}_\gamma)^{\mathrm{z}}. \tag{5.4.7}$$

With (3.7.6), we obtain

$$H_\alpha^{\bar{\mathrm{P}}\mathrm{T}} = \frac{\mathrm{i}}{4} \sum_{\substack{\beta\neq\gamma \\ \beta,\gamma\neq\alpha}} \frac{\eta_\alpha + \eta_\beta}{\eta_\alpha - \eta_\beta} \left(1 + b\, T_{\alpha\alpha}^{0}\right)(\boldsymbol{S}_\beta \times \boldsymbol{S}_\gamma)^{\mathrm{z}}. \tag{5.4.8}$$

While the scalar component of (5.4.8) vanishes, the vector component does not. With (5.3.5) and (5.3.25), we write

$$\begin{aligned} 1 + b\, T_{\alpha\alpha}^{0} &= 1 + \frac{5}{(s+1)(2s+3)} \left[3S^{\mathrm{z}2} - s(s+1)\right] \\ &= \frac{15 S^{\mathrm{z}2}}{(s+1)(2s+3)} - \frac{3(s-1)}{2s+3}. \end{aligned} \tag{5.4.9}$$

With (D.3.8) and (D.3.3) we find for the vector component of the product of the z-components

$$5\left\{S_\alpha^{\mathrm{z}2}(\boldsymbol{S}_\beta \times \boldsymbol{S}_\gamma)^{\mathrm{z}}\right\}_1 = S_\alpha^{\mathrm{z}}\left(\boldsymbol{S}_\alpha(\boldsymbol{S}_\beta \times \boldsymbol{S}_\gamma)\right) + \left(\boldsymbol{S}_\alpha(\boldsymbol{S}_\beta \times \boldsymbol{S}_\gamma)\right)S_\alpha^{\mathrm{z}} + s(s+1)(\boldsymbol{S}_\beta \times \boldsymbol{S}_\gamma)^{\mathrm{z}}.$$

Substitution into (5.4.8) yields

$$\begin{aligned} \left\{H_\alpha^{\bar{\mathrm{P}}\mathrm{T}}\right\}_1 = \frac{\mathrm{i}}{4} \frac{3}{2s+3} \sum_{\substack{\beta\neq\gamma \\ \beta,\gamma\neq\alpha}} \frac{\eta_\alpha + \eta_\beta}{\eta_\alpha - \eta_\beta} \\ \cdot \left[(\boldsymbol{S}_\beta \times \boldsymbol{S}_\gamma)^{\mathrm{z}} + \frac{1}{s+1}\boldsymbol{S}_\alpha\left(\boldsymbol{S}_\alpha(\boldsymbol{S}_\beta \times \boldsymbol{S}_\gamma)\right) + \frac{1}{s+1}\left(\boldsymbol{S}_\alpha(\boldsymbol{S}_\beta \times \boldsymbol{S}_\gamma)\right)\boldsymbol{S}_\alpha\right]. \end{aligned} \tag{5.4.10}$$

With

$$\sum_{\substack{\gamma \\ \gamma\neq\alpha,\beta}} \boldsymbol{S}_\gamma = \boldsymbol{S}_{\mathrm{tot}} - \boldsymbol{S}_\alpha - \boldsymbol{S}_\beta,$$

$\boldsymbol{S}_\beta \times \boldsymbol{S}_\beta = i\boldsymbol{S}_\beta$, and (3.7.18), we find from (5.4.10) that $\left|\psi_0^S\right\rangle$ is also annihilated by

$$\frac{\mathrm{i}}{2} \sum_{\substack{\beta \\ \beta\neq\alpha}} \frac{\eta_\alpha + \eta_\beta}{\eta_\alpha - \eta_\beta} \left[(\boldsymbol{S}_\alpha \times \boldsymbol{S}_\beta) - i\boldsymbol{S}_\beta + \frac{1}{s+1}\left(\boldsymbol{S}_\alpha(\boldsymbol{S}_\beta \times \boldsymbol{S}_{\mathrm{tot}})\right)\boldsymbol{S}_\alpha\right]. \tag{5.4.11}$$

With (4.6.10 and 4.6.11), we rewrite the product of the four spin operators in the last term as

$$\begin{aligned}\big(\boldsymbol{S}_\alpha(\boldsymbol{S}_\beta\times\boldsymbol{S}_{\mathrm{tot}})\big)\boldsymbol{S}_\alpha &= \mathrm{i}\boldsymbol{S}_\alpha\big(\boldsymbol{S}_\alpha\boldsymbol{S}_\beta\big)-\mathrm{i}\boldsymbol{S}_\alpha^2\boldsymbol{S}_\beta\\ &\quad+\text{term which annihilates every spin singlet,}\end{aligned}\tag{5.4.12}$$

which holds for $\alpha \neq \beta$. The spin liquid state (5.1.2) is therefore also annihilated by

$$\begin{aligned}\boldsymbol{D}_\alpha^S &= \frac{1}{2}\sum_{\substack{\beta\\ \beta\neq\alpha}}\frac{\eta_\alpha+\eta_\beta}{\eta_\alpha-\eta_\beta}\left[\mathrm{i}(\boldsymbol{S}_\alpha\times\boldsymbol{S}_\beta)+(s+1)\,\boldsymbol{S}_\beta-\frac{1}{s+1}\boldsymbol{S}_\alpha\big(\boldsymbol{S}_\alpha\boldsymbol{S}_\beta\big)\right],\\ \boldsymbol{D}_\alpha^S\big|\psi_0^S\big\rangle &= 0\quad\forall\alpha.\end{aligned}\tag{5.4.13}$$

This is the generalization of the auxiliary operator (3.7.8) of the Haldane–Shastry model.

Equation (5.4.13) implies that the spin liquid state $\big|\psi_0^S\big\rangle$ is further annihilated by

$$\boldsymbol{\Lambda}^S=\sum_{\alpha=1}^{N}\boldsymbol{D}_\alpha^S=\frac{1}{2}\sum_{\alpha\neq\beta}\frac{\eta_\alpha+\eta_\beta}{\eta_\alpha-\eta_\beta}\left[\mathrm{i}(\boldsymbol{S}_\alpha\times\boldsymbol{S}_\beta)-\frac{1}{s+1}\boldsymbol{S}_\alpha\big(\boldsymbol{S}_\alpha\boldsymbol{S}_\beta\big)\right],\tag{5.4.14}$$

where we have used (B.16). This is the analog of the rapidity operator (2.2.8) or (3.7.9) of the Haldane–Shastry model. In contrast to the Haldane–Shastry model, however, the operator (5.4.14) does not commute with the Hamiltonian (5.3.40). The model is hence not likely to share the integrability structure of the Haldane–Shastry model. It is possible, however, that the model is integrabel in the thermodynamic limit $N\to\infty$.

5.4.2 Annihilation Operators Which Transform Odd Under T

Finally, we consider annihilation operators we can construct from (5.4.2), and which transform odd under T,

$$H_\alpha^{\bar{\mathrm{T}}}=\frac{1}{2}\left(H_\alpha-\Theta H_\alpha\Theta\right)=H_\alpha^{\bar{\mathrm{T}}=}+H_\alpha^{\bar{\mathrm{T}}\neq},\tag{5.4.15}$$

where

$$\begin{aligned}H_\alpha^{\bar{\mathrm{T}}=} &= \frac{1}{2}\sum_{\substack{\beta\\ \beta\neq\alpha}}\omega_{\alpha\beta\beta}\left[\left(a\,V_\alpha^0+c\,W_{\alpha\alpha\alpha}^0\right)\frac{1}{2}\{S_\beta^+,S_\beta^-\}+\left(1+b\,T_{\alpha\alpha}^0\right)\frac{1}{2}[S_\beta^+,S_\beta^-]\right]\\ &= \frac{1}{2}\sum_{\substack{\beta\\ \beta\neq\alpha}}\omega_{\alpha\beta\beta}\left[\left(a\,S_\alpha^{\mathrm{z}}+c\,W_{\alpha\alpha\alpha}^0\right)\left(\frac{2s(s+1)}{3}-\frac{1}{\sqrt{6}}\,T_{\beta\beta}^0\right)+\left(1+b\,T_{\alpha\alpha}^0\right)S_\beta^{\mathrm{z}}\right],\end{aligned}\tag{5.4.16}$$

$$H_\alpha^{\bar{\mathrm{T}}\neq} = \frac{1}{2} \sum_{\substack{\beta\neq\gamma \\ \beta,\gamma\neq\alpha}} \omega_{\alpha\beta\gamma} \left(a\, S_\alpha^{\mathrm{z}} + c\, W_{\alpha\alpha\alpha}^0\right) \left(\frac{2}{3} \boldsymbol{S}_\beta \boldsymbol{S}_\gamma - \mathrm{i}(\boldsymbol{S}_\beta \times \boldsymbol{S}_\gamma)^{\mathrm{z}} - \frac{1}{\sqrt{6}}\, T_{\beta\gamma}^0\right). \tag{5.4.17}$$

Let us first look at the component which transforms odd under P,

$$H_\alpha^{\bar{\mathrm{P}}\bar{\mathrm{T}}} = \frac{1}{2}\left(H_\alpha^{\bar{\mathrm{T}}} - \Pi H_\alpha^{\bar{\mathrm{T}}} \Pi\right) = H_\alpha^{\bar{\mathrm{P}}\bar{\mathrm{T}}=} + H_\alpha^{\bar{\mathrm{P}}\bar{T}\neq}, \tag{5.4.18}$$

where

$$H_\alpha^{\bar{\mathrm{P}}\bar{\mathrm{T}}=} = 0, \quad H_\alpha^{\bar{\mathrm{P}}\bar{\mathrm{T}}\neq} = -\frac{\mathrm{i}}{2} \sum_{\substack{\beta\neq\gamma \\ \beta,\gamma\neq\alpha}} \omega_{\alpha\beta\gamma} \left(a\, S_\alpha^{\mathrm{z}} + c\, W_{\alpha\alpha\alpha}^0\right) (\boldsymbol{S}_\beta \times \boldsymbol{S}_\gamma)^{\mathrm{z}}. \tag{5.4.19}$$

This operator has no vector component. With (D.2.5) and (5.3.25), we obtain the scalar component

$$\left\{H_\alpha^{\bar{\mathrm{T}}}\right\}_0 = -\frac{\mathrm{i}}{2(s+1)} \sum_{\substack{\beta\neq\gamma \\ \beta,\gamma\neq\alpha}} \frac{\boldsymbol{S}_\alpha (\boldsymbol{S}_\beta \times \boldsymbol{S}_\gamma)}{(\bar{\eta}_\alpha - \bar{\eta}_\beta)(\eta_\alpha - \eta_\gamma)}. \tag{5.4.20}$$

This is identical to (3.7.16) in the Haldane–Shastry model, and annihilates every spin singlet by the line of reasoning pursued in (3.7.17). We will not consider it further.

We will now turn the component which transforms even under P,

$$H_\alpha^{\mathrm{P}\bar{\mathrm{T}}} = \frac{1}{2}\left(H_\alpha^{\bar{\mathrm{T}}} + \Pi H_\alpha^{\bar{\mathrm{T}}} \Pi\right) = H_\alpha^{\mathrm{P}\bar{\mathrm{T}}=} + H_\alpha^{\mathrm{P}\bar{\mathrm{T}}\neq}, \tag{5.4.21}$$

where

$$\begin{aligned} H_\alpha^{\mathrm{P}\bar{\mathrm{T}}=} &= H_\alpha^{\bar{\mathrm{T}}=}, \\ &= \frac{1}{2} \sum_{\substack{\beta \\ \beta\neq\alpha}} \omega_{\alpha\beta\beta} \left[\left(a\, S_\alpha^{\mathrm{z}} + c\, W_{\alpha\alpha\alpha}^0\right) \left(\frac{2s(s+1)}{3} - \frac{1}{\sqrt{6}}\, T_{\beta\beta}^0\right) + \left(1 + b\, T_{\alpha\alpha}^0\right) S_\beta^{\mathrm{z}} \right]. \end{aligned} \tag{5.4.22}$$

$$H_\alpha^{\mathrm{P}\bar{\mathrm{T}}\neq} = \frac{1}{2} \sum_{\substack{\beta\neq\gamma \\ \beta,\gamma\neq\alpha}} \omega_{\alpha\beta\gamma} \left(a\, S_\alpha^{\mathrm{z}} + c\, W_{\alpha\alpha\alpha}^0\right) \left(\frac{2}{3} \boldsymbol{S}_\beta \boldsymbol{S}_\gamma - \frac{1}{\sqrt{6}}\, T_{\beta\gamma}^0\right). \tag{5.4.23}$$

which has no scalar, but a vector component. The vector components of (5.4.22) and (5.4.23) are given by

$$\{H_\alpha^{\mathrm{P}\bar{\mathrm{T}}=}\}_1 = \frac{1}{2}\sum_{\substack{\beta\\ \beta\neq\alpha}} \omega_{\alpha\beta\beta}\Big[\frac{a}{2}\{S_\alpha^{\mathrm{z}}(S_\beta^+ S_\beta^- + S_\beta^- S_\beta^+)\}_1 - \frac{c}{\sqrt{6}}\{W_{\alpha\alpha\alpha}^0\, T_{\beta\beta}^0\}_1 + \{(1+b\,T_{\alpha\alpha}^0)S_\beta^{\mathrm{z}}\}_1\Big], \tag{5.4.24}$$

$$\{H_\alpha^{\mathrm{P}\bar{\mathrm{T}}\neq}\}_1 = \frac{1}{2}\sum_{\substack{\beta\neq\gamma\\ \beta,\gamma\neq\alpha}} \omega_{\alpha\beta\gamma}\Big[\frac{a}{2}\{S_\alpha^{\mathrm{z}}(S_\beta^+ S_\gamma^- + S_\beta^- S_\gamma^+)\}_1 - \frac{c}{\sqrt{6}}\{W_{\alpha\alpha\alpha}^0\, T_{\beta\gamma}^0\}_1\Big], \tag{5.4.25}$$

where we have rewritten the first term in the way we originally obtained it. For $S=\frac{1}{2}$ or $S=1$, these expressions simplify significantly as $W_{\alpha\alpha\alpha}^0 = 0$, which follows directly from $W_{\alpha\alpha\alpha}^3 = -(S_\alpha^+)^3 = 0$ for $S<\frac{3}{2}$. For general $S=s$, however, we have to evaluate $\{W_{\alpha\alpha\alpha}^0\, T_{\beta\gamma}^0\}_1$.

5.4.3 Evaluation of $\{\mathbf{W}_{\alpha\alpha\alpha}^0\, \mathbf{T}_{\beta\gamma}^0\}_1$

We evaluate the vector component of the tensor product of $W_{\alpha\alpha\alpha}^0$ and $T_{\beta\gamma}^0$ with $\alpha\neq\beta,\gamma$ using (3.5.20),

$$\{W_{\alpha\alpha\alpha}^0 T_{\beta\gamma}^0\}_1 = \langle 1,0|3,0;2,0\rangle \sum_{m=-2}^{2} W_{\alpha\alpha\alpha}^m T_{\beta\gamma}^{-m}\,\langle 3,m;2,-m|1,0\rangle\,. \tag{5.4.26}$$

From either (D.3.6) or directly from (3.5.9), we obtain

$$\begin{aligned}
W_{\alpha\alpha\alpha}^3 &= -S_\alpha^+ S_\alpha^+ S_\alpha^+,\\
W_{\alpha\alpha\alpha}^2 &= \frac{1}{\sqrt{6}}\Big[S_\alpha^-, W_{\alpha\alpha\alpha}^3\Big] = \sqrt{6}\,S_\alpha^+ S_\alpha^+(S_\alpha^{\mathrm{z}}+1),\\
W_{\alpha\alpha\alpha}^1 &= \frac{1}{\sqrt{10}}\Big[S_\alpha^-, W_{\alpha\alpha\alpha}^2\Big] = -\sqrt{\frac{3}{5}}\,S_\alpha^+\Big[5S_\alpha^{\mathrm{z}}(S_\alpha^{\mathrm{z}}+1) - s(s+1)+2\Big],\\
W_{\alpha\alpha\alpha}^0 &= \frac{1}{\sqrt{12}}\Big[S_\alpha^-, W_{\alpha\alpha\alpha}^1\Big] = \frac{2}{\sqrt{5}}\Big[5S_\alpha^{\mathrm{z}\,2} - 3s(s+1)+1\Big]S_\alpha^{\mathrm{z}}\\
W_{\alpha\alpha\alpha}^{-1} &= \frac{1}{\sqrt{12}}\Big[S_\alpha^-, W_{\alpha\alpha\alpha}^0\Big] = \sqrt{\frac{3}{5}}\,S_\alpha^-\Big[5S_\alpha^{\mathrm{z}}(S_\alpha^{\mathrm{z}}-1) - s(s+1)+2\Big],\\
W_{\alpha\alpha\alpha}^{-2} &= \frac{1}{\sqrt{10}}\Big[S_\alpha^-, W_{\alpha\alpha\alpha}^{-1}\Big] = \sqrt{6}\,S_\alpha^- S_\alpha^-(S_\alpha^{\mathrm{z}}-1),\\
W_{\alpha\alpha\alpha}^{-3} &= \frac{1}{\sqrt{6}}\Big[S_\alpha^-, W_{\alpha\alpha\alpha}^{-2}\Big] = S_\alpha^- S_\alpha^- S_\alpha^-.
\end{aligned} \tag{5.4.27}$$

With (D.2.3) and the Clebsch–Gordan coefficients

$$\langle 3, m; 2, -m | 1, 0 \rangle = \frac{1}{\sqrt{7 \cdot 5}} \begin{cases} \sqrt{5} & \text{for } m = \pm 2, \\ -2\sqrt{2} & \text{for } m = \pm 1, \\ 3 & \text{for } m = 0, \end{cases} \tag{5.4.28}$$

we obtain

$$\begin{aligned} \frac{7 \cdot 5}{3} \sqrt{\frac{5}{6}} \{ W^0_{\alpha\alpha\alpha} T^0_{\beta\gamma} \}_1 = & \, 5\, S^+_\alpha S^z_\alpha S^+_\alpha S^-_\beta S^-_\gamma \\ & + \left[5 \big(S^+_\alpha S^z_\alpha S^z_\alpha + S^z_\alpha S^z_\alpha S^+_\alpha \big) - \big(2s(s+1) + 1 \big) S^+_\alpha \right] \\ & \cdot \big(S^z_\beta S^-_\gamma + S^-_\beta S^z_\gamma \big) \\ & + \left[5 S^{z\,2}_\alpha - 3s(s+1) + 1 \right] S^z_\alpha \left[4 S^z_\beta S^z_\gamma - S^+_\beta S^-_\gamma - S^-_\beta S^+_\gamma \right] \\ & + \left[5 \big(S^-_\alpha S^z_\alpha S^z_\alpha + S^z_\alpha S^z_\alpha S^-_\alpha \big) - \big(2s(s+1) + 1 \big) S^-_\alpha \right] \\ & \cdot \big(S^z_\beta S^+_\gamma + S^+_\beta S^z_\gamma \big) \\ & + 5\, S^-_\alpha S^z_\alpha S^-_\alpha S^+_\beta S^+_\gamma . \end{aligned} \tag{5.4.29}$$

We wish to write this in a more convenient form, which directly displays that it transforms as a vector under spin rotations.

Let us consider first the case $\beta \neq \gamma$, and try an Ansatz of the form[1]

$$\begin{aligned} & 2a \left[(\boldsymbol{S}_\alpha \boldsymbol{S}_\beta) S^z_\alpha (\boldsymbol{S}_\alpha \boldsymbol{S}_\gamma) + (\boldsymbol{S}_\alpha \boldsymbol{S}_\gamma) S^z_\alpha (\boldsymbol{S}_\alpha \boldsymbol{S}_\beta) \right] \\ & \quad + 2b \left[S^z_\beta (\boldsymbol{S}_\alpha \boldsymbol{S}_\gamma) + (\boldsymbol{S}_\alpha \boldsymbol{S}_\beta) S^z_\gamma \right] + 2c\, S^z_\alpha (\boldsymbol{S}_\beta \boldsymbol{S}_\gamma). \end{aligned} \tag{5.4.30}$$

Comparing the coefficients of the (five-spin) terms containing $S^-_\beta S^-_\gamma$ and $S^-_\beta S^+_\gamma$ yields $a = 5$. Comparing the coefficients of the three-spin terms containing $S^z_\beta S^-_\gamma$, $S^-_\beta S^z_\gamma$, $S^z_\beta S^+_\gamma$, and $S^+_\beta S^z_\gamma$ yields $b = -2s(s+1) - 1$. If compare the coefficients of both the three-spin and the five-spin terms containing $S^+_\beta S^-_\gamma$ and $S^-_\beta S^+_\gamma$ terms,

$$\begin{aligned} -\big(5 S^{z\,2}_\alpha - 3s(s+1) + 1 \big) S^z_\alpha &= \frac{a}{2} \big(S^+_\alpha S^z_\alpha S^-_\alpha + S^-_\alpha S^z_\alpha S^+_\alpha \big) + c S^z_\alpha \\ &= \frac{5}{2} \big(S^+_\alpha S^-_\alpha (S^z_\alpha - 1) + S^-_\alpha S^+_\alpha (S^z_\alpha + 1) \big) + c S^z_\alpha \\ &= \frac{5}{2} \big((2s(s+1) - 2 S^{z\,2}_\alpha) S^z_\alpha - 2 S^z_\alpha \big) + c S^z_\alpha \\ &= -5 S^{z\,3}_\alpha + \big(5s(s+1) - 5 \big) S^z_\alpha + c S^z_\alpha , \end{aligned}$$

we obtain $c = -2s(s+1) + 4$. With these choices, the coefficients of both the three-spin and the five-spin terms containing $S^z_\beta S^z_\gamma$ agree as well,

[1] Note that there is no relation between these coefficients and those introduced in (5.3.7).

$$\big(20S_\alpha^{z\,2} - 12s(s+1) + 4\big)S_\alpha^z = 4a\, S_\alpha^z S_\alpha^z S_\alpha^z + (4b+2c)S_\alpha^z.$$

Finally, the coefficients of the five-spin terms containing $S_\beta^z S_\gamma^-$, $S_\beta^- S_\gamma^z$, $S_\beta^z S_\gamma^+$, and $S_\beta^+ S_\gamma^z$, in (5.4.30),

$$\begin{aligned} a\Big[& S_\alpha^z S_\beta^z S_\alpha^z \big(S_\alpha^+ S_\gamma^- + S_\alpha^- S_\gamma^+\big) + S_\alpha^z S_\gamma^z S_\alpha^z \big(S_\alpha^+ S_\beta^- + S_\alpha^- S_\beta^+\big) \\ & + \big(S_\alpha^+ S_\beta^- + S_\alpha^- S_\beta^+\big) S_\alpha^z S_\alpha^z S_\gamma^z + \big(S_\alpha^+ S_\gamma^- + S_\alpha^- S_\gamma^+\big) S_\alpha^z S_\alpha^z S_\beta^z \Big], \end{aligned}$$

agree with those in (5.4.29).

For the equivalence to hold for the case $\beta = \gamma$ as well, we need to order the spin operators in all terms in (5.4.30) such that the S_β's are to the left of the S_γ's, as this is the order of the spin operators in (5.4.29). We hence have to replace the second term in the first bracket in (5.4.30) by

$$\boldsymbol{S}_\alpha \big(S_\alpha^z (\boldsymbol{S}_\alpha \boldsymbol{S}_\beta)\big) \boldsymbol{S}_\gamma \equiv S_\alpha^i S_\alpha^z S_\alpha^j S_\beta^j S_\gamma^i,$$

or equivalently add a term

$$\begin{aligned} & \boldsymbol{S}_\alpha \big(S_\alpha^z (\boldsymbol{S}_\alpha \boldsymbol{S}_\beta)\big) \boldsymbol{S}_\gamma - (\boldsymbol{S}_\alpha \boldsymbol{S}_\gamma) S_\alpha^z (\boldsymbol{S}_\alpha \boldsymbol{S}_\beta) \\ & = S_\alpha^i S_\alpha^z S_\alpha^j \big[S_\beta^j, S_\gamma^i\big] \\ & = -\delta_{\beta\gamma}\, \mathrm{i}\varepsilon^{ijk} S_\alpha^i S_\alpha^z S_\alpha^j S_\beta^k \\ & \doteq -\delta_{\beta\gamma} \big(\mathrm{i}\varepsilon^{ijk} S_\alpha^z S_\alpha^i S_\alpha^j S_\beta^k + \mathrm{i}\varepsilon^{ijk} \big[S_\alpha^i, S_\alpha^z\big] S_\alpha^j S_\beta^k\big) \\ & = -\delta_{\beta\gamma} \big(S_\alpha^z\, \mathrm{i}(\boldsymbol{S}_\alpha \times \boldsymbol{S}_\alpha) \boldsymbol{S}_\beta - \varepsilon^{ijk}\varepsilon^{izl} S_\alpha^l S_\alpha^j S_\beta^k\big) \\ & = -\delta_{\beta\gamma} \big(- S_\alpha^z (\boldsymbol{S}_\alpha \boldsymbol{S}_\beta) - \big(\delta^{jz}\delta^{kl} - \delta^{jl}\delta^{kz}\big) S_\alpha^l S_\alpha^j S_\beta^k\big) \\ & = \delta_{\beta\gamma} \big(S_\alpha^z (\boldsymbol{S}_\alpha \boldsymbol{S}_\beta) + (\boldsymbol{S}_\alpha \boldsymbol{S}_\beta) S_\alpha^z - s(s+1) S_\beta^z\big). \end{aligned} \tag{5.4.31}$$

Taking all the terms together, we finally obtain

$$\begin{aligned} \frac{7\cdot 5}{2\cdot 3}\sqrt{\frac{5}{6}}\, \big\{W_{\alpha\alpha\alpha}^0 T_{\beta\gamma}^0\big\}_1 = {} & 5\big[(\boldsymbol{S}_\alpha \boldsymbol{S}_\beta) S_\alpha^z (\boldsymbol{S}_\alpha \boldsymbol{S}_\gamma) + (\boldsymbol{S}_\alpha \boldsymbol{S}_\gamma) S_\alpha^z (\boldsymbol{S}_\alpha \boldsymbol{S}_\beta)\big] \\ & - \big(2s(s+1)+1\big)\big[S_\beta^z (\boldsymbol{S}_\alpha \boldsymbol{S}_\gamma) + (\boldsymbol{S}_\alpha \boldsymbol{S}_\beta) S_\gamma^z\big] \\ & - \big(2s(s+1)-4\big)\, S_\alpha^z (\boldsymbol{S}_\beta \boldsymbol{S}_\gamma) \\ & + 5\,\delta_{\beta\gamma}\big[S_\alpha^z (\boldsymbol{S}_\alpha \boldsymbol{S}_\beta) + (\boldsymbol{S}_\alpha \boldsymbol{S}_\beta) S_\alpha^z - s(s+1) S_\beta^z\big]. \end{aligned} \tag{5.4.32}$$

5.4.4 Annihilation Operators Which Transform Odd Under T (Continued)

Substitution of (5.4.32) into (5.4.24) yields with (5.3.25), (D.3.9), (5.4.9) and (D.3.8)

$$
\begin{aligned}
\{H_\alpha^{\mathrm{P\bar{T}=}}\}_1 &= \frac{1}{2}\sum_{\substack{\beta\\ \beta\neq\alpha}} \omega_{\alpha\beta\beta}\Bigg[\frac{3}{2(s+1)}\{S_\alpha^z(S_\beta^+ S_\beta^- + S_\beta^- S_\beta^+)\}_1 \\
&\qquad - \frac{7}{2(s+1)(2s+3)(s+2)}\sqrt{\frac{5}{6}}\,\{W^0_{\alpha\alpha\alpha}\,T^0_{\beta\beta}\}_1 \\
&\qquad + \frac{15}{(s+1)(2s+3)}\{S_\alpha^{z\,2} S_\beta^z\}_1 - \frac{3(s-1)}{2s+3}S_\beta^z\Bigg] \\
&= \frac{3}{2(s+1)}\sum_{\substack{\beta\\ \beta\neq\alpha}} \omega_{\alpha\beta\beta}\Bigg[\frac{4}{5}s(s+1)S_\alpha^z - \frac{1}{5}S_\beta^z(\boldsymbol{S}_\alpha\boldsymbol{S}_\beta) - \frac{1}{5}(\boldsymbol{S}_\alpha\boldsymbol{S}_\beta)S_\beta^z \\
&\qquad - \frac{1}{5(2s+3)(s+2)}\Big[10(\boldsymbol{S}_\alpha\boldsymbol{S}_\beta)S_\alpha^z(\boldsymbol{S}_\alpha\boldsymbol{S}_\beta) \\
&\qquad - \big(2s(s+1)+1\big)\big[S_\beta^z(\boldsymbol{S}_\alpha\boldsymbol{S}_\beta) + (\boldsymbol{S}_\alpha\boldsymbol{S}_\beta)S_\beta^z\big] \\
&\qquad - \big(2s(s+1)-4\big)\,s(s+1)\,S_\alpha^z \\
&\qquad + 5\big[S_\alpha^z(\boldsymbol{S}_\alpha\boldsymbol{S}_\beta) + (\boldsymbol{S}_\alpha\boldsymbol{S}_\beta)S_\alpha^z - s(s+1)S_\beta^z\big]\Big] \\
&\qquad + \frac{1}{2s+3}\Big[S_\alpha^z(\boldsymbol{S}_\alpha\boldsymbol{S}_\beta) + (\boldsymbol{S}_\alpha\boldsymbol{S}_\beta)S_\alpha^z + s(s+1)S_\beta^z\Big] - \frac{s^2-1}{2s+3}S_\beta^z\Bigg] \\
&= \frac{3}{2(2s+3)(s+2)}\sum_{\substack{\beta\\ \beta\neq\alpha}} \omega_{\alpha\beta\beta}\Bigg[-\frac{2}{s+1}(\boldsymbol{S}_\alpha\boldsymbol{S}_\beta)S_\alpha^z(\boldsymbol{S}_\alpha\boldsymbol{S}_\beta) \\
&\qquad + \big[S_\alpha^z(\boldsymbol{S}_\alpha\boldsymbol{S}_\beta) + (\boldsymbol{S}_\alpha\boldsymbol{S}_\beta)S_\alpha^z\big] \\
&\qquad - \big[S_\beta^z(\boldsymbol{S}_\alpha\boldsymbol{S}_\beta) + (\boldsymbol{S}_\alpha\boldsymbol{S}_\beta)S_\beta^z\big] \\
&\qquad + 2s(s+1)(s+2)\,S_\alpha^z + 2(s+1)\,S_\beta^z\Bigg] \\
&= \frac{3}{(2s+3)(s+2)}\sum_{\substack{\beta\\ \beta\neq\alpha}} \omega_{\alpha\beta\beta}\Bigg[-\frac{1}{s+1}(\boldsymbol{S}_\alpha\boldsymbol{S}_\beta)S_\alpha^z(\boldsymbol{S}_\alpha\boldsymbol{S}_\beta) \\
&\qquad + S_\alpha^z(\boldsymbol{S}_\alpha\boldsymbol{S}_\beta) - (\boldsymbol{S}_\alpha\boldsymbol{S}_\beta)S_\beta^z \\
&\qquad + s(s+1)(s+2)\,S_\alpha^z + (s+1)\,S_\beta^z\Bigg].
\end{aligned}
\tag{5.4.33}
$$

Similarly, substitution of (5.4.32) into (5.4.25) yields with (5.3.25) and (D.3.9)

$$
\begin{aligned}
\left\{H_{\alpha}^{\mathrm{P\bar{T}}\neq}\right\}_1 &= \frac{1}{2}\sum_{\substack{\beta\neq\gamma\\ \beta,\gamma\neq\alpha}} \omega_{\alpha\beta\gamma}\left[\frac{3}{2(s+1)}\left\{S_{\alpha}^{z}(S_{\beta}^{+}S_{\gamma}^{-}+S_{\beta}^{-}S_{\gamma}^{+})\right\}_1\right.\\
&\qquad\left.-\frac{7}{2(s+1)(2s+3)(s+2)}\sqrt{\frac{5}{6}}\left\{W_{\alpha\alpha\alpha}^{0}\,T_{\beta\gamma}^{0}\right\}_1\right]\\
&= \frac{3}{2(s+1)}\sum_{\substack{\beta\neq\gamma\\ \beta,\gamma\neq\alpha}} \omega_{\alpha\beta\gamma}\left[\frac{4}{5}S_{\alpha}^{z}(\boldsymbol{S}_{\beta}\boldsymbol{S}_{\gamma})-\frac{1}{5}S_{\beta}^{z}(\boldsymbol{S}_{\alpha}\boldsymbol{S}_{\gamma})-\frac{1}{5}(\boldsymbol{S}_{\alpha}\boldsymbol{S}_{\beta})S_{\gamma}^{z}\right.\\
&\qquad-\frac{1}{5(2s+3)(s+2)}\Big[5\left[(\boldsymbol{S}_{\alpha}\boldsymbol{S}_{\beta})S_{\alpha}^{z}(\boldsymbol{S}_{\alpha}\boldsymbol{S}_{\gamma})+(\boldsymbol{S}_{\alpha}\boldsymbol{S}_{\gamma})S_{\alpha}^{z}(\boldsymbol{S}_{\alpha}\boldsymbol{S}_{\beta})\right]\\
&\qquad-\big(2s(s+1)+1\big)\left[S_{\beta}^{z}(\boldsymbol{S}_{\alpha}\boldsymbol{S}_{\gamma})+(\boldsymbol{S}_{\alpha}\boldsymbol{S}_{\beta})S_{\gamma}^{z}\right]\\
&\qquad\left.-\big(2s(s+1)-4\big)\,S_{\alpha}^{z}(\boldsymbol{S}_{\beta}\boldsymbol{S}_{\gamma})\Big]\right]\\
&= \frac{3}{2(2s+3)(s+2)}\sum_{\substack{\beta\neq\gamma\\ \beta,\gamma\neq\alpha}} \omega_{\alpha\beta\gamma}\left[-\frac{(\boldsymbol{S}_{\alpha}\boldsymbol{S}_{\beta})S_{\alpha}^{z}(\boldsymbol{S}_{\alpha}\boldsymbol{S}_{\gamma})+(\boldsymbol{S}_{\alpha}\boldsymbol{S}_{\gamma})S_{\alpha}^{z}(\boldsymbol{S}_{\alpha}\boldsymbol{S}_{\beta})}{s+1}\right.\\
&\qquad+2(s+2)\,S_{\alpha}^{z}(\boldsymbol{S}_{\beta}\boldsymbol{S}_{\gamma})\\
&\qquad\left.-S_{\beta}^{z}(\boldsymbol{S}_{\alpha}\boldsymbol{S}_{\gamma})-(\boldsymbol{S}_{\alpha}\boldsymbol{S}_{\beta})S_{\gamma}^{z}\right].
\end{aligned}
\tag{5.4.34}
$$

Combining (5.4.33) and (5.4.34), we finally obtain the vector annihilation operator

$$
\begin{aligned}
\boldsymbol{A}_{\alpha}^{S} &\equiv \frac{2(2s+3)(s+2)}{3}\Big(\left\{H_{\alpha}^{\mathrm{P\bar{T}}=}\right\}_{\boldsymbol{1}}+\left\{H_{\alpha}^{\mathrm{P\bar{T}}\neq}\right\}_{\boldsymbol{1}}\Big)\\
&= \sum_{\substack{\beta\\ \beta\neq\alpha}} \frac{\boldsymbol{S}_{\alpha}(\boldsymbol{S}_{\alpha}\boldsymbol{S}_{\beta})+(\boldsymbol{S}_{\alpha}\boldsymbol{S}_{\beta})\boldsymbol{S}_{\alpha}+2(s+1)\,\boldsymbol{S}_{\beta}}{|\eta_{\alpha}-\eta_{\beta}|^{2}}\\
&\quad+\sum_{\substack{\beta,\gamma\\ \beta,\gamma\neq\alpha}} \frac{1}{(\bar{\eta}_{\alpha}-\bar{\eta}_{\beta})(\eta_{\alpha}-\eta_{\gamma})}\left[-\frac{(\boldsymbol{S}_{\alpha}\boldsymbol{S}_{\beta})\boldsymbol{S}_{\alpha}(\boldsymbol{S}_{\alpha}\boldsymbol{S}_{\gamma})+(\boldsymbol{S}_{\alpha}\boldsymbol{S}_{\gamma})\boldsymbol{S}_{\alpha}(\boldsymbol{S}_{\alpha}\boldsymbol{S}_{\beta})}{s+1}\right.\\
&\quad\left.+2(s+2)\,\boldsymbol{S}_{\alpha}(\boldsymbol{S}_{\beta}\boldsymbol{S}_{\gamma})-\boldsymbol{S}_{\beta}(\boldsymbol{S}_{\alpha}\boldsymbol{S}_{\gamma})-(\boldsymbol{S}_{\alpha}\boldsymbol{S}_{\beta})\boldsymbol{S}_{\gamma}\right],\\
\boldsymbol{A}_{\alpha}^{S}\left|\psi_{0}^{S}\right\rangle &= 0 \quad \forall\alpha.
\end{aligned}
\tag{5.4.35}
$$

This operator is even more complicated than the corresponding operator (4.6.27) for $S=1$, and only simplifies moderately if we sum over α. From (5.4.33), we obtain

Table 5.1 Annihilation operators for the general spin liquid ground state

Annihilation operators for $\lvert\psi_0^S\rangle$					
Operator	Equation	Symmetry transformation properties			
		T	P	Order of tensor	Transl. inv.
$\boldsymbol{S}_{\text{tot}}$	(2.2.6)	$-$	$+$	Vector	Yes
Ω_α^S	(5.2.1)	No	No	$2s+1$	No
$H^S - E_0^S$	(5.3.40)	$+$	$+$	Scalar	Yes
$\boldsymbol{D}_\alpha^S$	(5.4.13)	$+$	$-$	Vector	No
$\boldsymbol{\Lambda}^S$	(5.4.14)	$+$	$-$	Vector	Yes
$\boldsymbol{A}_\alpha^S$	(5.4.35)	$-$	$+$	Vector	No
$\boldsymbol{\Upsilon}^S$	(5.4.36)	$-$	$+$	Vector	Yes

With the exception of the defining operator Ω_α^S, which is the $m = 2s+1$ component of a tensor of order $2s+1$, we have only included scalar and vector annihilation operators

$$\begin{aligned}
&\frac{2(2s+3)(s+2)}{3}\sum_\alpha \left\{H_\alpha^{\mathrm{P\bar{T}}=}\right\}_1 \\
&= -\frac{2}{s+1}\sum_{\alpha\neq\beta}\omega_{\alpha\beta\beta}\,(\boldsymbol{S}_\alpha\boldsymbol{S}_\beta)S_\alpha^{\mathrm{z}}(\boldsymbol{S}_\alpha\boldsymbol{S}_\beta) \;+\; 2(s+1)^3\sum_\alpha \boldsymbol{S}_\alpha \sum_{\substack{\beta\\ \beta\neq\alpha}}\omega_{\alpha\beta\beta} \\
&= -\frac{2}{s+1}\sum_{\alpha\neq\beta}\omega_{\alpha\beta\beta}\,(\boldsymbol{S}_\alpha\boldsymbol{S}_\beta)S_\alpha^{\mathrm{z}}(\boldsymbol{S}_\alpha\boldsymbol{S}_\beta) \;+\; (s+1)^3\,\frac{N^2-1}{6}\boldsymbol{S}_{\text{tot}}.
\end{aligned}$$

This implies that $\lvert\psi_0^S\rangle$ is also annihilated by

$$\begin{aligned}
\boldsymbol{\Upsilon}^S = &-\frac{1}{s+1}\sum_{\substack{\alpha,\beta,\gamma\\ \beta,\gamma\neq\alpha}}\frac{(\boldsymbol{S}_\alpha\boldsymbol{S}_\beta)\boldsymbol{S}_\alpha(\boldsymbol{S}_\alpha\boldsymbol{S}_\gamma)+(\boldsymbol{S}_\alpha\boldsymbol{S}_\gamma)\boldsymbol{S}_\alpha(\boldsymbol{S}_\alpha\boldsymbol{S}_\beta)}{(\bar{\eta}_\alpha-\bar{\eta}_\beta)(\eta_\alpha-\eta_\gamma)} \\
&+\sum_{\substack{\alpha,\beta,\gamma\\ \alpha\neq\beta\neq\gamma\neq\alpha}}\frac{2(s+2)\,\boldsymbol{S}_\alpha(\boldsymbol{S}_\beta\boldsymbol{S}_\gamma)-\boldsymbol{S}_\beta(\boldsymbol{S}_\alpha\boldsymbol{S}_\gamma)-(\boldsymbol{S}_\alpha\boldsymbol{S}_\beta)\boldsymbol{S}_\gamma}{(\bar{\eta}_\alpha-\bar{\eta}_\beta)(\eta_\alpha-\eta_\gamma)}.
\end{aligned} \tag{5.4.36}$$

Whether this operator is of any practical use for further study of the model, however, remains an open question. The derivation of it concludes our study of non-trivial scalar and vector operators we can obtain from the defining condition (5.2.1) for the critical spin liquid state (5.1.3). These operators are summarized in Table 5.1.

5.5 Scalar Operators Constructed from Vectors

We see from Table 5.1 that there are two simple ways of constructing translationally, parity, and time reversal invariant scalar operators which annihilate $\lvert\psi_0^S\rangle$ from vector operators. These operators are

$$\sum_{\alpha} \boldsymbol{D}_{\alpha}^{S\dagger} \boldsymbol{D}_{\alpha}^{S} \quad \text{and} \quad \sum_{\alpha} \boldsymbol{S}_{\alpha} \boldsymbol{A}_{\alpha}^{S}. \tag{5.5.1}$$

These could potentially lead to alternative parent Hamiltonians for $\left|\psi_0^S\right\rangle$. If we just recover (5.3.40), the evaluation of the first operator will show that $H^S - E_0$ is positive semi-definite, or in other words, that $\left|\psi_0^S\right\rangle$ is a ground state of H^S.

5.5.1 Factorization of the Hamiltonian

In this section, we will evaluate

$$\sum_{\alpha} \boldsymbol{D}_{\alpha}^{S\dagger} \boldsymbol{D}_{\alpha}^{S},$$

with $\boldsymbol{D}_{\alpha}^{S}$ given by (5.4.13), or explicitly

$$\boldsymbol{D}_{\alpha}^{S\dagger} = \frac{1}{2} \sum_{\substack{\beta \\ \beta \neq \alpha}} \frac{\eta_\alpha + \eta_\beta}{\eta_\alpha - \eta_\beta} \left[\mathrm{i}(\boldsymbol{S}_\alpha \times \boldsymbol{S}_\beta) - (s+1)\boldsymbol{S}_\beta + \frac{1}{s+1} (\boldsymbol{S}_\alpha \boldsymbol{S}_\beta) \boldsymbol{S}_\alpha \right],$$

$$\boldsymbol{D}_{\alpha}^{S} = \frac{1}{2} \sum_{\substack{\gamma \\ \gamma \neq \alpha}} \frac{\eta_\alpha + \eta_\gamma}{\eta_\alpha - \eta_\gamma} \left[\mathrm{i}(\boldsymbol{S}_\alpha \times \boldsymbol{S}_\gamma) + (s+1)\boldsymbol{S}_\gamma - \frac{1}{s+1} \boldsymbol{S}_\alpha (\boldsymbol{S}_\alpha \boldsymbol{S}_\gamma) \right].$$

With $\alpha \neq \beta, \gamma$ and

$$\begin{aligned}
\mathrm{i}(\boldsymbol{S}_\alpha \times \boldsymbol{S}_\beta)\mathrm{i}(\boldsymbol{S}_\alpha \times \boldsymbol{S}_\gamma) &= \varepsilon^{ijk}\varepsilon^{ilm} S_\beta^j S_\alpha^k S_\alpha^l S_\gamma^m \\
&= \delta^{jl}\delta^{km} \left(S_\beta^j S_\alpha^l S_\alpha^k S_\gamma^m - S_\beta^j [S_\alpha^l, S_\alpha^k] S_\gamma^m \right) - \delta^{jm}\delta^{kl} S_\beta^j S_\alpha^k S_\alpha^l S_\gamma^m \\
&= (\boldsymbol{S}_\alpha \boldsymbol{S}_\beta)(\boldsymbol{S}_\alpha \boldsymbol{S}_\gamma) - \mathrm{i}\boldsymbol{S}_\alpha (\boldsymbol{S}_\beta \times \boldsymbol{S}_\gamma) - s(s+1)\boldsymbol{S}_\beta \boldsymbol{S}_\gamma,
\end{aligned} \tag{5.5.2}$$

we obtain for the product of the two square brackets

$$\begin{aligned}
&(\boldsymbol{S}_\alpha \boldsymbol{S}_\beta)(\boldsymbol{S}_\alpha \boldsymbol{S}_\gamma) - \mathrm{i}\boldsymbol{S}_\alpha (\boldsymbol{S}_\beta \times \boldsymbol{S}_\gamma) - s(s+1)\boldsymbol{S}_\beta \boldsymbol{S}_\gamma \\
&\quad + 2(s+1)\mathrm{i}\boldsymbol{S}_\alpha (\boldsymbol{S}_\beta \times \boldsymbol{S}_\gamma) - \frac{2}{s+1} (\boldsymbol{S}_\alpha \boldsymbol{S}_\beta)(\boldsymbol{S}_\alpha \boldsymbol{S}_\gamma) \\
&\quad - (s+1)^2 \boldsymbol{S}_\beta \boldsymbol{S}_\gamma + 2(\boldsymbol{S}_\alpha \boldsymbol{S}_\beta)(\boldsymbol{S}_\alpha \boldsymbol{S}_\gamma) - \frac{s}{s+1} (\boldsymbol{S}_\alpha \boldsymbol{S}_\beta)(\boldsymbol{S}_\alpha \boldsymbol{S}_\gamma) \\
&\quad = (2s+1) \left[\frac{1}{s+1} (\boldsymbol{S}_\alpha \boldsymbol{S}_\beta)(\boldsymbol{S}_\alpha \boldsymbol{S}_\gamma) + \mathrm{i}\boldsymbol{S}_\alpha (\boldsymbol{S}_\beta \times \boldsymbol{S}_\gamma) - (s+1)\boldsymbol{S}_\beta \boldsymbol{S}_\gamma \right].
\end{aligned} \tag{5.5.3}$$

The product of the prefactors is given by

$$
\begin{aligned}
&\frac{\eta_\alpha+\eta_\beta}{\eta_\alpha-\eta_\beta}\cdot\frac{\eta_\alpha+\eta_\gamma}{\eta_\alpha-\eta_\gamma}\\
&=\frac{1}{2}\left[1+\frac{2\eta_\beta}{\eta_\alpha-\eta_\beta}\right]\left[-1+\frac{2\eta_\alpha}{\eta_\alpha-\eta_\gamma}\right]+\frac{1}{2}\left[-1+\frac{2\eta_\alpha}{\eta_\alpha-\eta_\beta}\right]\left[1+\frac{2\eta_\gamma}{\eta_\alpha-\eta_\gamma}\right]\\
&=-1+1+1+2\frac{\eta_\alpha\eta_\beta}{(\eta_\alpha-\eta_\beta)(\eta_\alpha-\eta_\gamma)}+2\frac{\eta_\alpha\eta_\gamma}{(\eta_\alpha-\eta_\beta)(\eta_\alpha-\eta_\gamma)}\\
&=1-2(\omega_{\alpha\beta\gamma}+\omega_{\alpha\gamma\beta}).
\end{aligned}
\tag{5.5.4}
$$

We now define

$$
\begin{aligned}
\boldsymbol{B}_\alpha^S &\equiv \frac{\mathrm{i}}{2}\sum_{\substack{\gamma\\ \gamma\neq\alpha}}\left[\mathrm{i}(\boldsymbol{S}_\alpha\times\boldsymbol{S}_\gamma)+(s+1)\boldsymbol{S}_\gamma-\frac{1}{s+1}\boldsymbol{S}_\alpha\left(\boldsymbol{S}_\alpha\boldsymbol{S}_\gamma\right)\right]\\
&=\frac{\mathrm{i}}{2}\left[\mathrm{i}\left(\boldsymbol{S}_\alpha\times\boldsymbol{S}_{\mathrm{tot}}\right)+(s+1)\boldsymbol{S}_{\mathrm{tot}}-\frac{\boldsymbol{S}_\alpha\left(\boldsymbol{S}_\alpha\boldsymbol{S}_{\mathrm{tot}}\right)}{s+1}\right],
\end{aligned}
\tag{5.5.5}
$$

and its Hermitian conjugate,

$$
\boldsymbol{B}_\alpha^{S\dagger}=\frac{\mathrm{i}}{2}\sum_{\substack{\beta\\ \beta\neq\alpha}}\left[\mathrm{i}(\boldsymbol{S}_\alpha\times\boldsymbol{S}_\beta)-(s+1)\boldsymbol{S}_\beta+\frac{1}{s+1}\left(\boldsymbol{S}_\alpha\boldsymbol{S}_\beta\right)\boldsymbol{S}_\alpha\right].
$$

Obviously, $\boldsymbol{B}_\alpha^S$ annihilates every spin singlet, and $\left|\psi_0^S\right\rangle$ in particular. With (5.5.5), we may write

$$
\begin{aligned}
&\frac{1}{2s+1}\sum_\alpha\left(\boldsymbol{D}_\alpha^{S\dagger}\boldsymbol{D}_\alpha^S+\boldsymbol{B}_\alpha^{S\dagger}\boldsymbol{B}_\alpha^S\right)\\
&\quad=-\sum_{\substack{\alpha,\beta,\gamma\\ \beta,\gamma\neq\alpha}}\frac{\omega_{\alpha\beta\gamma}+\omega_{\alpha\gamma\beta}}{2}\left[\frac{(\boldsymbol{S}_\alpha\boldsymbol{S}_\beta)(\boldsymbol{S}_\alpha\boldsymbol{S}_\gamma)}{s+1}+\mathrm{i}\boldsymbol{S}_\alpha(\boldsymbol{S}_\beta\times\boldsymbol{S}_\gamma)-(s+1)\boldsymbol{S}_\beta\boldsymbol{S}_\gamma\right]\\
&\quad=-\sum_{\alpha\neq\beta}\omega_{\alpha\beta\beta}\left[\frac{(\boldsymbol{S}_\alpha\boldsymbol{S}_\beta)(\boldsymbol{S}_\alpha\boldsymbol{S}_\gamma)}{s+1}-\boldsymbol{S}_\alpha\boldsymbol{S}_\beta-s(s+1)^2\right]\\
&\quad\quad-\sum_{\substack{\alpha,\beta,\gamma\\ \alpha\neq\beta\neq\gamma\neq\alpha}}\omega_{\alpha\beta\gamma}\left[\frac{(\boldsymbol{S}_\alpha\boldsymbol{S}_\beta)(\boldsymbol{S}_\alpha\boldsymbol{S}_\gamma)+(\boldsymbol{S}_\alpha\boldsymbol{S}_\gamma)(\boldsymbol{S}_\alpha\boldsymbol{S}_\beta)}{2(s+1)}-(s+1)\boldsymbol{S}_\beta\boldsymbol{S}_\gamma\right]\\
&\quad=(2s+3)\sum_{\alpha\neq\beta}\omega_{\alpha\beta\beta}\boldsymbol{S}_\alpha\boldsymbol{S}_\beta-\sum_{\substack{\alpha,\beta,\gamma\\ \beta,\gamma\neq\alpha}}\omega_{\alpha\beta\gamma}\frac{(\boldsymbol{S}_\alpha\boldsymbol{S}_\beta)(\boldsymbol{S}_\alpha\boldsymbol{S}_\gamma)+(\boldsymbol{S}_\alpha\boldsymbol{S}_\gamma)(\boldsymbol{S}_\alpha\boldsymbol{S}_\beta)}{2(s+1)}\\
&\quad\quad+s(s+1)^2\frac{N(N^2+5)}{12}-\frac{s+1}{2}\boldsymbol{S}_{\mathrm{tot}}^2,
\end{aligned}
\tag{5.5.6}
$$

where we have used (B.15) and (B.20). With the Hamiltonian (5.3.40) and the ground state energy (5.3.41) derived in Sect. 5.3, we may write

$$\begin{aligned}
&\frac{1}{(2s+1)(2s+3)}\sum_{\alpha}\left(\boldsymbol{D}_{\alpha}^{S\dagger}\boldsymbol{D}_{\alpha}^{S}+\boldsymbol{B}_{\alpha}^{S\dagger}\boldsymbol{B}_{\alpha}^{S}\right)+\frac{s+1}{2(2s+3)}\boldsymbol{S}_{\text{tot}}^{2}\\
&=\sum_{\alpha\neq\beta}\frac{\boldsymbol{S}_{\alpha}\boldsymbol{S}_{\beta}}{|\eta_{\alpha}-\eta_{\beta}|^{2}}-\frac{1}{2(s+1)(2s+3)}\sum_{\substack{\alpha,\beta,\gamma\\ \alpha\neq\beta,\gamma}}\frac{(\boldsymbol{S}_{\alpha}\boldsymbol{S}_{\beta})(\boldsymbol{S}_{\alpha}\boldsymbol{S}_{\gamma})+(\boldsymbol{S}_{\alpha}\boldsymbol{S}_{\gamma})(\boldsymbol{S}_{\alpha}\boldsymbol{S}_{\beta})}{(\bar{\eta}_{\alpha}-\bar{\eta}_{\beta})(\eta_{\alpha}-\eta_{\gamma})}\\
&\quad+\frac{s(s+1)^{2}}{2s+3}\frac{N(N^{2}+5)}{12}\\
&=\frac{N^{2}}{2\pi^{2}}\left[H^{S}-E_{0}^{S}\right].
\end{aligned} \tag{5.5.7}$$

Since all the operators on the left hand side of (5.5.7) are positive semi-definite, i.e., have only non-negative eigenvalues, the operator $H^S - E_0^S$ on the right has to be positive semi-definite as well. Furthermore, since all the operators on the left annihilate $\left|\psi_0^S\right\rangle$ we have shown that $\left|\psi_0^S\right\rangle$ is a zero energy ground state of $H^S - E_0^S$. Exact diagonalization studies [2] carried out numerically for up to $N=18$ sites for the $S=1$ model and for up to $N=12$ sites for the $S=\frac{1}{2}$ model further show that $\left|\psi_0^{S=1}\right\rangle$ and $\left|\psi_0^{S=\frac{3}{2}}\right\rangle$ are the unique ground states of (5.3.40). We assume this property to hold for general spin S, but are not aware of any method to prove this analytically.

Note that the derivation using the operators $\boldsymbol{D}_\alpha^S$ is actually the simplest derivation of (5.3.40) we are aware of. As compared to our original derivation in Sect. 5.3, it has the advantage that, except for the tensor decomposition of $(S^+)^{2s}(S^-)^{2s}$ spelled out in Sect. 5.3.2, we only needed the formula (D.3.8) for the vector content of $S_1^z S_2^z S_3^z$, but not the significantly more complicated formula (4.5.22) for the scalar component of $T_{\alpha\alpha}^0 T_{\beta\gamma}^0$ derived in Sect. 4.5.3. That we have arrived at the same model twice using different methods gives us some confidence in the uniqueness of the final Hamiltonian (5.3.40).

5.5.2 A Variation of the Model

The analysis in the previous section suggests that another, closely related Hamiltonian is positive semi-definite as well. Writing the product of prefactors (5.5.4) as

$$\frac{\eta_\alpha+\eta_\beta}{\eta_\alpha-\eta_\beta}\cdot\frac{\eta_\alpha+\eta_\gamma}{\eta_\alpha-\eta_\gamma}=-2\left(\omega_{\alpha\beta\gamma}+\omega_{\alpha\gamma\beta}-\frac{1}{2}\right), \tag{5.5.8}$$

we can derive a model directly from

$$\sum_\alpha \boldsymbol{D}_\alpha^{S\dagger}\boldsymbol{D}_\alpha^S,$$

without any need to introduce the operators $\boldsymbol{B}_\alpha^S$ and $\boldsymbol{B}_\alpha^{S^\dagger}$. This yields

$$\begin{aligned}
&\frac{1}{2s+1}\sum_\alpha \boldsymbol{D}_\alpha^{S^\dagger}\boldsymbol{D}_\alpha^S\\
&\quad= -\sum_{\alpha\neq\beta}\left(\omega_{\alpha\beta\beta}-\frac{1}{4}\right)\left[\frac{(\boldsymbol{S}_\alpha\boldsymbol{S}_\beta)(\boldsymbol{S}_\alpha\boldsymbol{S}_\gamma)}{s+1}-\boldsymbol{S}_\alpha\boldsymbol{S}_\beta - s(s+1)^2\right]\\
&\quad\quad - \sum_{\substack{\alpha,\beta,\gamma\\ \alpha\neq\beta\neq\gamma\neq\alpha}}\left(\omega_{\alpha\beta\gamma}-\frac{1}{4}\right)\left[\frac{(\boldsymbol{S}_\alpha\boldsymbol{S}_\beta)(\boldsymbol{S}_\alpha\boldsymbol{S}_\gamma)+(\boldsymbol{S}_\alpha\boldsymbol{S}_\gamma)(\boldsymbol{S}_\alpha\boldsymbol{S}_\beta)}{2(s+1)} - (s+1)\boldsymbol{S}_\beta\boldsymbol{S}_\gamma\right]\\
&\quad= (2s+3)\sum_{\alpha\neq\beta}\left(\omega_{\alpha\beta\beta}-\frac{1}{4}\right)\boldsymbol{S}_\alpha\boldsymbol{S}_\beta\\
&\quad\quad - \sum_{\substack{\alpha,\beta,\gamma\\ \beta,\gamma\neq\alpha}}\left(\omega_{\alpha\beta\gamma}-\frac{1}{4}\right)\frac{(\boldsymbol{S}_\alpha\boldsymbol{S}_\beta)(\boldsymbol{S}_\alpha\boldsymbol{S}_\gamma)+(\boldsymbol{S}_\alpha\boldsymbol{S}_\gamma)(\boldsymbol{S}_\alpha\boldsymbol{S}_\beta)}{2(s+1)}\\
&\quad\quad + s(s+1)^2\sum_{\alpha\neq\beta}\omega_{\alpha\beta\beta} - \frac{1}{4}(s+1)\sum_{\substack{\alpha,\beta,\gamma\\ \beta,\gamma\neq\alpha}}\boldsymbol{S}_\beta\boldsymbol{S}_\gamma\\
&\quad= (2s+3)\sum_{\alpha\neq\beta}\left(\omega_{\alpha\beta\beta}-\frac{1}{4}\right)\boldsymbol{S}_\alpha\boldsymbol{S}_\beta\\
&\quad\quad - \sum_{\substack{\alpha,\beta,\gamma\\ \beta,\gamma\neq\alpha}}\left(\omega_{\alpha\beta\gamma}-\frac{1}{4}\right)\frac{(\boldsymbol{S}_\alpha\boldsymbol{S}_\beta)(\boldsymbol{S}_\alpha\boldsymbol{S}_\gamma)+(\boldsymbol{S}_\alpha\boldsymbol{S}_\gamma)(\boldsymbol{S}_\alpha\boldsymbol{S}_\beta)}{2(s+1)}\\
&\quad\quad + s(s+1)^2\frac{N(N^2-4)}{12} - \frac{(s+1)(N-2)}{4}\boldsymbol{S}_{\text{tot}}^2,
\end{aligned} \tag{5.5.9}$$

where we have used (B.15) and (B.20). If we now define the alternative model

$$\begin{aligned}
\tilde{H}^S \equiv \frac{2\pi^2}{N^2}\Bigg[&\sum_{\alpha\neq\beta}\left(\frac{1}{|\eta_\alpha-\eta_\beta|^2}-\frac{1}{4}\right)\boldsymbol{S}_\alpha\boldsymbol{S}_\beta\\
&-\sum_{\substack{\alpha,\beta,\gamma\\ \alpha\neq\beta,\gamma}}\left(\frac{1}{(\bar{\eta}_\alpha-\bar{\eta}_\beta)(\eta_\alpha-\eta_\gamma)}-\frac{1}{4}\right)\frac{(\boldsymbol{S}_\alpha\boldsymbol{S}_\beta)(\boldsymbol{S}_\alpha\boldsymbol{S}_\gamma)+(\boldsymbol{S}_\alpha\boldsymbol{S}_\gamma)(\boldsymbol{S}_\alpha\boldsymbol{S}_\beta)}{2(s+1)(2s+3)}\Bigg]
\end{aligned} \tag{5.5.10}$$

with energy eigenvalue

$$\tilde{E}_0^S = -\frac{2\pi^2}{N^2}\frac{s(s+1)^2}{2s+3}\frac{N(N^2-4)}{12} = -\frac{\pi^2}{6}\frac{s(s+1)^2}{2s+3}\left(N-\frac{4}{N}\right), \tag{5.5.11}$$

we may rewrite (5.5.9) as

$$\frac{2\pi^2}{N^2}\left[\frac{1}{(2s+1)(2s+3)}\sum_{\alpha} \boldsymbol{D}_\alpha^{S\dagger}\boldsymbol{D}_\alpha^S + \frac{(s+1)(N-2)}{4(2s+3)}\boldsymbol{S}_{\text{tot}}^2\right] = \tilde{H}^S - \tilde{E}_0^S. \tag{5.5.12}$$

This implies that $|\psi_0^S\rangle$ is also a ground state of $\tilde{H}^S$ with energy $\tilde{E}_0^S$, as defined in (5.5.10) and (5.5.11), respectively.

Since the maximal distance of η_α and η_β on the unit circle is 2, the shift in the coefficients in (5.5.10) (as compared to (5.3.40)) effects that these coefficients go to zero as the sites η_α and η_β, η_γ are maximally separated on the unit circle. The alternative model (5.5.10) is hence more local than the original model (5.3.40). It is possible that the alternative model (5.5.10) possesses symmetries (or even an integrability structure) the original model does not share.

5.5.3 The Third Derivation

Finally, another translationally, parity, and time reversal invariant scalar operator which annihilates $|\psi_0^S\rangle$ is given by

$$\sum_{\alpha} \boldsymbol{S}_\alpha \boldsymbol{A}_\alpha^S, \tag{5.5.13}$$

where $\boldsymbol{A}_\alpha^S$ is given by (5.4.35),

$$\begin{aligned}\boldsymbol{A}_\alpha^S = &\sum_{\substack{\beta\\ \beta\neq\alpha}} \omega_{\alpha\beta\beta}\left[\boldsymbol{S}_\alpha(\boldsymbol{S}_\alpha\boldsymbol{S}_\beta) + (\boldsymbol{S}_\alpha\boldsymbol{S}_\beta)\boldsymbol{S}_\alpha + 2(s+1)\boldsymbol{S}_\beta\right]\\ &+ \sum_{\substack{\beta,\gamma\\ \beta,\gamma\neq\alpha}} \omega_{\alpha\beta\gamma}\Bigg[-\frac{(\boldsymbol{S}_\alpha\boldsymbol{S}_\beta)\boldsymbol{S}_\alpha(\boldsymbol{S}_\alpha\boldsymbol{S}_\gamma) + (\boldsymbol{S}_\alpha\boldsymbol{S}_\gamma)\boldsymbol{S}_\alpha(\boldsymbol{S}_\alpha\boldsymbol{S}_\beta)}{s+1}\\ &\qquad + 2(s+2),\, \boldsymbol{S}_\alpha(\boldsymbol{S}_\beta\boldsymbol{S}_\gamma) - \boldsymbol{S}_\beta(\boldsymbol{S}_\alpha\boldsymbol{S}_\gamma) - (\boldsymbol{S}_\alpha\boldsymbol{S}_\beta)\boldsymbol{S}_\gamma\Bigg].\end{aligned}$$

With

$$\begin{aligned}\boldsymbol{S}_\alpha(\boldsymbol{S}_\alpha\boldsymbol{S}_\beta)\boldsymbol{S}_\alpha &= \boldsymbol{S}_\alpha\left[\boldsymbol{S}_\alpha(\boldsymbol{S}_\alpha\boldsymbol{S}_\beta) + i\boldsymbol{S}_\alpha\times\boldsymbol{S}_\beta\right]\\ &= (s(s+1)-1)\,\boldsymbol{S}_\alpha\boldsymbol{S}_\beta,\\ \boldsymbol{S}_\alpha(\boldsymbol{S}_\alpha\boldsymbol{S}_\beta)\boldsymbol{S}_\gamma &= \boldsymbol{S}_\alpha\left[\boldsymbol{S}_\gamma(\boldsymbol{S}_\alpha\boldsymbol{S}_\beta) - i\delta_{\beta\gamma}\boldsymbol{S}_\alpha\times\boldsymbol{S}_\beta\right]\\ &= (\boldsymbol{S}_\alpha\boldsymbol{S}_\gamma)(\boldsymbol{S}_\alpha\boldsymbol{S}_\beta) + \delta_{\beta\gamma}\boldsymbol{S}_\alpha\boldsymbol{S}_\beta,\end{aligned}$$

which follows from (4.6.23, 4.6.24) and holds for $\alpha \neq \beta, \gamma$, we obtain

$$\sum_{\alpha} \boldsymbol{S}_\alpha \boldsymbol{A}_\alpha^S = \sum_{\alpha\neq\beta} \omega_{\alpha\beta\beta} \left[2s(s+1) - 1 + 2(s+1) - 1\right] \boldsymbol{S}_\alpha \boldsymbol{S}_\beta$$
$$+ \sum_{\substack{\alpha,\beta,\gamma \\ \alpha\neq\beta,\gamma}} \omega_{\alpha\beta\gamma} \Bigg[-(s(s+1)-1) \frac{(\boldsymbol{S}_\alpha \boldsymbol{S}_\beta)(\boldsymbol{S}_\alpha \boldsymbol{S}_\gamma) + (\boldsymbol{S}_\alpha \boldsymbol{S}_\gamma)(\boldsymbol{S}_\alpha \boldsymbol{S}_\beta)}{s+1}$$
$$+ 2s(s+1)(s+2) \boldsymbol{S}_\beta \boldsymbol{S}_\gamma - (\boldsymbol{S}_\alpha \boldsymbol{S}_\beta)(\boldsymbol{S}_\alpha \boldsymbol{S}_\gamma) - (\boldsymbol{S}_\alpha \boldsymbol{S}_\gamma)(\boldsymbol{S}_\alpha \boldsymbol{S}_\beta) \Bigg]. \tag{5.5.14}$$

With (B.15) and (B.20), we find

$$\sum_{\substack{\alpha,\beta,\gamma \\ \alpha\neq\beta,\gamma}} \omega_{\alpha\beta\gamma} \boldsymbol{S}_\beta \boldsymbol{S}_\gamma = s(s+1) \frac{N(N^2+5)}{12} + 2 \sum_{\alpha\neq\beta} \omega_{\alpha\beta\beta} \boldsymbol{S}_\alpha \boldsymbol{S}_\beta - \frac{1}{2} \boldsymbol{S}_{\text{tot}}^2,$$

and therewith

$$\sum_{\alpha} \boldsymbol{S}_\alpha \boldsymbol{A}_\alpha^S = 2s(s+2)(2s+3) \sum_{\alpha\neq\beta} \omega_{\alpha\beta\beta} \boldsymbol{S}_\alpha \boldsymbol{S}_\beta$$
$$- \frac{s(s+2)}{s+1} \sum_{\substack{\alpha,\beta,\gamma \\ \alpha\neq\beta,\gamma}} \omega_{\alpha\beta\gamma} \left[(\boldsymbol{S}_\alpha \boldsymbol{S}_\beta)(\boldsymbol{S}_\alpha \boldsymbol{S}_\gamma) + (\boldsymbol{S}_\alpha \boldsymbol{S}_\gamma)(\boldsymbol{S}_\alpha \boldsymbol{S}_\beta) \right]$$
$$+ 2s(s+2)s(s+1)^2 \frac{N(N^2+5)}{12} - s(s+1)(s+2) \boldsymbol{S}_{\text{tot}}^2. \tag{5.5.15}$$

We may rewrite this

$$\frac{1}{2s(s+2)(2s+3)} \sum_{\alpha} \boldsymbol{S}_\alpha \boldsymbol{A}_\alpha^S + \frac{s+1}{2(2s+3)} \boldsymbol{S}_{\text{tot}}^2$$
$$= \sum_{\alpha\neq\beta} \frac{\boldsymbol{S}_\alpha \boldsymbol{S}_\beta}{|\eta_\alpha - \eta_\beta|^2} - \frac{1}{2(s+1)(2s+3)} \sum_{\substack{\alpha,\beta,\gamma \\ \alpha\neq\beta,\gamma}} \frac{(\boldsymbol{S}_\alpha \boldsymbol{S}_\beta)(\boldsymbol{S}_\alpha \boldsymbol{S}_\gamma) + (\boldsymbol{S}_\alpha \boldsymbol{S}_\gamma)(\boldsymbol{S}_\alpha \boldsymbol{S}_\beta)}{(\bar{\eta}_\alpha - \bar{\eta}_\beta)(\eta_\alpha - \eta_\gamma)}$$
$$+ \frac{s(s+1)^2}{2s+3} \frac{N(N^2+5)}{12}. \tag{5.5.16}$$

In other words, we obtain the model Hamiltonian (5.3.40) for a third time. The present derivation is the most complicated one, and does not yield any new insights, except that it further strengthens the case that there is a certain uniqueness to our Hamiltonian.

5.6 The Case $S = \frac{1}{2}$ Once More

Finally, we wish to demonstrate that the general spin S model introduced and derived in this section includes the Haldane–Shastry model as the special case $S = \frac{1}{2}$.

For $S = \frac{1}{2}$, the higher order interaction terms in the Hamiltonian (5.3.40) simplify, as

$$(\boldsymbol{S}_\alpha \boldsymbol{S}_\beta)^2 = -\frac{1}{2}\boldsymbol{S}_\alpha \boldsymbol{S}_\beta + \frac{3}{16}, \quad \alpha \neq \beta, \tag{5.6.1}$$

and

$$(\boldsymbol{S}_\alpha \boldsymbol{S}_\beta)(\boldsymbol{S}_\alpha \boldsymbol{S}_\gamma) + (\boldsymbol{S}_\alpha \boldsymbol{S}_\gamma)(\boldsymbol{S}_\alpha \boldsymbol{S}_\beta) = \frac{1}{2}\boldsymbol{S}_\beta \boldsymbol{S}_\gamma, \quad \alpha \neq \beta \neq \gamma \neq \alpha. \tag{5.6.2}$$

We can verify (5.6.1) and (5.6.2) with

$$\begin{aligned}(\boldsymbol{S}_\alpha \boldsymbol{S}_\beta)(\boldsymbol{S}_\alpha \boldsymbol{S}_\gamma) &= S_\beta^i S_\alpha^i S_\alpha^j S_\gamma^j \\ &= S_\beta^i \left(\frac{1}{4}\delta^{ij} + \frac{1}{2}\varepsilon^{ijk} S_\alpha^k\right) S_\gamma^j \\ &= \frac{1}{4}\boldsymbol{S}_\beta \boldsymbol{S}_\gamma + \frac{1}{2}\boldsymbol{S}_\alpha(\boldsymbol{S}_\beta \times \boldsymbol{S}_\gamma),\end{aligned} \tag{5.6.3}$$

which holds only for $S = \frac{1}{2}$ and $\alpha \neq \beta, \gamma$. Alternatively, since $S_\alpha^{+2} = 0$ for $S = \frac{1}{2}$, $T_{\alpha\alpha}^m = 0$ for all m, and (4.5.22) reduces to

$$-\boldsymbol{S}_\beta \boldsymbol{S}_\gamma + 2\left[(\boldsymbol{S}_\alpha \boldsymbol{S}_\beta)(\boldsymbol{S}_\alpha \boldsymbol{S}_\gamma) + (\boldsymbol{S}_\alpha \boldsymbol{S}_\gamma)(\boldsymbol{S}_\alpha \boldsymbol{S}_\beta)\right] + 2\delta_{\beta\gamma} \boldsymbol{S}_\alpha \boldsymbol{S}_\beta = 0. \tag{5.6.4}$$

For $\beta = \gamma$ and $\beta \neq \gamma$, this yields (5.6.1) and (5.6.2), respectively.

Substitution of (5.6.1, 5.6.2), and $s = \frac{1}{2}$ into the general Hamiltonian (5.3.40) yields

$$\begin{aligned}H^{S=\frac{1}{2}} &= \frac{2\pi^2}{N^2}\left[\sum_{\alpha\neq\beta} \frac{1}{|\eta_\alpha - \eta_\beta|^2}\left(\boldsymbol{S}_\alpha \boldsymbol{S}_\beta + \frac{1}{12}\boldsymbol{S}_\alpha \boldsymbol{S}_\beta - \frac{1}{32}\right)\right. \\ &\qquad \left. - \frac{1}{24}\sum_{\substack{\alpha,\beta,\gamma \\ \alpha\neq\beta\neq\gamma\neq\alpha}} \frac{\boldsymbol{S}_\beta \boldsymbol{S}_\gamma}{(\bar{\eta}_\alpha - \bar{\eta}_\beta)(\eta_\alpha - \eta_\gamma)}\right] \\ &= \frac{2\pi^2}{N^2}\left[\sum_{\alpha\neq\beta} \frac{\boldsymbol{S}_\alpha \boldsymbol{S}_\beta}{|\eta_\alpha - \eta_\beta|^2} - \frac{1}{32}\frac{N(N^2-1)}{12} + \frac{1}{48}\boldsymbol{S}_{\text{tot}}^2 - \frac{N}{64}\right] \\ &= H^{\text{HS}} - \frac{1}{32}\frac{N(N^2+5)}{12} + \frac{1}{48}\boldsymbol{S}_{\text{tot}}^2.\end{aligned} \tag{5.6.5}$$

The energy of the Haldane–Shastry ground state $\left|\psi_0^{\mathrm{HS}}\right\rangle$ is hence with (5.3.41) given by

$$E_0^{\mathrm{HS}} = E_0^{S=\frac{1}{2}} + \frac{2\pi^2}{N^2}\frac{1}{32}\frac{N(N^2+5)}{12} = -\frac{2\pi^2}{N^2}\frac{N(N^2+5)}{48},$$

which agrees with (2.2.5). Note that as the derivation in (5.6.5) stands, we have lost the information that $H^{\mathrm{HS}} - E_0^{\mathrm{HS}}$ is positive semi-definite, due to the $\boldsymbol{S}_{\mathrm{tot}}^2$ term. This information, however, can be recovered if we take the last term on the left-hand side of (5.5.7) into account. The spin S model we have derived here hence includes the Haldane–Shastry model as the special case $S = \frac{1}{2}$.

References

1. N. Read, E. Rezayi, Beyond paired quantum Hall states: parafermions and incompressible states in the first excited Landau level. Phys. Rev. B **59**, 8084 (1999)
2. R. Thomale, S. Rachel, P. Schmitteckert, M. Greiter, manuscript in preparation

Chapter 6
Conclusions and Unresolved Issues

The model—In this monograph, we have presented an exact model of a critical spin chain with spin S. The Hamiltonian is given by

$$H^S = \frac{2\pi^2}{N^2}\left[\sum_{\alpha\neq\beta}^{N} \frac{\boldsymbol{S}_\alpha \boldsymbol{S}_\beta}{|\eta_\alpha - \eta_\beta|^2} - \frac{1}{2(S+1)(2S+3)} \sum_{\substack{\alpha,\beta,\gamma \\ \alpha\neq\beta,\gamma}}^{N} \frac{(\boldsymbol{S}_\alpha \boldsymbol{S}_\beta)(\boldsymbol{S}_\alpha \boldsymbol{S}_\gamma) + (\boldsymbol{S}_\alpha \boldsymbol{S}_\gamma)(\boldsymbol{S}_\alpha \boldsymbol{S}_\beta)}{(\bar{\eta}_\alpha - \bar{\eta}_\beta)(\eta_\alpha - \eta_\gamma)}\right], \tag{6.1}$$

where $\eta_\alpha = e^{\mathrm{i}\frac{2\pi}{N}\alpha}$, $\alpha = 1, \ldots, N$, are the coordinates of N sites on a unit circle embedded in the complex plane. If we write the ground state of the Haldane–Shastry model [1, 2], which is equivalent to the Gutzwiller state obtained by projection of filled bands [3–5], in terms of Schwinger bosons,

$$\begin{aligned} \left|\psi_0^{\mathrm{HS}}\right\rangle &= \sum_{\{z_1,\ldots,z_M;w_1,\ldots,w_M\}} \psi_0^{\mathrm{HS}}(z_1,\ldots,z_M)\, a_{z_1}^{\dagger}\cdots a_{z_M}^{\dagger} b_{w_1}^{\dagger}\cdots b_{w_M}^{\dagger}|0\rangle \\ &\equiv \Psi_0^{\mathrm{HS}}[a^{\dagger}, b^{\dagger}]|0\rangle, \end{aligned} \tag{6.2}$$

where $M = \frac{N}{2}$ and the w_k's are those coordinates on the unit circle which are not occupied by any of the z_i's, then the exact ground state of our model Hamiltonian (6.1) is given by

$$\left|\psi_0^{S=1}\right\rangle = \left(\Psi_0^{\mathrm{HS}}\left[a^{\dagger}, b^{\dagger}\right]\right)^S |0\rangle. \tag{6.3}$$

The ground state energy is

$$E_0^S = -\frac{2\pi^2}{N^2}\frac{S(S+1)^2}{2S+3}\frac{N(N^2+5)}{12}. \tag{6.4}$$

M. Greiter, *Mapping of Parent Hamiltonians*, Springer Tracts in Modern Physics 244,
DOI: 10.1007/978-3-642-24384-4_6, © Springer-Verlag Berlin Heidelberg 2011

For $S = \frac{1}{2}$, the model (6.1) reduces to the Haldane–Shastry model. Since the model describes a critical spin chain with spin S, the low energy effective field theory is given by the SU(2) level $k=2S$ Wess–Zumino–Witten model [6, 7].

The Hamiltonian was constructed from the condition

$$\Omega_\alpha^S = \sum_{\substack{\beta=1\\ \beta\neq\alpha}}^{N} \frac{1}{\eta_\alpha - \eta_\beta}(S_\alpha^-)^{2S} S_\beta^-, \qquad \Omega_\alpha^S |\psi_0^S\rangle = 0 \quad \forall\alpha, \tag{6.5}$$

which we obtained for $S = \frac{1}{2}$ and for $S = 1$ from the two- and three-body parent Hamiltonians of bosonic Laughlin and Moore–Read states in quantum Hall systems, respectively, and then generalized to arbitrary spin.

Uniqueness and the quest for integrability—Starting with the defining condition (6.5), we constructed a total of three translationally, parity, and time reversal invariant scalar annihilation operator for the state (6.3)—one directly, and two by taking the scalar products of vector operators. All three operators yielded the parent Hamiltonian (6.1). This attests are certain uniqueness to the model.

Nonetheless, it is clear that the model is not completely unique. First, the ground state (6.3) is trivially annihilated by all terms which annihilate every spin singlet. For example, we could add the term

$$\sum_\alpha (\boldsymbol{S}_\alpha \boldsymbol{S}_{\text{tot}})^2 \tag{6.6}$$

with an arbitrary coefficient to (6.1). Then (6.3) would remain the ground state as long as the operator $H^S - E_0^S$ were to remain positive semi-definite. (This ambivalence was exploited in Sect. 5.5.2, when we derived the alternative Hamiltonian (5.5.10).) Another three-spin term which annihilates every spin singlet is given by (5.4.20), even though this term is not suitable as it violates both parity and time reversal symmetry. If we allow for four-spin interactions, there is a plethora of parity and time reversal invariant scalar operators we could add.

Second, we could construct another parent Hamiltonian from the annihilation operator

$$\Xi_\alpha = \sum_{\substack{\beta,\gamma=1\\ \beta,\gamma\neq\alpha}}^{N} \frac{S_\alpha^- S_\beta^- S_\gamma^-}{(\eta_\alpha - \eta_\beta)(\eta_\alpha - \eta_\gamma)} - \sum_{\substack{\beta=1\\ \beta\neq\alpha}}^{N} \frac{(S_\alpha^-)^2 S_\beta^-}{(\eta_\alpha - \eta_\beta)^2}, \quad \Xi_\alpha |\psi_0^{S=1}\rangle = 0\,\forall\alpha, \tag{6.7}$$

which we derived in Sect. 4.4.2. This Hamiltonian will presumably contain five-spin interactions.

The issue of uniqueness of the model is relevant to the question of whether the model, or a closely related model, is integrable. Preliminary numerical work [8] indicates that the model (6.1) is not integrable for finite system sizes, while the data are consistent with integrability in the thermodynamic limit.

Momentum spacings and topological degeneracies—The other highly important, unresolved issue regarding the model concerns the momentum spacings of the spinon excitations. In Sect. 2.4.5, we proposed that the spacings for the $S=1$ model would alternate between being odd multiples of $\frac{\pi}{N}$ and being either odd or even multiples of $\frac{\pi}{N}$. (Recall that odd multiples of $\frac{\pi}{N}$ correspond to half-fermions in one dimension, while even multiples represent either fermions or bosons.) Whenever we have a choice between even and odd, this choice represents a topological quantum number, which is insensitive to local perturbations. These topological quantum numbers span an internal or topological Hilbert space of dimension 2^L when $2L$ spinons are present. All the states in this space are degenerate in the thermodynamic limit. This topological Hilbert space is the one-dimensional analog of the topological Hilbert space spanned by the Majorana fermion states [9–12] in the vortex cores of the Moore–Read state [13–15] or the non-abelian chiral spin liquid [16]. In Sect. 2.4.7, we generalized these conditions for the momentum spacings to the models with arbitrary spin $S > 1$.

The first unresolved issue with regard to our proposal is whether it is correct. In view of the established momentum spacings [17] for the spinons in the Haldane–Shastry model, the construction of the state suggests that it is. Since the model (6.1) is presumably not integrable, however, the momenta of the individual spinons will not be good quantum numbers when more than one spinon is present. (This is always the case, as the minimal number of spinons for the models with $S \geq 1$ is two.) Nonetheless, the topological shifts can still be good quantum numbers. In this regard, the situation is similar to the Moore–Read state, where, when long-ranged interaction are present, the state vectors in the internal Hilbert space are degenerate in the thermodynamic limit only.

Assuming that our assignment of the momentum spacings is correct, the next question to ask is whether the picture applies only to the exact model we have constructed in this monograph, or to a whole range of critical spin chain models with $S=1$. If it applies to a range of models, as we believe, the topological space spanned by the spinons may be useful in applications as protected cubits. The internal state vector can probably be manipulated though measurements involving several spinons simultaneously, but it is far from clear how to do so efficiently.

To study the spinon excitations systematically, it would be highly desirable to apply the method reviewed in Sect. 2.2.4 for the Haldane–Shastry model to the general model (6.1). Unfortunately, this does not appear straightforward. The problem arises when we write out the $S_\beta^+ S_\alpha^- \, S_\gamma^+ S_\alpha^-$ term along the lines of (2.2.36–2.2.40). When we evaluated the $S_\alpha^+ S_\beta^-$ term in the Haldane–Shastry model, we used the Taylor series expansion (2.2.39) to shift the variable η_β in the function

$$\frac{\psi(z_1, \ldots, z_{j-1}, \eta_\beta, z_{j+1}, \ldots, z_M)}{\eta_\beta}$$

in (2.2.38) to z_j. When we evaluate the action of the $S_\beta^+ S_\alpha^- \, S_\gamma^+ S_\alpha^-$ term in the $S=1$ model, we need to shift “two variables” η_α in the function

$$\frac{\psi(z_1, \ldots, z_{j-1}, \eta_\alpha, z_{j+1}, \ldots, z_{k-1}, \eta_\alpha, z_{k+1}, \ldots, z_M)}{\eta_\alpha^2}$$

via Taylor expansions, one to z_j and one to z_k. This yields for $z_j \neq z_k$

$$\frac{\psi(z_1, \ldots, \eta_\alpha, \ldots, \eta_\alpha, \ldots, z_M)}{\eta_\alpha^2}$$
$$= \sum_{l=0}^{N-1} \frac{(\eta_\alpha - z_j)^l}{l!} \sum_{m=0}^{N-1} \frac{(\eta_\alpha - z_k)^m}{m!} \frac{\partial^l}{\partial z_j^l} \frac{\partial^m}{\partial z_k^m} \frac{\psi(z_1, \ldots, z_M)}{z_j z_k}. \tag{6.8}$$

The sum over α we need to evaluate is hence

$$\sum_{\substack{\alpha=1 \\ \eta_\alpha \neq z_j z_k}}^{N} \frac{\eta_\alpha^2 (\eta_\alpha - z_j)^l (\eta_\alpha - z_k)^m}{(\eta_\alpha - z_j)(\bar{\eta}_\alpha - \bar{z}_k)}, \quad 0 \leq l, m \leq N-1. \tag{6.9}$$

In the Haldane–Shastry model, the corresponding sum (2.2.40) is non-zero only for l=0,1, and 2. In the present case, however, further terms arise for $l + m + 1 = N$. These yield terms with very high derivatives when substituted in (6.8). It is not clear whether an analysis along these lines is feasible.

Static spin correlations—Another open issue is the static spin correlation functions of the ground state (6.3). We conjecture that it can be evaluated via a generalization of the method employed by Metzner and Vollhardt [4] for the Gutzwiller wave function.

Generalization to symmetric representations of SU(n)—The generalization of the model to symmetric representations of SU(n), like the representations **6** or **10** of SU(3), appears to follow without incident. If we write the SU(3) Gutzwiller or Haldane–Shastry ground state [18, 19] in terms of SU(3) Schwinger bosons $b^\dagger$, $r^\dagger$, $g^\dagger$ (for blue, red, and green; see e.g. [20]),

$$\left|\psi_0^{\text{HS}}\right\rangle \equiv \Psi_0^{\text{HS}}\big[b^\dagger, r^\dagger, g^\dagger\big]|0\rangle, \tag{6.10}$$

the generalizations to the SU(3) representation **6** and **10** are given by

$$\left|\psi_0^k\right\rangle = \left(\Psi_0^{\text{HS}}\big[b^\dagger, r^\dagger, g^\dagger\big]\right)^k |0\rangle, \tag{6.11}$$

with $k=2$ and $k=3$, respectively. The generalization of the defining condition (6.5) is

$$\Omega_\alpha^k = \sum_{\substack{\beta=1 \\ \beta\neq\alpha}}^{N} \frac{1}{\eta_\alpha - \eta_\beta} (I_\alpha^-)^k I_\beta^-, \quad \Omega_\alpha^k \left|\psi_0^k\right\rangle = 0 \quad \forall \alpha, \tag{6.12}$$

where $I^- \equiv r^\dagger b$ is one of the three "lowering" operators for the SU(3) spins. We assume that the construction of a parent Hamiltonian along the lines of Chaps. 4 and 5 will proceed without incident. The momentum spacings in the SU(n) models are likely to follow patterns which have no analog in quantum Hall systems, and have hence not been studied before.

Generalization to include mobile holes—It appears likely that the model can be generalized to include mobile holes as well, a task which has been accomplished for the $S = \frac{1}{2}$ model by Kuramoto and Yokoyama [21].

Conclusion—We have introduced an exact model of critical spin chains with arbitrary spin S. For $S = \frac{1}{2}$, the model reduces to one previously discovered by Haldane [1] and Shastry [2]. The spinon excitations obey non-abelian statistics for $S \geq 1$, with the internal Hilbert space spanned by topological spacings of the single spinon momenta. There is a long list of unresolved issues, including the quest for integrability and the viability of potential applications as protected cubits in quantum computation.

References

1. F.D.M. Haldane, Exact Jastrow–Gutzwiller resonant-valence-bond ground state of the spin-$\frac{1}{2}$ antiferromagnetic Heisenberg chain with $1/r^2$ exchange. Phys. Rev. Lett. **60**, 635 (1988)
2. B.S. Shastry, Exact solution of an $S = \frac{1}{2}$ Heisenberg antiferromagnetic chain with long-ranged interactions. Phys. Rev. Lett. **60**, 639 (1988)
3. M.C. Gutzwiller, Effect of correlation on the ferromagnetism of transition metals. Phys. Rev. Lett. **10**, 159 (1963)
4. W. Metzner, D. Vollhardt, Ground-state properties of correlated fermions: exact analytic results for the Gutzwiller wave function. Phys. Rev. Lett. **59**, 121 (1987)
5. F. Gebhard, D. Vollhardt, Correlation functions for Hubbard-type models: the exact results for the Gutzwiller wave function in one dimension. Phys. Rev. Lett. **59**, 1472 (1987)
6. J. Wess, B. Zumino, Consequences of anomalous ward identities. Phys. Lett. **37**, 95 (1971)
7. E. Witten, Non-Abelian bosonization in two dimensions. Commun. Math. Phys. **92**, 455 (1984)
8. R. Thomale, S. Rachel, P. Schmitteckert, M. Greiter, manuscript in preparation
9. N. Read, D. Green, Paired states of fermions in two dimensions with breaking of parity and time-reversal symmetries and the fractional quantum Hall effect. Phys. Rev. B **61**, 10267 (2000)
10. C. Nayak, F. Wilczek, $2n$-quasihole states realize 2^{n-1}-dimensional spinor braiding statistics in paired quantum Hall states. Nucl. Phys. B **479**, 529 (1996)
11. D.A. Ivanov, Non-Abelian statistics of half-quantum vortices in p-wave superconductors. Phys. Rev. Lett. **86**, 268 (2001)
12. A. Stern, von F. Oppen, E. Mariani, Geometric phases and quantum entanglement as building blocks for non-Abelian quasiparticle statistics. Phys. Rev. B **70**, 205338 (2004)
13. G. Moore, N. Read, Nonabelions in the fractional quantum Hall effect. Nucl. Phys. B **360**, 362 (1991)
14. M. Greiter, X.G. Wen, F. Wilczek, Paired Hall state at half filling. Phys. Rev. Lett. **66**, 3205 (1991)
15. M. Greiter, X.G. Wen, F. Wilczek, Paired Hall states. Nucl. Phys. B **374**, 567 (1992)
16. M. Greiter, R. Thomale, Non-Abelian statistics in a quantum antiferromagnet. Phys. Rev. Lett. **102**, 207203 (2009)
17. M. Greiter, D. Schuricht, Many-spinon states and the secret significance of Young tableaux. Phys. Rev. Lett. **98**, 237202 (2007)
18. N. Kawakami, Asymptotic Bethe-ansatz solution of multicomponent quantum systems with $1/r^2$ long-range interaction. Phys. Rev. B **46**, 1005 (1992)

19. N. Kawakami, SU(*N*) generalization of the Gutzwiller–Jastrow wave function and its critical properties in one dimension. Phys. Rev. B **46**, 3191 (1992)
20. M. Greiter, S. Rachel, Valence bond solids for SU(*n*) spin chains: exact models, spinon confinement, and the Haldane gap. Phys. Rev. B **75**, 184441 (2007)
21. Y. Kuramoto, H. Yokoyama, Exactly soluble supersymmetric *t*–*J*-type model with a long-range exchange and transfer. Phys. Rev. Lett. **67**, 1338 (1991)

Appendix A
Spherical Coordinates

The formalism for Landau level quantization on the sphere developed in Sect. 2.1.5 requires vector analysis in spherical coordinates. In this appendix, we will briefly review the conventions. Vectors and vector fields are given by

$$\boldsymbol{r} = r\boldsymbol{e}_{\mathrm{r}}, \tag{A.1}$$

$$\boldsymbol{v}(\boldsymbol{r}) = v_r\boldsymbol{e}_{\mathrm{r}} + v_\theta\boldsymbol{e}_\theta + v_\varphi\boldsymbol{e}_\varphi, \tag{A.2}$$

with

$$\boldsymbol{e}_r = \begin{pmatrix} \cos\varphi\,\sin\theta \\ \sin\varphi\,\sin\theta \\ \cos\theta \end{pmatrix}, \quad \boldsymbol{e}_\theta = \begin{pmatrix} \cos\varphi\,\cos\theta \\ \sin\varphi\,\cos\theta \\ -\sin\theta \end{pmatrix}, \quad \boldsymbol{e}_\varphi = \begin{pmatrix} -\sin\varphi \\ \cos\varphi \\ 0 \end{pmatrix}. \tag{A.3}$$

where $\varphi \in [0, 2\pi[$ and $\theta \in [0, \pi]$. This implies

$$\boldsymbol{e}_{\mathrm{r}} \times \boldsymbol{e}_\theta = \boldsymbol{e}_\varphi, \quad \boldsymbol{e}_\theta \times \boldsymbol{e}_\varphi = \boldsymbol{e}_{\mathrm{r}}, \quad \boldsymbol{e}_\varphi \times \boldsymbol{e}_{\mathrm{r}} = \boldsymbol{e}_\theta, \tag{A.4}$$

and

$$\begin{aligned} &\frac{\partial \boldsymbol{e}_r}{\partial\theta} = \boldsymbol{e}_\theta, && \frac{\partial \boldsymbol{e}_\theta}{\partial\theta} = -\boldsymbol{e}_r, && \frac{\partial \boldsymbol{e}_\varphi}{\partial\theta} = 0, \\ &\frac{\partial \boldsymbol{e}_r}{\partial\varphi} = \sin\theta\,\boldsymbol{e}_\varphi, && \frac{\partial \boldsymbol{e}_\theta}{\partial\varphi} = \cos\theta\,\boldsymbol{e}_\varphi, && \frac{\partial \boldsymbol{e}_\varphi}{\partial\varphi} = -\sin\theta\,\boldsymbol{e}_r - \cos\theta\,\boldsymbol{e}_\theta. \end{aligned} \tag{A.5}$$

With

$$\nabla = \boldsymbol{e}_{\mathrm{r}}\frac{\partial}{\partial r} + \boldsymbol{e}_\theta\frac{1}{r}\frac{\partial}{\partial\theta} + \boldsymbol{e}_\varphi\frac{1}{r\sin\theta}\frac{\partial}{\partial\varphi} \tag{A.6}$$

we obtain

$$\nabla\boldsymbol{v} = \frac{1}{r^2}\frac{\partial(r^2 v_r)}{\partial r} + \frac{1}{r\sin\theta}\frac{\partial(\sin\theta v_\theta)}{\partial\theta} + \frac{1}{r\sin\theta}\frac{\partial v_\varphi}{\partial\varphi}, \tag{A.7}$$

M. Greiter, *Mapping of Parent Hamiltonians*, Springer Tracts in Modern Physics 244,
DOI: 10.1007/978-3-642-24384-4,

$$\begin{aligned}\nabla \times \boldsymbol{v} = \boldsymbol{e}_{\mathrm{r}} \frac{1}{r \sin\theta}\left(\frac{\partial(\sin\theta v_\varphi)}{\partial\theta} - \partial\frac{\boldsymbol{v}_\theta}{\partial\varphi}\right) \\ + \boldsymbol{e}_\theta\left(\frac{1}{r\sin\theta}\frac{\partial v_r}{\partial\varphi} - \frac{1}{r}\frac{\partial(r v_\varphi)}{\partial r}\right) \\ + \boldsymbol{e}_\varphi\left(\frac{1}{r}\frac{\partial(r v_\theta)}{\partial r} - \frac{1}{r}\frac{\partial v_r}{\partial\theta}\right),\end{aligned} \tag{A.8}$$

$$\nabla^2 = \frac{1}{r^2}\frac{\partial}{\partial r}\left(r^2\frac{\partial}{\partial r}\right) + \frac{1}{r^2\sin\theta}\frac{\partial}{\partial\theta}\left(\sin\theta\frac{\partial}{\partial\theta}\right) + \frac{1}{r^2\sin^2\theta}\frac{\partial^2}{\partial\varphi^2}. \tag{A.9}$$

Appendix B
Fourier Sums for One-Dimensional Lattices

In this appendix we collect and proof some useful formulas for the explicit calculations of the Haldane–Shastry model. In particular, we provide the Fourier sums required for the evaluation of the coefficients A_l in (2.2.40) using two different methods, first by contour integration loosely following Laughlin et al. [1], and second by Feynmanesque algebra.

For $\eta_\alpha = e^{i\frac{2\pi}{N}\alpha}$ with $\alpha = 1, \ldots, N$ the following hold:

(a)

$$\eta_\alpha^N = 1. \tag{B.1}$$

(b)

$$\sum_{\alpha=1}^{N} \eta_\alpha^m = N\delta_{m,0} \mod N. \tag{B.2}$$

(c)

$$\prod_{\alpha=1}^{N} (\eta - \eta_\alpha) = \eta^N - 1. \tag{B.3}$$

Proof The η_α are by definition roots of 1. □

(d)

$$\sum_{\alpha=1}^{N} \frac{1}{\eta - \eta_\alpha} = \frac{N\eta^{N-1}}{\eta^N - 1}. \tag{B.4}$$

Proof Take $\frac{\partial}{\partial\eta}$ of (B.3) and divide both sides by $\eta^N - 1$. □

M. Greiter, *Mapping of Parent Hamiltonians*, Springer Tracts in Modern Physics 244, DOI: 10.1007/978-3-642-24384-4, © Springer-Verlag Berlin Heidelberg 2011

(e)

$$\sum_{\alpha=1}^{N} \frac{\eta_\alpha}{\eta - \eta_\alpha} = \frac{N}{\eta^N - 1}. \tag{B.5}$$

Proof Substitute $\eta_\alpha \to \frac{1}{\eta_\alpha}$, $\eta \to \frac{1}{\eta}$ in (B.4) and divide by $(-\eta)$. □

(f)

$$\sum_{\substack{\alpha,\beta,\gamma=1 \\ \alpha\neq\beta\neq\gamma\neq\alpha}}^{N} \frac{\eta_\gamma^2}{(\eta_\alpha - \eta_\gamma)(\eta_\beta - \eta_\gamma)} = \frac{N(N-1)(N-2)}{3}. \tag{B.6}$$

Proof Use the algebraic identity

$$\frac{a^2}{(a-b)(a-c)} + \frac{b^2}{(b-a)(b-c)} + \frac{c^2}{(c-a)(c-b)} = 1. \tag{B.7}$$

□

(g)

$$\sum_{\substack{\alpha,\beta=1 \\ \alpha\neq\beta}}^{N-1} \frac{1}{(\eta_\alpha - 1)(\eta_\beta - 1)} = \frac{(N-1)(N-2)}{3}. \tag{B.8}$$

Proof Substitute $\eta_\alpha \to \eta_\alpha \eta_\gamma, \eta_\beta \to \eta_\beta \eta_\gamma$ in (B.6)

(h)

$$\sum_{\alpha=1}^{N-1} \frac{\eta_\alpha^m}{\eta_\alpha - 1} = \frac{N+1}{2} - m, \quad 1 \leq m \leq N \tag{B.9}$$

Proof by contour integration Use Cauchy's theorem [2] for the function

$$f(z) = \frac{z^{m-1}}{z-1}, \quad N \geq 2,$$

with the contours shown in Fig. B.1 yields

$$\begin{aligned}
\sum_{\alpha=1}^{N-1} \frac{\eta_\alpha^m}{\eta_\alpha - 1} &= \frac{1}{2\pi \mathrm{i}} \sum_{\alpha=1}^{N-1} \oint_C \frac{z^{m-1}}{z-1} \frac{\eta_\alpha}{z - \eta_\alpha} \mathrm{d}z \\
&= \frac{N}{2\pi \mathrm{i}} \oint_C \frac{z^{m-1}}{(z-1)(z^N - 1)} \mathrm{d}z \\
&= -\frac{N}{2\pi \mathrm{i}} \oint_{C'} \underbrace{\frac{z^{m-1}}{(z-1)(z^N-1)}}_{=f(z)} \mathrm{d}z
\end{aligned}$$

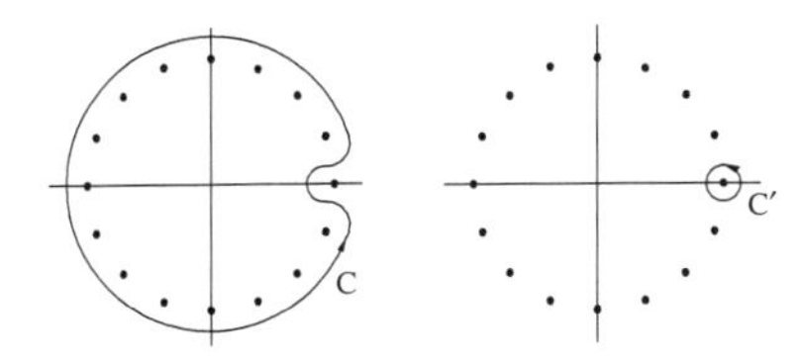

Fig. B.1 Contours for integrations

where we have first used (B.5) and then deformed the contour C such that the radius of circle goes to infinity, used that the circle at infinity does not contribute to the integral as the integrand falls off as at least $1/z^2$ for $m \leq N$, and finally reversed the direction of integration to replace C by C′.
Since $f(z)$ has a pole of second order at $z = 1$, the residue is given by

$$c_{-1} = \lim_{z \to 1} \frac{\mathrm{d}}{\mathrm{d}z} \underbrace{(z-1)^2 f(z)}_{=1/g(z)} = -\lim_{z \to 1} \frac{g'(z)}{g^2(z)}.$$

With

$$g(z) = \frac{1}{z^{m-1}} \frac{z^N - 1}{z - 1} = \sum_{k=1}^{N} z^{k-m} \stackrel{z \to 1}{\longrightarrow} N, \tag{B.10}$$

$$g'(z) = \sum_{k=1}^{N} (k-m) z^{k-m-1} \stackrel{z \to 1}{\longrightarrow} \frac{N(N+1)}{2} - mN, \tag{B.11}$$

we obtain

$$\sum_{\alpha=1}^{N-1} \frac{\eta_\alpha^m}{\eta_\alpha - 1} = -Nc_{-1} = \frac{(N+1)}{2} - m.$$

□

Proof by algebra With the definition

$$S_m \equiv \sum_{\alpha=1}^{N-1} \frac{\eta_\alpha^m}{\eta_\alpha - 1},$$

we find

$$S_{m+1} - S_m = \sum_{\alpha=1}^{N-1} \eta_\alpha^m = \begin{cases} -1, & 1 \leq m \leq N-1, \\ N-1, & m = 0. \end{cases}$$

and

$$S_0 = \sum_{\alpha=1}^{N-1} \frac{1}{\eta_\alpha - 1} = -\sum_{\alpha=1}^{N-1} \frac{\eta_\alpha}{\eta_\alpha - 1} = -S_1,$$

where we substituted $\eta_\alpha \to \frac{1}{\eta_\alpha}$. This directly implies

$$S_1 = -S_0 = \frac{N-1}{2} \quad \text{and} \quad S_m = \frac{(N+1)}{2} - m, \quad 1 \le m \le N.$$

(i)

$$\sum_{\alpha=1}^{N-1} \frac{1}{\eta_\alpha - 1} = -\frac{N-1}{2}. \tag{B.12}$$

Proof Use (B.9) with $m = N$. □

(j)

$$\sum_{\alpha=1}^{N-1} \frac{\eta_\alpha^m}{(\eta_\alpha - 1)^2} = -\frac{N^2-1}{12} + \frac{(m-1)(N-m+1)}{2}, \quad 1 \le m \le N. \tag{B.13}$$

Proof by contour integration In analogy to the proof of (B.9) we write

$$\begin{aligned}
\sum_{\alpha=1}^{N-1} \frac{\eta_\alpha^m}{(\eta_\alpha - 1)^2} &= \frac{1}{2\pi \mathrm{i}} \sum_{\alpha=1}^{N-1} \oint_C \frac{z^{m-1}}{(z-1)^2} \frac{\eta_\alpha}{z - \eta_\alpha} \mathrm{d}z \\
&= \frac{N}{2\pi \mathrm{i}} \oint_C \frac{z^{m-1}}{(z-1)^2 (z^N - 1)} \mathrm{d}z \\
&= -\frac{N}{2\pi \mathrm{i}} \oint_{C'} \underbrace{\frac{z^{m-1}}{(z-1)^2 (z^N - 1)}}_{=h(z)} \mathrm{d}z
\end{aligned}$$

where we have again used (B.5) and replaced the contour C by C′. As $h(z)$ has a now pole of third order at $z = 1$, the residue is given by

$$c_{-1} = \frac{1}{2} \lim_{z \to 1} \frac{\mathrm{d}^2}{\mathrm{d}z^2} \underbrace{(z-1)^3 h(z)}_{=1/g(z)} = \lim_{z \to 1} \left(-\frac{g''(z)}{2g^2(z)} + \frac{(g'(z))^2}{g^3(z)} w \right).$$

With $g(1)$ and $g'(1)$ as given by (B.10) and (B.11) and

$$\begin{aligned}
g''(z) &= \sum_{k=1}^{N} (k-m)(k-m-1) z^{k-m-2} \\
&\xrightarrow{z \to 1} \frac{N(N+1)(2N+1)}{6} - (2m+1)\frac{N(N+1)}{2} + m(m+1),
\end{aligned}$$

we find after some algebra that $-Nc_{-1}$ equals the expression on the right of (B.13). □

Proof by algebra With the definition

$$R_m \equiv \sum_{\alpha=1}^{N-1} \frac{\eta_\alpha^m}{(\eta_\alpha - 1)^2},$$

we find

$$R_{m+1} - R_m = \sum_{\alpha=1}^{N-1} \frac{\eta_\alpha^m}{(\eta_\alpha - 1)} = S_m$$

and

$$\begin{aligned} R_0 &= \sum_{\alpha=1}^{N-1} \frac{1}{(\eta_\alpha - 1)^2} \\ &= \sum_{\alpha=1}^{N-1}\sum_{\beta=1}^{N-1} \frac{1}{(\eta_\alpha - 1)(\eta_\beta - 1)} - \sum_{\substack{\alpha,\beta=1 \\ (\alpha\neq\beta)}}^{N-1} \frac{1}{(\eta_\alpha - 1)(\eta_\beta - 1)} \\ &= -\frac{(N-1)(N-5)}{12}, \end{aligned}$$

where we have used (B.12) and (B.8). This implies

$$\begin{aligned} R_{m+1} &= R_0 + S_0 + \sum_{n=1}^{m} S_n \\ &= -\frac{N^2-1}{12} + \sum_{n=1}^{m} \left(\frac{(N+1)}{2} - n \right) \\ &= -\frac{N^2-1}{12} + \frac{m(N-m)}{2}. \end{aligned}$$

for $1 \leq m \leq N$. □

(k)

$$\sum_{\alpha=1}^{N-1} \frac{1}{(\eta_\alpha - 1)^2} = -\frac{(N-1)(N-5)}{12}. \tag{B.14}$$

Proof Use (B.13) with $m = N$. □

(l)

$$\sum_{\alpha=1}^{N-1} \frac{\eta_\alpha^m}{|\eta_\alpha - 1|^2} = \frac{N^2-1}{12} - \frac{m(N-m)}{2}, \quad 0 \leq m \leq N. \tag{B.15}$$

Proof Use (B.13) with $m \to m + 1$. □

(m)

$$\sum_{\substack{\alpha=1 \\ \alpha\neq\beta}}^{N} \frac{\eta_\alpha + \eta_\beta}{\eta_\alpha - \eta_\beta} = 0. \tag{B.16}$$

Proof Substitute $\eta_\alpha \to \frac{1}{\eta_\alpha}, \eta_\beta \to \frac{1}{\eta_\beta}$ in one of the terms or use (B.9) and (B.12).

□

(n)

$$\sum_{\substack{\alpha=1\\ \alpha\neq\beta,\gamma\\ \beta\neq\gamma}}^{N} \frac{1}{\bar{\eta}_\alpha - \bar{\eta}_\beta} \frac{1}{\eta_\alpha - \eta_\gamma} = -\frac{\eta_\beta}{\eta_\beta - \eta_\gamma} + \frac{2}{|\eta_\beta - \eta_\gamma|^2}. \tag{B.17}$$

Proof With

$$\frac{1}{(\eta_\alpha - \eta_\beta)(\eta_\alpha - \eta_\gamma)} = \frac{1}{\eta_\beta - \eta_\gamma}\left(\frac{1}{\eta_\alpha - \eta_\beta} - \frac{1}{\eta_\alpha - \eta_\gamma}\right) \tag{B.18}$$

and

$$\sum_{\substack{\alpha=1\\ \alpha\neq\beta,\gamma\\ \beta\neq\gamma}}^{N} \frac{\eta_\alpha}{\eta_\alpha - \eta_\beta} = \frac{N-1}{2} - \frac{\eta_\gamma}{\eta_\gamma - \eta_\beta}, \tag{B.19}$$

which follows directly from (B.9), we write

$$\begin{aligned}
\sum_{\substack{\alpha=1\\ \alpha\neq\beta,\gamma\\ \beta\neq\gamma}}^{N} \frac{1}{\bar{\eta}_\alpha - \bar{\eta}_\beta} \frac{1}{\eta_\alpha - \eta_\gamma} &= -\sum_{\substack{\alpha=1\\ \alpha\neq\beta,\gamma\\ \beta\neq\gamma}}^{N} \frac{\eta_\beta}{\eta_\beta - \eta_\gamma}\left(\frac{\eta_\alpha}{\eta_\alpha - \eta_\beta} - \frac{\eta_\alpha}{\eta_\alpha - \eta_\gamma}\right)\\
&= -\frac{\eta_\beta}{\eta_\beta - \eta_\gamma}\left(-\frac{\eta_\gamma}{\eta_\gamma - \eta_\beta} + \frac{\eta_\beta}{\eta_\beta - \eta_\gamma}\right)\\
&= -\frac{\eta_\beta}{\eta_\beta - \eta_\gamma}\left(1 + \frac{2\eta_\gamma}{\eta_\beta - \eta_\gamma}\right)\\
&= -\frac{\eta_\beta}{\eta_\beta - \eta_\gamma} + \frac{2}{|\eta_\beta - \eta_\gamma|^2}.
\end{aligned}$$

□

(o) For symmetric operators $A_{\beta\gamma} = A_{\gamma\beta}$ it holds:

$$\sum_{\substack{\alpha,\beta,\gamma=1\\ \alpha\neq\beta\neq\gamma\neq\alpha}}^{N} \frac{A_{\beta\gamma}}{(\bar{\eta}_\alpha - \bar{\eta}_\beta)(\eta_\alpha - \eta_\gamma)} = \sum_{\beta\neq\gamma} \frac{2A_{\beta\gamma}}{|\eta_\beta - \eta_\gamma|^2} - \frac{1}{2}\sum_{\beta\neq\gamma} A_{\beta\gamma}. \tag{B.20}$$

Proof Use (B.17). □

Appendix C
Angular Momentum Algebra

In this appendix, we review a few very well known relations for angular momentum operators [3, 4]. The components of the angular momentum operator $\boldsymbol{J}$ obey the SU(2) Lie algebra

$$\left[J^a, J^b\right] = \mathrm{i}\varepsilon^{abc} J^c \quad \text{for} \quad a, b, c = \mathrm{x}, \mathrm{y}, \mathrm{z}. \tag{C.1}$$

Since $\left[\boldsymbol{J}^2, J^{\mathrm{z}}\right] = 0$, we can choose a basis of simultaneous eigenstates of $\boldsymbol{J}^2$ and J^{z},

$$\begin{aligned} \boldsymbol{J}^2 |j,m\rangle &=^2 j(j+1)|j,m\rangle, \\ J^{\mathrm{z}}|j,m\rangle &= m|j,m\rangle, \end{aligned} \tag{C.2}$$

where $m = -j, \ldots, j$. With $J^{\pm} \equiv J^{\mathrm{x}} \pm \mathrm{i} J^{\mathrm{y}}$, we have

$$\left[J^{\mathrm{z}}, J^{\pm}\right] = \pm J^{\pm}. \tag{C.3}$$

We further have

$$\begin{aligned} J^+ J^- &= (J^{\mathrm{x}})^2 + (J^{\mathrm{y}})^2 - \mathrm{i}[J^{\mathrm{x}}, J^{\mathrm{y}}] = \boldsymbol{J}^2 - (J^{\mathrm{z}})^2 + J^{\mathrm{z}}, \\ J^- J^+ &= \boldsymbol{J}^2 - (J^{\mathrm{z}})^2 - J^{\mathrm{z}}, \end{aligned} \tag{C.4}$$

and therefore

$$[J^+, J^-] = 2J^{\mathrm{z}}. \tag{C.5}$$

Equations (C.3) and (C.4) further imply

$$J^{\pm}|j,m\rangle = \sqrt{j(j+1) - m(m \pm 1)}\,|j, m-1\rangle, \tag{C.6}$$

where we have chosen the phases between $J^-|j,m\rangle$ and $|j,m-1\rangle$ real.

M. Greiter, *Mapping of Parent Hamiltonians*, Springer Tracts in Modern Physics 244,
DOI: 10.1007/978-3-642-24384-4, © Springer-Verlag Berlin Heidelberg 2011

Appendix D
Tensor Decompositions of Spin Operators

In this appendix, we will write out the tensor components [3, 4] of all the tensors of different order we can form from one, two, or three spins operators.

D.1 One Spin Operator

A single spin $\boldsymbol{S}$ transforms as a vector under rotations, which we normalize such that the $m = 0$ component equals S^z (see (3.5.10) in Sect. 3.5). The components of V^m are

$$\begin{aligned} V^1 &= -\frac{1}{\sqrt{2}}S^+, \\ V^0 &= \frac{1}{\sqrt{2}}\left[S^-, V^1\right] = S^z, \\ V^{-1} &= \frac{1}{\sqrt{2}}\left[S^-, V^0\right] = \frac{1}{\sqrt{2}}S^-. \end{aligned} \tag{D.1.1}$$

D.2 Two Spin Operators

Since each spin operator transforms as a vector, and the representation content of four vectors is given by

$$\mathbf{1} \otimes \mathbf{1} = \mathbf{0} \oplus \mathbf{1} \oplus \mathbf{2},$$

we can form one scalar, one vector, and one tensor of second order from two spin operators $\boldsymbol{S}_1$ and $\boldsymbol{S}_2$. The scalar is given by

$$U_{12} = \boldsymbol{S}_1\boldsymbol{S}_2 = \frac{1}{2}\left(S_1^+S_2^- + S_1^-S_2^+\right) + S_1^zS_2^z \tag{D.2.1}$$

M. Greiter, *Mapping of Parent Hamiltonians*, Springer Tracts in Modern Physics 244,
DOI: 10.1007/978-3-642-24384-4, © Springer-Verlag Berlin Heidelberg 2011

and the vector by $-\mathrm{i}(\boldsymbol{S}_1 \times \boldsymbol{S}_2)$. Written out in components, we obtain

$$\begin{aligned} V_{12}^{1} &= \frac{\mathrm{i}}{\sqrt{2}}(\boldsymbol{S}_1 \times \boldsymbol{S}_2)^{+} = \frac{1}{\sqrt{2}}\left(S_1^{+}S_2^{z} - S_1^{z}S_2^{+}\right), \\ V_{12}^{0} &= -\mathrm{i}(\boldsymbol{S}_1 \times \boldsymbol{S}_2)^{z} = \frac{1}{2}\left(S_1^{+}S_2^{-} - S_1^{-}S_2^{+}\right), \\ V_{12}^{-1} &= -\frac{\mathrm{i}}{\sqrt{2}}(\boldsymbol{S}_1 \times \boldsymbol{S}_2)^{-} = \frac{1}{\sqrt{2}}\left(S_1^{-}S_2^{z} - S_1^{z}S_2^{-}\right). \end{aligned} \tag{D.2.2}$$

With regard to the 2nd order tensor, note that $S_1^{+}S_2^{+}$ is the only operator we can construct with two spin operators which raises the S_{tot}^{z} quantum number by two. It must hence be proportional to the $m = 2$ component of the 2nd order tensor. As there is no particularly propitious way to normalize this tensor, we simply set the $m = 2$ component equal to $S_1^{+}S_2^{+}$, and then obtain the other components using (3.5.9). This yields[1]

$$\begin{aligned} T_{12}^{2} &= S_1^{+}S_2^{+}, \\ T_{12}^{1} &= \frac{1}{2}\left[S_1^{-} + S_2^{-}, T_{12}^{2}\right] = -S_1^{z}S_2^{+} - S_1^{+}S_2^{z}, \\ T_{12}^{0} &= \frac{1}{\sqrt{6}}\left[S_1^{-} + S_2^{-}, T_{12}^{1}\right] = \frac{1}{\sqrt{6}}\left(4S_1^{z}S_2^{z} - S_1^{+}S_2^{-} - S_1^{-}S_2^{+}\right), \\ T_{12}^{-1} &= \frac{1}{\sqrt{6}}\left[S_1^{-} + S_2^{-}, T_{12}^{0}\right] = S_1^{z}S_2^{-} + S_1^{-}S_2^{z}, \\ T_{12}^{-2} &= \frac{1}{2}\left[S_1^{-} + S_2^{-}, T_{12}^{-1}\right] = S_1^{-}S_2^{-}. \end{aligned} \tag{D.2.3}$$

Equations (D.2.1) and (D.2.3) imply

$$\frac{1}{2}\left(S_1^{+}S_2^{-} + S_1^{-}S_2^{+}\right) = \frac{2}{3}\boldsymbol{S}_1\boldsymbol{S}_2 - \frac{1}{\sqrt{6}}T_{12}^{0}, \tag{D.2.4}$$

$$S_1^{z}S_2^{z} = \frac{1}{3}\boldsymbol{S}_1\boldsymbol{S}_2 + \frac{1}{\sqrt{6}}T_{12}^{0}. \tag{D.2.5}$$

Combining (D.2.4) with (D.2.2) yields

$$\begin{aligned} S_1^{+}S_2^{-} &= \frac{2}{3}\boldsymbol{S}_1\boldsymbol{S}_2 - \mathrm{i}(\boldsymbol{S}_1 \times \boldsymbol{S}_2)^{z} - \frac{1}{\sqrt{6}}T_{12}^{0}, \\ S_1^{-}S_2^{+} &= \frac{2}{3}\boldsymbol{S}_1\boldsymbol{S}_2 + \mathrm{i}(\boldsymbol{S}_1 \times \boldsymbol{S}_2)^{z} - \frac{1}{\sqrt{6}}T_{12}^{0}. \end{aligned} \tag{D.2.6}$$

[1] We denote general tensors of order j with $T^{(j)}$ and 2nd order tensors with T.

For $\boldsymbol{S}_1 = \boldsymbol{S}_2$, (D.2.6) reduces with $\boldsymbol{S}_1 \times \boldsymbol{S}_1 = \mathrm{i}\boldsymbol{S}_1$ to

$$\begin{aligned} S_1^+ S_1^- &= \frac{2}{3} \boldsymbol{S}_1^2 + S_1^z - \frac{1}{\sqrt{6}} T_{11}^0, \\ S_1^- S_1^+ &= \frac{2}{3} \boldsymbol{S}_1^2 - S_1^z - \frac{1}{\sqrt{6}} T_{11}^0. \end{aligned} \tag{D.2.7}$$

D.3 Three Spin Operators

Since

$$\mathbf{1} \otimes \mathbf{1} \otimes \mathbf{1} = \mathbf{0} \oplus 3 \cdot \mathbf{1} \oplus 2 \cdot \mathbf{2} \oplus \mathbf{3},$$

we can form one scalar, three vectors, two tensors of second order, and one tensor of third order, from three spin operators $\boldsymbol{S}_1$, $\boldsymbol{S}_2$, and $\boldsymbol{S}_3$.

The scalar is given by

$$\begin{aligned} U_{123} &= -\mathrm{i}\boldsymbol{S}_1(\boldsymbol{S}_2 \times \boldsymbol{S}_3) \\ &= \frac{1}{2} S_1^z (S_2^+ S_3^- - S_2^- S_3^+) + 2 \text{ cyclic permutations} \\ &= \frac{1}{2}(S_1^z S_2^+ S_3^- + S_1^+ S_2^- S_3^z + S_1^- S_2^z S_3^+ \\ &\quad - S_1^z S_2^- S_3^+ - S_1^- S_2^+ S_3^z - S_1^+ S_2^z S_3^-). \end{aligned} \tag{D.3.1}$$

The three vectors are given by

$$\boldsymbol{S}_1(\boldsymbol{S}_2\boldsymbol{S}_3), \quad \boldsymbol{S}_1(\boldsymbol{S}_2)\boldsymbol{S}_3, \quad \text{and} \quad (\boldsymbol{S}_1\boldsymbol{S}_2)\boldsymbol{S}_3, \tag{D.3.2}$$

where the scalar product in the second expression is understood to contract $\boldsymbol{S}_1$ and $\boldsymbol{S}_3$. The components for each m are according to the conventions specified in (D.1.1). For later purposes, we write for the $m = 0$ components,

$$\begin{aligned} V_{a,123}^0 &= S_1^z(\boldsymbol{S}_2\boldsymbol{S}_3) = \frac{1}{2}(S_1^z S_2^+ S_3^- + S_1^z S_2^- S_3^+) + S_1^z S_2^z S_3^z, \\ V_{b,123}^0 &= \boldsymbol{S}_1(S_2^z)\boldsymbol{S}_3 = \frac{1}{2}(S_1^- S_2^z S_3^+ + S_1^+ S_2^z S_3^-) + S_1^z S_2^z S_3^z, \\ V_{c,123}^0 &= (\boldsymbol{S}_1\boldsymbol{S}_2)S_3^z = \frac{1}{2}(S_1^+ S_2^- S_3^z + S_1^- S_2^+ S_3^z) + S_1^z S_2^z S_3^z. \end{aligned} \tag{D.3.3}$$

To obtain a tensor operator of second order, or more precisely the $m = 2$ component of it, all we need to do is to form the product of the $m = 1$ components of two vector operators constructed out of the three spins, like $\boldsymbol{S}_1$ and $-\mathrm{i}(\boldsymbol{S}_2 \times \boldsymbol{S}_3)$ or $-\mathrm{i}(\boldsymbol{S}_1 \times \boldsymbol{S}_2)$ and $\boldsymbol{S}_3$. In this way, we construct the tensor operators of second order

$$\begin{aligned} T^2_{a,123} &= -\mathrm{i}S_1^+(\boldsymbol{S}_2 \times \boldsymbol{S}_3)^+, \\ T^2_{b,123} &= -\mathrm{i}(\boldsymbol{S}_1 \times \boldsymbol{S}_2)^+ S_3^+. \end{aligned} \tag{D.3.4}$$

The other components are obtained as in (D.2.3). As we are primarily interested in the $m = 0$ component, we may use (D.2.3) directly to write

$$\begin{aligned} T^0_{a,123} &= -\frac{\mathrm{i}}{\sqrt{6}}[4S_1^z(\boldsymbol{S}_2 \times \boldsymbol{S}_3)^z - S_1^+(\boldsymbol{S}_2 \times \boldsymbol{S}_3)^- - S_1^-(\boldsymbol{S}_2 \times \boldsymbol{S}_3)^+] \\ &= \frac{1}{\sqrt{6}}[2S_1^z(S_2^+S_3^- - S_2^-S_3^+) - S_1^+S_2^-S_3^z + S_1^+S_2^zS_3^- \\ &\qquad + S_1^-S_2^+S_3^z - S_1^-S_2^zS_3^+], \end{aligned} \tag{D.3.5}$$

and similarly for $T^0_{b,123}$, which can be obtained from $T^0_{b,123}$ by a cyclical permutation of the superscripts $+, -, z$. Note that there is no third tensor of this kind, as the sum of the three tensors obtained from (D.3.5) by cyclic permutations of the superscripts equals zero.

We obtain the tensor of third order with the method we used to obtain the second order tensor (D.2.3) formed by two spins:

$$\begin{aligned} W^3_{123} &= -S_1^+S_2^+S_3^+, \\ W^2_{123} &= \frac{1}{\sqrt{6}}\left[S_1^- + S_2^- + S_3^-, W^3_{123}\right] \\ &= -\frac{1}{\sqrt{6}}[S_1^-, S_1^+]S_2^+S_3^+ + 2 \text{ cycl. permutations} \\ &= \sqrt{\frac{2}{3}}S_1^zS_2^+S_3^+ + 2 \text{ cycl. permutations}, \\ W^1_{123} &= \frac{1}{\sqrt{10}}[S_1^- + S_2^- + S_3^-, W^2_{123}] \\ &= \frac{1}{\sqrt{15}}([S_1^-, S_1^z]S_2^+S_3^+ + S_1^z[S_2^- + S_3^-, S_2^+S_3^+]) + 2 \text{ cycl. permutations}, \\ &= \frac{1}{\sqrt{15}}(S_1^-S_2^+S_3^+ - 4S_1^zS_2^zS_3^+) + 2 \text{ cycl. permutations}, \end{aligned}$$

$$
\begin{aligned}
W_{123}^{0} &= \frac{1}{\sqrt{12}}\left[S_1^- + S_2^- + S_3^-, W_{123}^{1}\right] \\
&= \frac{1}{6\sqrt{5}}\Big(S_1^-[S_2^- + S_3^-, S_2^+ S_3^+] - 4[S_1^- + S_2^-, S_1^z S_2^z] S_3^+ \\
&\quad - 4 S_1^z S_2^z [S_3^-, S_3^+]\Big) + 2 \text{ cycl. permutations} \\
&= -\frac{1}{\sqrt{5}}(S_1^- S_2^+ S_3^z + \text{ 5 permutations}) + \frac{4}{\sqrt{5}} S_1^z S_2^z S_3^z,
\end{aligned}
\tag{D.3.6}
$$

$$
\begin{aligned}
W_{123}^{-1} &= \frac{1}{\sqrt{12}}\left[S_1^- + S_2^- + S_3^-, W_{123}^{0}\right] \\
&= -\frac{1}{\sqrt{15}}(S_1^- S_2^- S_3^+ - 4 S_1^- S_2^z S_3^z) + 2 \text{ cycl. permutations},
\end{aligned}
$$

$$
\begin{aligned}
W_{123}^{-2} &= \frac{1}{\sqrt{10}}\left[S_1^- + S_2^- + S_3^-, W_{123}^{-1}\right] \\
&= \sqrt{\frac{2}{3}} S_1^- S_2^- S_3^z + 2 \text{ cycl. permutations}, \\
W_{123}^{-3} &= \frac{1}{\sqrt{6}}\left[S_1^- + S_2^- + S_3^-, W_{123}^{-2}\right] \\
&= S_1^- S_2^- S_3^- .
\end{aligned}
$$

The permutations here always refer to permutations of the superscripts $+, -, z$, as otherwise we would have to assume again that none of the three spin operators are identical. In particular, writing out the $m = 0$ yields

$$
\begin{aligned}
W_{123}^{0} = -\frac{1}{\sqrt{5}}(&S_1^- S_2^+ S_3^z + S_1^+ S_2^z S_3^- + S_1^z S_2^- S_3^+ \\
&+ S_1^+ S_2^- S_3^z + S_1^- S_2^z S_3^+ + S_1^z S_2^+ S_3^-) + \frac{4}{\sqrt{5}} S_1^z S_2^z S_3^z.
\end{aligned}
\tag{D.3.7}
$$

Combining (D.3.3) and (D.3.7), we obtain

$$
S_1^z S_2^z S_3^z = \frac{1}{5}\left(V_{a,123}^{0} + V_{b,123}^{0} + V_{c,123}^{0}\right) + \frac{1}{2\sqrt{5}} W_{123}^{0},
\tag{D.3.8}
$$

and hence

$$
\begin{aligned}
\frac{1}{2} S_1^z (S_2^+ S_3^- + S_2^- S_3^+) &= V_{a,123}^{0} - S_1^z S_2^z S_3^z \\
&= \frac{4}{5} V_{a,123}^{0} - \frac{1}{5} V_{b,123}^{0} - \frac{1}{5} V_{c,123}^{0} - \frac{1}{2\sqrt{5}} W_{123}^{0}.
\end{aligned}
\tag{D.3.9}
$$

From (D.3.1) and (D.3.5) we obtain

$$\frac{1}{2}S_1^z\left(S_2^+S_3^- - S_2^-S_3^+\right) = \frac{1}{3}\,U_{123} + \frac{1}{\sqrt{6}}\,T^0_{a,123}. \tag{D.3.10}$$

Combining (D.3.9) and (D.3.10) we finally obtain

$$\begin{aligned} S_1^zS_2^+S_3^- &= +\frac{1}{3}U_{123} + \frac{1}{5}\left(4\,V^0_{a,123} - V^0_{b,123} - V^0_{c,123}\right) + \frac{1}{\sqrt{6}}\,T^0_{a,123} - \frac{1}{2\sqrt{5}}\,W^0_{123} \\ &= +\frac{1}{3}\boldsymbol{S}_1(\boldsymbol{S}_2\times\boldsymbol{S}_3) + \frac{1}{5}\left[4\,S_1^z(\boldsymbol{S}_2\boldsymbol{S}_3) - \boldsymbol{S}_1(S_2^z)\boldsymbol{S}_3 - (\boldsymbol{S}_1\boldsymbol{S}_2)S_3^z\right] \\ &\quad + \frac{1}{\sqrt{6}}\,T^0_{a,123} - \frac{1}{2\sqrt{5}}\,W^0_{123}, \end{aligned} \tag{D.3.11}$$

$$\begin{aligned} S_1^zS_2^-S_3^+ &= -\frac{1}{3}\,U_{123} + \frac{1}{5}\left(4\,V^0_{a,123} - V^0_{b,123} - V^0_{c,123}\right) - \frac{1}{\sqrt{6}}\,T^0_{a,123} - \frac{1}{2\sqrt{5}}\,W^0_{123} \\ &= -\frac{1}{3}\boldsymbol{S}_1(\boldsymbol{S}_2\times\boldsymbol{S}_3) + \frac{1}{5}\left[4\,S_1^z(\boldsymbol{S}_2\boldsymbol{S}_3) - \boldsymbol{S}_1(S_2^z)\boldsymbol{S}_3 - (\boldsymbol{S}_1\boldsymbol{S}_2)S_3^z\right] \\ &\quad - \frac{1}{\sqrt{6}}\,T^0_{a,123} - \frac{1}{2\sqrt{5}}\,W^0_{123}. \end{aligned} \tag{D.3.12}$$

References

1. R.B. Laughlin, D. Giuliano, R. Caracciolo, O.L. White, Quantum number fractionalization in antiferromagnets. In: G. Morandi, P. Sodano, A. Tagliacozzo, V. Tognetti (eds) *Field Theories for Low-Dimensional Condensed Matter Systems* (Springer, Berlin, 2000)
2. S. Lang, *Complex Analysis* (Springer, New York, 1985)
3. K. Gottfried, *Quantum mechanics,volume I: Fundamentals* (Benjamin/Addison Wesley, New York, 1966)
4. G. Baym, *Lectures on Quantum Mechanics* (Benjamin/Addison Wesley, New York, 1969)

PGSTL